AIRCRAFT CONTROL ALLOCATION

Aerospace Series

The aerospace sector draws upon many specialist areas of expertise; mechanical, electrical/electronic design, systems and computer engineering, planning, logistics, psychology (the human-machine interface), communications, risk analysis, and economics. Each title in the Aerospace Series explores a different aspect or cohesive topic that those working in or aspiring to work in the aerospace sector will find useful and informative. Books in the series range from high level student texts, explanation and dissemination of technology and good practice, through to interesting and important research that is immediately relevant to industrial development or practice.

Morphing Aerospace Vehicles and Structures by John Valasek (Editor)

Spacecraft Systems Engineering, 4th Edition by Peter Fortescue (Editor), Graham Swinerd (Editor), John Stark (Editor)

Basic Helicopter Aerodynamics, 3rd Edition by John M. Seddon, Simon Newman

Gas Turbine Propulsion Systems by Bernie MacIsaac, Roy Langton

System Health Management: with Aerospace Applications by Stephen B Johnson (Editor), Thomas Gormley (Co-Editor), Seth Kessler (Co-Editor), Charles Mott (Co-Editor), Ann Patterson-Hine (Co-Editor), Karl Reichard (Co-Editor), Philip Scandura, Jr. (Co-Editor)

Advanced Control of Aircraft, Spacecraft and Rockets by Ashish Tewari

Cooperative Path Planning of Unmanned Aerial Vehicles by Antonios Tsourdos, Brian White, Madhavan Shanmugavel

Principles of Flight for Pilots by Peter J. Swatton

Air Travel and Health: A Systems Perspective by Allan Seabridge, Shirley Morgan

Unmanned Aircraft Systems: UAVS Design, Development and Deployment by Reg Austin

Introduction to Antenna Placement and Installation by Thereza Macnamara

Principles of Flight Simulation by David Allerton

Aircraft Fuel Systems by Roy Langton, Chuck Clark, Martin Hewitt, Lonnie Richards

Computational Modelling and Simulation of Aircraft and the Environment, Volume 1: Platform Kinematics and Synthetic Environment by Dominic J. Diston

Handbook of Space Technology by Wilfried Ley (Editor), Klaus Wittmann (Editor), Willi Hallmann (Editor)

Aircraft Performance Theory and Practice for Pilots, 2nd Edition by Peter J. Swatton

Aircraft Systems: Mechanical, Electrical and Avionics Subsystems Integration, 3rd Edition by Ian Moir, Allan Seabridge

Stability and Control of Aircraft Systems: Introduction to Classical Feedback Control by Roy Langton

Military Avionics Systems by Ian Moir, Allan Seabridge, Malcolm Jukes (Contributions by)

Aircraft Loading and Structural Layout by Denis Howe

Aircraft Display Systems by Malcolm Jukes

Aircraft Conceptual Design Synthesis by Denis Howe

AIRCRAFT CONTROL ALLOCATION

Wayne Durham

Virginia Polytechnic Institute and State University, USA

Kenneth A. Bordignon

Embry-Riddle Aeronautical University, USA

Roger Beck

Dynamic Concepts, Inc., USA

This edition first published 2017
© 2017 John Wiley & Sons, Ltd

Registered office John Wiley & Sons Ltd, The Atrium, Southern Gate, Chichester, West Sussex, PO19 8SQ, United Kingdom

For details of our global editorial offices, for customer services and for information about how to apply for permission to reuse the copyright material in this book please see our website at www.wiley.com.

Library of Congress Cataloging-in-Publication Data

Names: Durham, Wayne, 1941- author. | Bordignon, Kenneth A., author. | Beck,
   Roger, 1977- author.
Title: Aircraft control allocation / Wayne Durham, Kenneth A. Bordignon,
   Roger Beck.
Description: Chichester, West Sussex, United Kingdom : John Wiley & Sons,
   Inc., [2017] | Includes bibliographical references and index.
Identifiers: LCCN 2016024806| ISBN 9781118827796 (cloth) | ISBN 9781118827772
   (epub) | ISBN 9781118827765 (Adobe PDF)
Subjects: LCSH: Airplanes–Control systems–Mathematics. | Airplanes–Control
   systems–Design and construction.
Classification: LCC TL678 .D87 2017 | DDC 629.135–dc23 LC record available at
https://lccn.loc.gov/2016024806

A catalogue record for this book is available from the British Library.

Cover photo: NASA's NF-15B research aircraft is shown during an early yaw-vectoring mission in the ACTIVE flight research project. Credits: NASA Photo

Typeset in 10/12pt TimesLTStd by SPi Global, Chennai, India

10 9 8 7 6 5 4 3 2 1

Contents

Dedication

For Craig Steidle, Bob Hanley, and John Foster. Thanks guys.

Series Preface

The field of aerospace is multi-disciplinary and wide ranging, covering a large variety of products, disciplines and domains, not merely in engineering but in many related supporting activities. These combine to enable the aerospace industry to produce innovative and techno-logically advanced vehicles. The wealth of knowledge and experience that has been gained by expert practitioners in the various aerospace fields needs to be passed onto others working in the industry and also researchers, teachers and the student body in universities.

The *Aerospace Series* aims to be a practical, topical and relevant series of books aimed at people working in the aerospace industry, including engineering professionals and opera-tors, engineers in academia, and allied professions such commercial and legal executives. The range of topics is intended to be wide ranging, covering design and development, manufacture, operation and support of aircraft, as well as topics such as infrastructure operations and current advances in research and technology.

Modern aircraft are designed with multiple control surfaces, and possibly other control effectors e.g. thrust vectoring, and therefore problems can arise as to how to combine these control devices in an optimal manner, particularly in low-speed flight regimes where the aero-dynamic surfaces lose their effectiveness.

This book, *Aircraft Control Allocation*, provides a detailed explanation of some selected topics relating to the aircraft control allocation problem. After providing some background material in aircraft control and control laws, a number of approaches that can be used to solve the control allocation problem are illustrated and the influence that they have on control law design discussed. Of particular note is the chapter describing some of the lessons learnt whilst designing the X-35 Flight Control System.

Peter Belobaba, Jonathan Cooper and Alan Seabridge

Glossary

$(\dot{\ })$ Dot over quantity: the derivative with respect to time of the contents of the parentheses ( ).

$(\hat{\ })$ Hat over quantity: the contents of the parentheses ( ) are approximate.

α Angle-of-attack: the aerodynamic angle between the projection of the relative wind onto the airplane's plane of symmetry and a suitably defined body fixed x-axis.

β Sideslip angle: The aerodynamic angle between the velocity vector and the airplane's plane of symmetry.

$\ell_1, \ell_2, \ell_\infty$ Vector norms: ℓ_2 is the square root of the sum of the squares of the entries in the vector. It appears everywhere. ℓ_1 is the sum of the absolute values of the entries and ℓ_∞ is the greatest absolute value. ℓ_1 and ℓ_∞ frequently appear in linear programming problems.

Ω Either:
1. Every combination of control effector deflections that are admissible; in other words, that are within the limits of travel or deflection.
2. A normally diagonal matrix used to specify the dynamics in a dynamic-inversion control law.

Φ The effects, usually body-axis moments, moment coefficients, or angular accelerations, of every combination of control effector deflections in Ω, q.v. (sense 1). Sometimes called the AMS, for 'attainable moment set' or subset.

ϕ Bank angle: one of three angles that define a 3-2-1 (z-y-x) rotation from inertial to body-fixed reference frames.

Π Either:
1. (Primarily) A subset of the attainable moments (Φ) consisting of all the moments that are generated by a particular control allocation method.
2. A plane surface that arises in Banks' method of allocation for the three-moment problem.

ψ Heading angle: one of three angles that define a 3-2-1 (z-y-x) rotation from inertial to body-fixed reference frames.

Θ A subset of Ω: all admissible controls that a particular control-allocation method can return as solutions to a control-allocation problem.

θ Pitch attitude: one of three angles that define a 3-2-1 (z-y-x) rotation from inertial to body-fixed reference frames.

B One of the matrices of the linearized equations of motion: A is the system matrix, B is control effectiveness matrix, and C is the output matrix.

C_{x_y} The non-dimensional stability or control derivative of x with respect to y: it is the non-dimensional form of X_y.

Comp   Complementary: a superscript to certain dynamic responses.

Cont   Controllable: a superscript to certain dynamic responses.

d, *des*   Desired: a subscript to a dynamic response, or any other quantity.

F_B Body-fixed reference frames. The origin is at the airplane's center of mass. The axes x_B and z_B lie in the airplane's plane of symmetry. y_B completes the right-hand system. Once defined, a body-fixed reference system's orientation with respect to the body does not change. Two frequently used body-fixed reference frames are the principal axes and the stability-axis system.

F_H Local-horizontal reference frame. The axes x_H, y_H, and z_H are oriented north, east, and down, respectively. The earth is flat.

F_W Wind-axis system. The axis x_W lies in the direction of flight, opposite the relative wind. z_W is in the plane of symmetry, oriented downward.

g Either:
1. Acceleration of gravity, or
2. The non-dimensional units of load factor n, q.v.

I With subscripts; moment of inertia.

Kine   Kinematic: a superscript to certain dynamic responses.

L, C, D Lift, side force, and drag: wind-axis forces in the x_W-, y_W- and z_W-directions, respectively.

L, M, N Body-axis moments about the x_B axis (rolling), y_B axis (pitching), and z_B axis (yawing), respectively.

L Either:
1. Lift, or
2. Rolling moment, depending on context.

LD   Lateral-directional, meaning all motions, accelerations, forces, and so on, that are not longitudinal, q.v. Sometimes *lat-dir*.

Long   Longitudinal, meaning all motions, accelerations, forces, and so on, that take place in the airplane's plane of symmetry. Pitching moments, velocities, and accelerations are about the airplane's y_B-axis but the motion is in the x_B–z_B plane.

m The mass of the airplane.

n Load factor, the ratio of lift to weight, $n = L/W$. Measured in gs.

p, q, r Body-axis roll rate, pitch rate, and yaw rate, respectively.

P A generalized inverse of a matrix B: $BPB = B$ and $PBP = P$, with appropriate dimensions.

Ref   Subscript, 'evaluated in reference conditions'.

u   Vector of control effector variables.

$\mathbf{u}_{min}$, $\mathbf{u}_{max}$ Vector of control effector limits, minimum or maximum.

$\mathbf{u}_l$, $\mathbf{u}_u$ Vector of control effector limits, lower or upper. This notation seems preferred by linear programmers over $\mathbf{u}_{min}$, $\mathbf{u}_{max}$, q.v.

x_B, y_B, z_B Names of body-axes.

W A weighting matrix, generally diagonal and positive.

x_W, y_W, z_W Names of wind axes.

X_y Where X is a force or moment and y is a state or control, a dimensional derivative, $\partial X/\partial y$. It is the dimensional form of C_{x_y}, q.v. The definition does *not* include division by mass or moment of inertia. If y is a control effector the result is called a control derivative, otherwise it is called a stability derivative.

X, Y, Z Body-axis forces in the x-, y- and z-directions, respectively.

x, y, z Names of axes. With no subscripts usually taken to be body-axes.

ACTIVE   Advanced Control Technology for Integrated Vehicles. A research F-15 with differential canards, axisymmetric thrust vectoring, and other novel features.

ADMIRE   Aero-Data Model In a Research Environment, simulation code. See Appendix B.

Admissible   Of a control effector or suite of control effectors, those deflections that are within the physical limits of employment.

AMS   Attainable moment subset or set, Φ.

Angular accelerations   See Objectives.

ARI   Aileron-rudder interconnect. Normally used to reduce adverse yaw due to aileron deflection.

Attainable   Of moments or accelerations; that which can be generated by some admissible combination of control effectors. The term may be applied globally, meaning there is some theoretical combination, or locally, to a particular control allocation method, meaning those combinations of control effectors that the method will generate using its rules.

Basic feasible solution   Of linear programs, a basic solution to the equality constraints in a linear program that also solves the inequality constraints.

Basic solution   Of linear programs, a solution to the l linear equality constraints of a linear program in 'standard form' with $k - l$ of the decision variables at their bound.

CAS   Control augmentation system.

Control effectiveness   A measure of the effect of utilizing a control effector, either moment, moment coefficient, or angular acceleration.

Control authority   The aggregate effect of the effectiveness of all the control effectors in whatever combination.

Control power   Angular acceleration per unit of control deflection.

CHR   Cooper–Harper rating; sometimes HQR.

Constraint   Of a control effector, a limiting position, usually imposed by the hardware. It may also refer to a limit on the rate of travel. In linear programming, a constraint may refer to the position limits, but also of an equality that must be satisfied. Thus $\mathbf{u} \leq \mathbf{u}_{max}$ is an inequality constraint, and $B\mathbf{u} = \mathbf{m}_{des}$ is an equality constraint.

Control effector   The devices that directly effect control by changing forces or moments, such as ailerons or rudders. When we say 'the controls' with no qualification, we usually mean the control effectors. The sign convention for conventional flapping control effectors follows a right-hand rule, with the thumb along the axis about which the effector is designed to generate moments, and the curled fingers denoting the positive deflection of the trailing edge.

Control inceptor   Cockpit devices that control, through direct linkage or a flight-control system or computer, the control effectors. Positive control inceptor deflections correspond to positive deflections of the effectors they are connected to, barring such things as aileron–rudder interconnects (ARI, q.v.).

Cycling   Of a linear program, a condition in which a sequence of vertices is visited by a solver for which the objective function does not decrease, eventually returning to the starting point in the cycle. Cycling represents a failure to converge and must be addressed by choosing an exchange rule designed to prevent it.

Degenerate basic solution   Of linear programs, a basic solution to a linear program in which one of the l decision variables in the basis is at its bound in addition to the non-basic variables.

Decision variables   The set of unknown parameters being optimized in a linear program.

FBW   Fly by wire. The pilot flies the computer, the computer flies the airplane.

FQ   Flying qualities.

Ganged   Said of mechanical devices that are linked so that they move in fixed relation to each other, such as ailerons.

HARV   High angle-of-attack research vehicle.

HQ   Handling qualities.

HQR   Handling qualities rating.

Interior point method   One of a family of numerical methods that seek to find the optimal solution to a linear program by moving through the interior of the feasible set.

Intersection   Of two objects (q.v.), an object that is wholly contained in each of the two.

Lat-Dir   Lateral-directional.

LEU, LED   Leading-edge up, down. Terms used to describe the deflection of leading-edge control surfaces.

Linear programming   A problem, or the method of solving that problem, of optimization of an objective subject to linear equality and inequality constraints. To the purpose of this book, a method of allocating controls subject to position constraints.

Moments   See Objectives.

Moment coefficients   See Objectives.

Object   A generalization of any of the several polytopes that describe sets of admissible controls and attainable moments.

Objectives   Those which control effectors are intended to generate. Originally control allocation sought to find the control effectors that generated specified moments, or moment coefficients. Subsequently researchers have tended toward using angular accelerations as the objectives. We will generally speak of the objectives as being moments.

Object notation   A method of identifying objects (q.v.) using a 0 for a control at its lower limit, a 1 at its upper limit, and a 2 if it can be anywhere in between.

OBM   On-board model. A set of aerodynamic data for an aircraft stored in the aircraft's flight control computer.

Over-actuated control system   See Redundant controls.

Phase one/two program   Phase one of a linear programming solver solves a modified problem in order to locate an initial feasible solution for the phase two solver that will optimize the original problem.

PIO   Pilot-induced oscillation. There's a more politically correct term that removes the onus from the pilot.

PR   Pilot rating; sometimes HQR, q.v.

Preferred   Of a solution to the control allocation problem, a control effector configuration that is as close as possible to to one that is preferred. Minimum norm solutions are used as preferred solutions often.

Pseudo control   A combination of control effectors intended to create a certain effect, such as the excitation of a particular dynamic response mode of the airplane.

Redundant controls   Control effectors are seldom *redundant*, in the sense that the designer had no use for them in mind. The control effectors that are redundant in higher-speed flight may be critical in slow-speed flight. The term just means that there are more control effectors than objectives, q.v. As used in this book, it means there are more than three control effectors to generate the three moments or angular accelerations.

SAS   Stability augmentation system.

Simplex   An extension of a triangle (two-dimensional), or tetrahedron (three-dimensional), to an arbitrary number of dimensions. An n-dimensional simplex is defined by the convex hull of $n + 1$ vertices.

Simplex method   Either:

1. (Dantzig) Algorithms based on Dantzig's original numerical algorithm for the solution of linear programs, introduced in 1947. The simplex method moves between neighboring vertices, basic solutions, of the feasible set, decreasing the cost function until the optimum is found.
2. (Nelder Mead) Also known as downhill simplex. Numerical solution algorithm that iterates an n-dimensional simplex to minimize n-dimensional, non-linear, unconstrained optimization problems. Heuristic rules at each step govern how to modify the simplex.

Slack variable   Variables augmenting the decision variables in a linear program so that inequality constraints can be converted to equality constraints.

TEU, TED, TEL, TER Trailing-edge up, down, left, right. Terms used to describe the deflection of flapping control surfaces.

Union Of two objects (q.v.), the smallest object of which the two given objects are both members.

Warm start A heuristic method for initializing a linear program solver given a pre-existing optimal solution to a similar problem.

About the Companion Website

Don't forget to visit the companion website for this book:

www.wiley.com/go/durham/aircraft_control_allocation

There you will find valuable material designed to enhance your learning, including:

- **Simulation quick start guide**

Scan this QR code to visit the companion website

1

Introduction

The general theme of the book is to reproduce the research and insights that led the authors through their seminal studies into airplane control allocation. There is much research remaining to be done in the field of control allocation, and by following the thinking that preceded the fruitful directions taken by other researchers, new areas of inquiry will be opened.

The authors defend their geometrical approach to visualizing the problem as one that provides greater insight into the mechanisms of the methods of solution that exist or may be contemplated. This is particularly true when considering the processes of reconfiguring the controls following the identification of a failure.

It is emphasized that the primary interest of the authors and the focus of the book is airplanes. Thus, we stick to a relatively small number objectives in the allocation problem, corresponding primarily to the three rotational degrees-of-freedom of airplanes and secondarily to the linear degrees-of-freedom. We acknowledge that there are many other fields that have similar problems, and believe our research lays a sound basis for other researchers to modify our results to apply them to their particular interests.

With respect to rigorous mathematical proofs, none will be found here. The authors are not mathematicians, as will be readily confirmed by any real mathematician who picks up this book. We certainly never thought before embarking on this research that 'null space' and 'airplane' would ever be used in the same sentence. We typically began by sketching a two-dimensional figure on the blackboard, something that seemed 'intuitively obvious' to us, and wondering if that figure generalized to higher dimensions.

Most important results have been proved in other sources: the many technical papers, theses, and dissertations that arose from our research, or in textbooks, particularly books that deal with linear algebra. Many of these publications are presented in Appendix C. Here we will just make claims that we are pretty sure are true. For instance, rather than prove that convexity is preserved under the mappings we describe, we will just assert it and perhaps give compelling evidence of its truth.

1.1 Redundant Control Effectors

The origins of our research into airplane control allocation lay in earlier research into model-following and dynamic-inversion control laws. The nature of model-following and

---

Aircraft Control Allocation, First Edition. Wayne Durham, Kenneth A. Bordignon and Roger Beck.
© 2017 John Wiley & Sons, Ltd. Published 2017 by John Wiley & Sons, Ltd.
Companion website: www.wiley.com/go/durham/aircraft_control_allocation

dynamic-inversion algorithms is such that one is required to find a vector of control effector deflections that yield a desired moment, force, or acceleration. With three moments and three controls, the answer for a linear problem is a trivial matrix inversion. The physical limits of the control effectors does not affect the solution since the solution is unique. That is, for an airplane with ganged ailerons, a rudder, and an elevator, the combination of these effectors that will generate a specific moment vector is unique; if one or more saturates then the problem is not in the math but in the hardware.

Early problems arose when considering an airplane whose horizontal tails were not ganged to generate pitching moments only, but that could be displaced differentially as well to generate rolling moments and, unintentionally, yawing moments. By considering the left and right horizontal tails as independent we now have four control effectors for the three components of the moment vector to be generated. The linearized control effectiveness matrix (to be defined in Eq. (2.20)) is no longer square, but has three rows and four columns.

Figure 1.1 depicts a variety of control effector types. The airplane is USAF S/N 71-0290, the F-15 ACTIVE (Advanced Control Technology for Integrated Vehicles). The canards and horizontal tails are all-moving surfaces. The two vertical stabilizers have hinged rudders at their trailing edges. The wings have trailing-edge ailerons and flaps. Finally, both engines have axisymmetric thrust vectoring capabilities. Each of these various control effectors is capable of independent action.

Redundant control effectors are employed to extend an airplane's performance envelope, typically in the low-speed regime. Thrust vectoring generates moments long after conventional

Figure 1.1   F-15 ACTIVE

flapping control surfaces have lost effectiveness at low dynamic pressure. Thrust vectoring enhances the dog-fighting potential of the F-22 Raptor, and permits maneuvers such as Pugachev's Cobra to be performed in other aircraft.

Clever control allocation is not needed in high-speed flight, where more than ample forces and moments can be generated with small effector deflections. In low-speed flight aerodynamic effectors lose effectiveness and must be combined with other effectors (aerodynamic or propulsive). However, if there are more control effectors than moments or accelerations to be generated, methods of allocating these controls are needed.

We are now faced with a 'wide' control effectiveness matrix: more columns than rows. As we will see there is a simple mathematical way to 'invert' such matrices. The real problem arises when the physical limitations of the control effectors are considered. In other words, control effectors have hard deflection limits that cannot be exceeded. When simple mathematical solutions to the problem are used, it is possible for one or more effectors to be *unnecessarily* commanded past its limits, meaning that it will saturate. When a control effector is saturated, the assumptions on which the flight control system was based are no longer valid.

1.2   Overview

We will begin by discussing aircraft flight dynamics and control. This will consist of a very brief overview of flight dynamics and its nomenclature, offered to provide the reader with explanations for some of the terms used subsequently.

Next we will spend some time describing dynamic inversion control. This form of control lends itself naturally to the control allocation problem as we have posed it. We will briefly discuss 'conventional' control, and even more briefly mention model-following control. All three forms of control law determination have some need of control allocation.

After formulating the problem, we address the geometry of control allocation. We do this first considering two-moment problems. Two-moment problems have application in aircraft control, since often the lateral-directional problem (rolling and yawing) is treated separately from the longitudinal problem (pitching). Moreover, it is much easier to make figures on the page of two-dimensional objects than of three-dimensional ones.

The geometry of the three-moment problem is a natural extension of that of the two-moment problem, and it is discussed in detail. For each of the two- and three-moment problems a metric is offered that permits comparison of different control allocation methods for their effectiveness in solving the problem. This gives rise to the idea of a 'maximum set' of moments that can be generated using different control allocation schemes, and its importance is discussed.

A large section on solution methods follows. We explore all the allocation methods of which the authors have first-hand experience, and most are accompanied by numerical illustrations. One of the control allocation methods—linear programming—is briefly discussed. Because of the current interest among researchers in the subject of linear programming solutions, there is a separate section (Appendix A) that further explores linear programming in greater detail.

All the preceding has been based on a global problem: the total set of control deflections that yield the whole moment vector. Now we turn our attention to a local problem. Digital flight control computers solve the allocation problem scores or even hundreds of times a second. We look within one frame of the computer's operation and consider not just how far an effector can move, but how fast. This permits us to incorporate rate limits into the problem. This framewise allocation comes with a serious drawback sometimes called 'windup'. The remedy to windup is not hard and comes with some beneficial side effects.

Next we briefly explore control allocation and flight control system design. Example designs are given for a roll-rate command, pitch-rate command, and a sideslip controller. Finally, the consequences of using a non-optimal control allocation method are graphically displayed.

At the end of the text there is a chapter on some of the real-life applications of the previously described research. Lessons learned from the design of the X-35 control system are presented.

Throughout the book we occasionally make reference to simulation code. The MATLAB®/Simulink® based code is available at a companion website to this book, and it is fully explained in Appendix B. The simulation code offers different modules that implement the various control allocation methods described in the book. Readers are free to adapt and use this code to further explore the concepts of control allocation.

We feel that a common simulation source is essential to creating reproducible results. Many technical papers present simulation results with insufficient information about that simulation for the reader or reviewer to reproduce them. There are many assumptions inherent in one researcher's simulation code that cannot be conveyed in a brief paper but that may greatly affect the results one obtains.[1] The simulation code we provide came to us courtesy of the Swedish Defence Research Agency with their permissions. MATLAB®/Simulink® code is available in student and academic editions and should be very widely accessible.

The final appendix is an annotated bibliography, in which we clean out our files of control allocation and dynamic inversion papers and list them, along with their abstracts or other descriptive material. This appendix is a good place to look to see if anyone is pursuing interests related to yours. It is inevitable that some have been overlooked, either through our inattention or the fault of some search engine or other. Our loosely-enforced cut-off criterion for inclusion in this list was that the source be refereed. Conference papers were generally not included unless the material was unique and relevant.

Finally, we wish to emphasize that the content of this book reflects only topics with which the authors are personally familiar. It is not a survey of the control allocation literature. There are many very good and sound areas of control allocation research that we have not addressed, except as indicated in the annotated bibliography (Appendix C). There one will find works by Marc Bodson, Jim Buffington, Dave Doman, Dale Enns, Tony Page, and many others with whom the authors have had collegial exchanges.

In particular we appreciate a group familiarly known as 'Bull studs': John Bolling, Josh Durham, Michelle Glaze, Bob Grogan, Matt Hederstrom, Jeff Leedy, Bruce Munro, Mark Nelson, Tony Page, Kevin Scalera, and the others who toiled in the 'Sim Lab' to help make sense of all this. And lastly Fred Lutze, who pondered the various stages of our progress and occasionally said, 'You *could* do that, but it would be wrong.'

---

[1] For example, whether one begins the calculations in a given frame assuming that the last commanded controls have been achieved, or using the actual deflections that resulted. It can make a big difference. Most of what this statement means will be made clear in Chapter 7. See Bodson and Pohlchuck (1998) for some more insight.

References

Aerodata Model in Research Environment (ADMIRE), Ver. 3.4h, Swedish Defence Research Agency (FOI), Stockholm, Sweden, 2003.

Bodson, M and Pohlchuck, E 1998 'Command Limiting in Reconfigurable Flight Control' *AIAA J. Guidance, Control, and Dynamics*, **21**(4), 639–646.

2

Aircraft Control

This chapter provides a brief overview of the subject of flight dynamics and control. Readers who are knowledgeable in the field will probably need only to resolve any differences between their notation and usage and ours (see the glossary for most of these). Our practices are based largely on those of Etkin (1972) and Etkin and Reid (1995). In particular we differ from many other authors in our definitions of stability axes (body-*fixed*), and of stability and control derivatives (no mass or moments of inertia).

2.1 Flight Dynamics

The usual derivation of the equations of motion of an airplane assume that it is a rigid body, and that an earth-fixed reference frame is inertial (the flat-earth assumption). The flat-earth assumption is justified in the present treatment because all the aircraft motions that we will consider will be limited to maneuvers at low speeds and of short duration, wherein the effects of centripetal, coriolis, and tangential accelerations due to the earth's rotation are negligible.

2.1.1 Equations of Motion

2.1.1.1 Kinematic Equations

The kinematic equations relate the angular accelerations of the airplane's body-fixed axes to the rate of change of its orientation with respect to the local horizontal reference frame. This representation may be done using Euler angles, Euler parameters, direction cosines, or any suitable characterization. Here we use Euler angles. The derivation is straightforward, and results in Eqs. (2.1).

$$\begin{Bmatrix} \dot{\phi} \\ \dot{\theta} \\ \dot{\psi} \end{Bmatrix} = \begin{bmatrix} 1 & \sin\phi\tan\theta & \cos\phi\tan\theta \\ 0 & \cos\phi & -\sin\phi \\ 0 & \sin\phi\sec\theta & \cos\phi\sec\theta \end{bmatrix} \begin{Bmatrix} p \\ q \\ r \end{Bmatrix} \qquad (2.1)$$

Aircraft Control Allocation, First Edition. Wayne Durham, Kenneth A. Bordignon and Roger Beck.
© 2017 John Wiley & Sons, Ltd. Published 2017 by John Wiley & Sons, Ltd.
Companion website: www.wiley.com/go/durham/aircraft_control_allocation

The angles ψ, θ, and ϕ are the names of the Euler angles that are defined in a 3-2-1 ($z - y - x$, respectively) rotation sequence from an earth-fixed reference frame to a body-fixed reference frame. They are called the heading angle ψ, pitch angle θ, and bank angle ϕ. Note that with the flat-earth assumption these angles are the same as in transforming from a local horizontal system. Also note that Eqs. (2.1) are invalid if $\theta = \pm\pi/2$ (pointed straight up or straight down).

The angular rates p, q, and r are the components of the inertial rotation rate of the body-axes, as represented in body-fixed coordinates. They are called the roll rate p, pitch rate q, and yaw rate r. The vector of inertial rotation rates is denoted $\boldsymbol{\omega}_B$, and that vector has components in the body-fixed coordinate system p, q, and r.

From a flight dynamics perspective, Eqs. (2.1) are needed to keep track of the orientation of the gravity vector with respect to the airplane.

2.1.1.2 Body-axis Force Equations

The body-axis force equations are the rates of change of the inertial components of velocity, as seen in the body-axes:

$$\{\dot{\mathbf{v}}_B\}_B = \frac{1}{m}\{\mathbf{F}\}_B - \{\Omega_B\}_B\{\mathbf{v}\}_B \tag{2.2}$$

In this expression $\mathbf{v}$ is the inertial velocity of the airplane's center of mass. The subscript B in $\{\dot{\mathbf{v}}_B\}$ means that the derivative with respect to time of $\mathbf{v}$ is taken with respect to the body-fixed coordinate system. Everything within the curly braces defines a vector, and the other subscript B outside the braces means this vector is represented, or has components in, the body-axis system.

$$\{\mathbf{v}\}_B \equiv \begin{Bmatrix} u \\ v \\ w \end{Bmatrix}, \{\dot{\mathbf{v}}_B\}_B = \begin{Bmatrix} \dot{u} \\ \dot{v} \\ \dot{w} \end{Bmatrix} \tag{2.3}$$

The parameter m is the mass of the aircraft. The vector $\{\mathbf{F}\}_B$ consists of forces $\{\mathbf{F}_A\}$ (aerodynamic), $\{\mathbf{T}\}$ (thrust), and $\{\mathbf{W}\}$ (the weight of the airplane) acting on the airplane.

$$\{\mathbf{F}\}_B = \{\mathbf{F}_A\}_B + \{\mathbf{W}\}_B + \{\mathbf{T}\}_B = \begin{Bmatrix} X \\ Y \\ Z \end{Bmatrix} + \begin{Bmatrix} -mg\,\sin\theta \\ mg\,\sin\phi\,\cos\theta \\ mg\,\cos\phi\,\cos\theta \end{Bmatrix} + \begin{Bmatrix} T_x \\ T_y \\ T_z \end{Bmatrix} \tag{2.4}$$

The components of $\{\mathbf{F}_A\}_B$ are named simply X, Y, and Z. The thrust vector $\{\mathbf{T}\}_B$ has been assumed to have components T_x, T_y, and T_z, to accommodate controlled thrust vectoring. The components of $\{\mathbf{W}\}_B$ arise from a simple transformation of the local-horizontal z-axis into body-fixed coordinates.

The body-axis forces are related to lift (L), drag (D), and side force (C) through a transformation

$$\begin{Bmatrix} X \\ Y \\ Z \end{Bmatrix} = \begin{bmatrix} \cos\alpha\cos\beta & -\cos\alpha\sin\beta & -\sin\alpha \\ \sin\beta & \cos\beta & 0 \\ \sin\alpha\cos\beta & -\sin\alpha\sin\beta & \cos\alpha \end{bmatrix} \begin{Bmatrix} -D \\ -C \\ -L \end{Bmatrix} \tag{2.5}$$

The matrix $\{\Omega_B\}_B$ replaces the cross-product operation $\{\omega_B\}_B\times$,

$$\{\Omega_B\}_B = \begin{bmatrix} 0 & -r & q \\ r & 0 & -p \\ -q & p & 0 \end{bmatrix} \tag{2.6}$$

As a result we may write the body-axis force equations,

$$\dot{u} = \frac{1}{m}(X + T_x) - g\sin\theta + rv - qw \tag{2.7a}$$

$$\dot{v} = \frac{1}{m}(Y + T_y) + g\sin\phi\cos\theta + pw - ru \tag{2.7b}$$

$$\dot{w} = \frac{1}{m}(Z + T_z) + g\cos\phi\cos\theta + qu - pv \tag{2.7c}$$

2.1.1.3 Body-axis Moment Equations

The body-axis moment equations are the equations of primary interest in control allocation. In terms of the rates of change of the inertial components of angular rotation, as seen in the body axes, these are

$$\{\dot{\omega}_B\}_B = I_B^{-1}[\{\mathbf{m}\}_B - \{\Omega_B\}_B I_B \{\omega_B\}_B] \tag{2.8}$$

The meaning of $\{\dot{\omega}_B\}_B$ is analogous to that for $\{\dot{\mathbf{v}}_B\}_B$, above. The derivative is taken with respect to the body-fixed reference frame, and the components of the resulting vector are represented in the body-fixed frame. I_B is the inertia matrix of the airplane.

The externally applied moments are those due to aerodynamics and thrust (gravity acts through the center of mass, the origin of our coordinate system, so there are no moments due to weight). On aircraft with controlled thrust vectoring, moments about all three axes may be produced. As a result the moments are

$$\{\mathbf{m}\}_B = \begin{Bmatrix} L + L_T \\ M + M_T \\ N + N_T \end{Bmatrix} \tag{2.9}$$

We assume a plane of symmetry, so in the inertia matrix the cross-products involving y become zero,

$$I_B = \begin{bmatrix} I_{xx} & 0 & -I_{xz} \\ 0 & I_{yy} & 0 \\ -I_{xz} & 0 & I_{zz} \end{bmatrix} \tag{2.10}$$

The inverse is then given by

$$I_B^{-1} = \frac{1}{I_D}\begin{bmatrix} I_{zz} & 0 & I_{xz} \\ 0 & I_D/I_{yy} & 0 \\ I_{xz} & 0 & I_{xx} \end{bmatrix} \tag{2.11}$$

Here, $I_D = I_{xx}I_{zz} - I_{xz}^2$. The resulting moment equations are given by Eqs. (2.12)

$$\dot{p} = \frac{I_{zz}}{I_D}[L + L_T + I_{xz}pq - (I_{zz} - I_{yy})qr] + \frac{I_{xz}}{I_D}[N + N_T - I_{xz}qr - (I_{yy} - I_{xx})pq] \qquad (2.12a)$$

$$\dot{q} = \frac{1}{I_{yy}}[M + M_T - (I_{xx} - I_{zz})pr - I_{xz}(p^2 - r^2)] \qquad (2.12b)$$

$$\dot{r} = \frac{I_{xz}}{I_D}[L + L_T + I_{xz}pq - (I_{zz} - I_{yy})qr] + \frac{I_{xx}}{I_D}[N + N_T - I_{xz}qr - (I_{yy} - I_{xx})pq] \qquad (2.12c)$$

If the problem is posed in principal axes, then Eq. (2.12) is greatly simplified:

$$\dot{p} = [L + L_T - (I_{zp} - I_{yp})qr]/I_{xp} \qquad (2.13a)$$

$$\dot{q} = [M + M_T - (I_{xp} - I_{zp})pr]/I_{yp} \qquad (2.13b)$$

$$\dot{r} = [N + N_T - (I_{yp} - I_{xp})pq]/I_{zp} \qquad (2.13c)$$

2.1.1.4 Body-axis Navigation Equations

The position of the aircraft relative to the Earth may be found by representing the velocity in Earth-fixed coordinates and integrating each component.

$$\begin{Bmatrix} \dot{x}_E \\ \dot{y}_E \\ \dot{z}_E \end{Bmatrix} = T_{H,B} \begin{Bmatrix} u \\ v \\ w \end{Bmatrix} \qquad (2.14)$$

We usually know the transformation from local horizontal to body, $T_{B,H}$, in terms of the roll angle ϕ, the pitch angle θ, and the heading angle ψ.

$$T_{B,H} = \begin{bmatrix} \cos\theta\cos\psi & \cos\theta\sin\psi & -\sin\theta \\ \begin{pmatrix} \sin\phi\sin\theta\cos\psi \\ -\cos\phi\sin\psi \end{pmatrix} & \begin{pmatrix} \sin\phi\sin\theta\sin\psi \\ +\cos\phi\cos\psi \end{pmatrix} & \sin\phi\cos\theta \\ \begin{pmatrix} \cos\phi\sin\theta\cos\psi \\ +\sin\phi\sin\psi \end{pmatrix} & \begin{pmatrix} \cos\phi\sin\theta\sin\psi \\ -\sin\phi\cos\psi \end{pmatrix} & \cos\phi\cos\theta \end{bmatrix} \qquad (2.15)$$

From the properties of direction cosine matrices we have $T_{B,H}^T = T_{B,H}^{-1} = T_{H,B}$. In other words, the transpose of $T_{B,H}$ is the transformation from body to local horizontal. Expanding the equations,

$$\dot{x}_E = u(\cos\theta\cos\psi) + v(\sin\phi\sin\theta\cos\psi - \cos\phi\sin\psi)$$
$$+ w(\cos\phi\sin\theta\cos\psi + \sin\phi\sin\psi) \qquad (2.16a)$$

$$\dot{y}_E = u(\cos\theta\sin\psi) + v(\sin\phi\sin\theta\sin\psi + \cos\phi\cos\psi)$$
$$+ w(\cos\phi\sin\theta\sin\psi - \sin\phi\cos\psi) \qquad (2.16b)$$

$$\dot{h} = -\dot{z}_E = u\sin\theta - v\sin\phi\cos\theta - w\cos\phi\cos\theta \qquad (2.16c)$$

2.1.1.5 Wind-axis Relationships

A complete set of differential equations may be derived using the wind axes, but the result is more complicated than that for body axes. However, if none of the forces and moments is functionally dependent on $\dot{\alpha}$ or $\dot{\beta}$ (a mathematical contrivance at best) then a somewhat simpler approach is possible. We may relate $\dot{V}$, $\dot{\alpha}$, and $\dot{\beta}$ directly to the body-axis force equations, based on the definitions of the wind-axis quantities V, α, and β.

$$\alpha \equiv \tan^{-1}\left(\frac{w}{u}\right) \Rightarrow \dot{\alpha} = \frac{u\dot{w} - w\dot{u}}{u^2 + w^2} \tag{2.17a}$$

$$V \equiv \sqrt{u^2 + v^2 + w^2} \Rightarrow \dot{V} = \frac{u\dot{u} + v\dot{v} + w\dot{w}}{\sqrt{u^2 + v^2 + w^2}} = \frac{u\dot{u} + v\dot{v} + w\dot{w}}{V} \tag{2.17b}$$

The $\dot{\beta}$ expression is best formulated in terms of V and $\dot{V}$

$$\beta \equiv \sin^{-1}\left(\frac{v}{V}\right) \Rightarrow \dot{\beta} = \frac{V\dot{v} - v\dot{V}}{V\sqrt{u^2 + w^2}} \tag{2.17c}$$

2.1.1.6 States of the Airplane

Each term in the equations of motion that appears as a time-derivative is a *state* of the airplane. Thus there are twelve states: three positional (x_E, y_E, and z_E), three angular (ϕ, θ, and ψ), three velocity (u, v, and w), and three angular velocity (p, q, and r). The wind-axis variables V, α, and β may be used in lieu of the body-axis velocities u, v, and w.

In the equations of motion there is a nonlinear ordinary differential equation for every state, expressed in terms of the states themselves, of forces and moments, and of many other variables not related to the control deflections.

2.1.2 Linearized Equations of Motion

The nonlinear equations of motion presented in Section 2.1.1 generally do not have analytical solutions. In the forms given they may be used in flight simulation using numerical integration.

Normally in a course in flight dynamics it is at this point that the nonlinear equations are linearized and placed in a form amenable to analytical solution. In this book we will not be concerned with such solutions, but we will have need of the linearized effects of the control effectors, so we will briefly describe the process of linearization.

Associated with the linearization process, there is usually a reference flight condition, and it is normally a steady condition. Linearization is accomplished by retaining just the first-order terms in a Taylor series expansion (or Maclaurin series with suitable definitions). The process can become somewhat convoluted, since there are explicit dependencies as well as assumed dependencies; that is, the dependencies of the forces and moments on the states and controls.

In this book we will have no need of the complete linearized equations of motion, but we will extensively use the linearized control effectiveness with respect to the three moments, or values related to the three moments. The quantities of primary interest are moments, moment coefficients, or angular accelerations in the kinematic equations. Unless we are talking about one or the other of these values specifically we will use a generic symbol m with a subscript to indicate the axis. Likewise, in a general discussion we will refer the effects simply as *moments*.

Moment coefficients are obtained from the moments by non-dimensionalization. This consists of division by the dynamic pressure, the wing area, and a characteristic length. The wing area is denoted by S. The dynamic pressure, $\bar{q}$, is a function of air density ρ and airspeed V,

$$\bar{q} = \frac{1}{2}\rho V^2 \tag{2.18}$$

The characteristic length is a suitably defined chord length $\bar{c}$ for the pitching moment, and the wing span b for the rolling and yawing moments. The non-dimensional coefficients are then

$$C_\ell = \frac{L}{\bar{q}Sb}$$

$$C_m = \frac{M}{\bar{q}S\bar{c}}$$

$$C_n = \frac{N}{\bar{q}Sb} \tag{2.19}$$

In Eqs. 2.12 and 2.13 the one-to-one correspondence between the angular accelerations and the moments is made clear. The moments and related effects are summarized in Table 2.1.

The control effectors are generically denoted u_i, $i = 1 \cdots m$ and it is assumed $m > 3$ (or $m > 2$ for the two-moment problem). As a vector they are $\mathbf{u}$. In a practical application these effectors may have specific names, such as the left horizontal tail or right aileron. In this case they will be denoted with an appropriately subscripted δ, such as δ_a for aileron. Then the entries in Eq. (2.20) will have the more familiar appearance of $L_{\delta_{LHT}}$ (dimensional control derivative) or $C_{L_{\delta_{LHT}}}$ (non-dimensional control derivative).

The *control effectiveness matrix,* using generic moments and generic controls, is given by the Jacobian:

$$B = \begin{bmatrix} \dfrac{\partial m_1}{\partial u_1} & \dfrac{\partial m_1}{\partial u_2} & \cdots & \dfrac{\partial m_1}{\partial u_m} \\[2mm] \dfrac{\partial m_2}{\partial u_1} & \dfrac{\partial m_2}{\partial u_2} & \cdots & \dfrac{\partial m_2}{\partial u_m} \\[2mm] \dfrac{\partial m_3}{\partial u_1} & \dfrac{\partial m_3}{\partial u_2} & \cdots & \dfrac{\partial m_3}{\partial u_m} \end{bmatrix} \tag{2.20}$$

From time to time we will have occasion to consider a smaller problem involving just roll and yaw. In that case all of the previous discussion should be taken using only the first and third rows of Eq. (2.20).

In general there will be no built-in ganging of any control effector pairs, in particular left and right. The horizontal tails and ailerons will constitute two control effectors each, with each effector capable of independent movement.

Table 2.1 Moments and related effects of control deflections

Axis	Moment	Moment coefficient	Angular acceleration	Generic
x_B	L	C_ℓ	$\dot{p}$	m_1
y_B	M	C_m	$\dot{q}$	m_2
z_B	N	C_n	$\dot{r}$	m_3

2.2   Control

2.2.1   General

Airplanes are maneuvered primarily through control of the angular accelerations, through control of the moments. That is, except for the use of thrust and a few special cases (direct lift control for example) there is usually no direct control of forces. The angular accelerations are used to orient the airplane with respect to the air mass, and with respect to an external object, such as another airplane or a runway. For example there is no control that exerts a horizontal force to turn an airplane; rather, roll control is used to attain a bank angle that creates a horizontal component of the lift vector, while pitch control is used to reorient the airplane to a higher angle of attack so that the vertical component balances the weight.

The presentation in this section is brief; for a full treatment of flight control systems see, for example, Pratt (2000).

2.2.1.1   Longitudinal Control

With respect to a given set of body-fixed axes, longitudinal pitching motion is about the y-axis. The pitch rate is designated by q. The pitch acceleration $\dot{q}$ is generated by pitching moments M. The pitching moment is created by aerodynamic forces, largely consisting of those that resist the rotary motion (pitch-rate damping, related to C_{m_q}), and by the flight control effectors that overcome them. Any control effector that has a non-negligible effect on pitching moment may contribute to the overall pitching moment. They are primarily things such as horizontal tails, canards, and maneuvering flaps.

Maximum-effort pitching maneuvers are relatively rare. That is, one seldom finds the longitudinal stick against the stops. Longitudinal control primarily is used to generate a particular angle of attack to regulate the turn rate. However, there are limits on load factor that must be observed at high speed to avoid structural damage; and at low speed, flight at high angle of attack bleeds energy rapidly due to induced drag. So far as nose-down longitudinal control is concerned, the only common occurrence of full-forward stick outside of air shows is during spin recovery.

2.2.1.2   Lateral Control

Lateral motion is about the x-axis, a rolling motion. The roll rate is designated by p. The roll acceleration $\dot{p}$ is generated by rolling moments L. The rolling moment is created by aerodynamic forces, largely consisting of those that resist the rotary motion of the wings (roll-rate damping, C_{ℓ_p}), and by the flight control effectors that overcome them. Any control effector that generates a rolling moment may be used, such as the left horizontal tail independent of the right.

During rapid, aggressive maneuvering, maximum lateral control effort is used frequently over a very wide range of airspeeds. Unlike earth-bound maneuvering in an automobile, there is seldom a penalty to be paid for application of maximum effort (as would be true of full deflection of the steering wheel of a car at high speed).

2.2.1.3 Directional Control

Directional motion is about the z-axis, a yawing motion. The yaw rate is designated by r. The yaw acceleration $\dot{r}$ is generated by yawing moments N. The yawing moment is created by aerodynamic forces, largely consisting of those that result from sideslip (C_{n_β}), and by the flight control effectors that overcome them. Any control effector that generates a yawing moment may be used, such as a rudder, but also almost every roll control effector as well.

In some aircraft with high engine torques, or torques generated by the airflow about the fuselage by propellers, yawing moments may be required to maintain directional control, especially during takeoff.

In multi-engine airplanes in which asymmetric thrust is obtainable (in particular, with engine failure) the aerodynamic yawing moment resulting from application of the rudder pedals may be used to balance the yawing moment created by the asymmetric thrust.

Occasionally sideslip may be used in a cross-wind landing, or as a means to generate higher drag during the approach to landing.

In aircraft used to perform air-show maneuvers, the pilot may intentionally generate large yawing moments to effect certain maneuvers, such as hammer-head stalls and the impressive Lomcevak (or Lomcovak).

Finally, in high-performance tactical aircraft being flown by aggressive pilots, large yawing moments may be used to perform defensive maneuvers to spoil an enemy aircraft's tracking solution. See Shaw (1985) for examples and descriptions.

Aside from these special applications, yawing moments are typically employed in combination with rolling moments to regulate sideslip. Slow-flying light aircraft, and all aircraft operated at high angle of attack, generally need coordination of rolling and yawing moments to keep sideslip low. Many aircraft are equipped with an aileron–rudder interconnect (ARI) or rolling-surface–rudder interconnect (RSRI) that serves this purpose.

2.2.1.4 Multi-axis Control

In general every maneuver performed in an airplane involves control of all three axes. As noted above, to simply turn the airplane, lateral control is used to tilt the lift vector, longitudinal control to increase the lift, and directional control to eliminate any sideslip caused by the roll. Only purely longitudinal maneuvers, such as a loop, are single-axis, and even they may require lateral and directional control if engine torque effects are considered.

One maneuver that may challenge a flight control system is the velocity vector roll (Durham *et al.* 1994). At a high angle-of-attack, a roll about the x_B-axis will generate large amounts of sideslip (simplistically, after 90° of roll all the α will become β). To avoid this a large amount of carefully coordinated directional control is required. The desired result is to keep the orientation of the velocity vector with respect to the airplane fixed throughout the roll, or constant angle of attack and zero sideslip, so that the airplane rolls about the velocity vector.

2.2.2 Aircraft Control Effectors

2.2.2.1 General

Control effectors are, for our purposes, any external devices that may be used to vary the forces with the primary purpose of changing the moments acting on the airplane. Control effector deflections are generally denoted by the Greek symbol δ with a subscript that specifies the effector in question. Thus one could see symbols such as δ_e for an elevator deflection, δ_r for the rudder. Further specification in the subscript is sometimes needed. If the left and right ailerons are not ganged (as they are conventionally) but are free to move independently, then δ_a as the ganged system would be replaced by δ_{aL} and δ_{aR} for left and right.

When we are dealing with a suite of control effectors mathematically, they will be denoted $u_i, i = 1 \cdots m$, where m is the number of effectors. These will be grouped together to form a column vector $\mathbf{u}$.

$$\mathbf{u} = \begin{Bmatrix} u_1 \\ u_2 \\ \vdots \\ u_m \end{Bmatrix} \tag{2.21}$$

2.2.2.2 Aerodynamic

Aerodynamic flight controls are generally devices that can change the camber of a larger aerodynamic surface, or vary the incidence of an aerodynamic surface. The purpose of the aerodynamic change is to create a change in force (through the local lift) on some surface. The force itself is not of primary interest to us here, but that the force is at a distance from the airplane center-of-gravity and thus produces a moment about that point.

The camber of an aerodynamic surface may be changed by a smaller flapping surface on either the leading or trailing edge of the larger surface. On the trailing edge are found the classical elevator on the horizontal tail, rudder on the vertical fin, and ailerons on the wing. Leading and trailing edge flaps are not usually thought of as primary flight controls, but they may be so used, especially if operated differentially, but also as pitch controls when operated symmetrically.

Control effectors that function by changing the incidence of an entire surface include all-moving horizontal tails, canards, and vertical fins. All-moving horizontal tails and canards may be operated symmetrically for pitch control, but also differentially for roll and yaw control.

Common devices that do not fit exactly into these categories are spoilers: devices that pop up from the surface of the wing to spoil the lift and create a change in force. The use of spoilers was important in early supersonic airplanes because the conventional ailerons lost effectiveness in the wake of the shock waves on the wing.

Sign conventions for moving-surface effectors vary in the literature, and special care must be taken to understand what is used in a particular application. In this text, the sign convention will be that that results from a right-hand rule, with the thumb along one of the three body-fixed axes. Thus, with the thumb in the direction of the positive x-axis, flapping surfaces on the right side of the airplane (say, the ailerons) are positive trailing-edge down (TED)[1] and vice versa on the left. With the thumb in the direction of the positive y-axis, flapping surfaces on the rear of the airplane (say, the elevators) are positive TED and vice versa on the forward part (say, the canards). With the thumb in the direction of the positive z-axis, flapping surfaces on the rear of the airplane (say, the rudder) are positive trailing-edge left (TEL).

There is a distinction between *primary* and *secondary* flight controls. Primary flight controls are used in maneuvering the airplane, while secondary flight controls are used to modify an

---

[1] More abbreviations for describing the deflection of moving surfaces are given in the glossary.

aerodynamic characteristic. Examples of the latter are speed brakes to increase drag and flaps to lower the landing speed. Leading- and trailing-edge flaps may also be used for maneuvering flight, and spoilers may be used to increase drag, so the distinction is not absolute.

2.2.2.3 Propulsive

Propulsive flight controls are generally thought of as those created by engine thrust vectoring, and these are normally effected by varying the geometry of exhaust nozzles. Some applications of thrust vectoring, such as vertical or short takeoff and landing, may also be found, but our concern in this text is with the use of vectoring to generate moments in maneuvering flight. Thus, any application of thrust that may be varied to modify the moments acting on the airplane, including reaction jets, is included.

Thrust vectoring may be designed to operate in two dimensions, normally pitch, and frequently accomplished by using paddles in the engine exhaust plume to deflect it away from axial. Three-dimensional systems can deflect the thrust at any angle, creating pitch and yaw. With two-engined aircraft the individual engines' vectoring controls may be employed differentially to create rolling moments.

There are many examples of research aircraft that employed thrust vectoring, such as the McDonnell Douglas F-15 ACTIVE (Figure 2.1), General Dynamics F-16 VISTA,

Figure 2.1   Variable nozzle thrust vectoring on the F-15 ACTIVE (*Source*: NASA)

Figure 2.2   Paddle thrust vectoring on the X-31 (*Source*: NASA)

Rockwell-MBB X-31 (Figure 2.2), and McDonnell Douglas F-18 HARV.[2] The Lockheed Martin F-22 Raptor and several Sukhoi aircraft are examples of operational aircraft employing thrust vectoring.

2.2.2.4   Zero Deflection

The position of zero deflection of the control effectors is completely arbitrary. The airplane does not care what you define as zero. On airplanes with all-moving horizontal tails there is often painted on the side of the fuselage, corresponding to the leading edge of the horizontal tail, a datum marked '0'. In an arc about this mark are others, corresponding to degrees of deflection from that datum. The zero-deflection mark was determined during the design phase, and is used when rigging the control system, much like the timing marks on an engine flywheel. There is nothing special about this zero except that everything mechanical about the horizontal tail actuation system is referenced to it.

Other control effectors may have some 'natural' zero deflection. A trailing-edge flapping device may have zero defined as that position which continues the natural camber of the surface it is attached to. Spoilers have an intuitively natural zero deflection, which is being flush with the surface of the wing.

On the other hand, zero deflection may have a more significant definition, such as a trim condition, or a condition of minimum control-induced drag, or the current position at the beginning of a frame of computations in the flight control computer.

The sense of zero deflection implied throughout this book is that of some reference flight condition, most often a trim condition. We will have occasion to apply a Taylor series

---

[2] ACTIVE: Advanced Control Technology for Integrated Vehicles; VISTA: Variable stability In-flight Simulator Test Aircraft; HARV: High Alpha (angle of attack) Research Vehicle.

expansion through the first-order terms in order to linearize the control effectiveness. Implicit in the definition of the series is a reference condition, to be applied to the evaluation of the partial derivatives in the series.

2.2.3 Aircraft Control Inceptors

2.2.3.1 Traditional Center Stick

Traditional control inceptors consist of rudder pedals and a center stick. The center stick moves fore and aft for longitudinal control; lateral control is accomplished through either lateral movement of the stick, or through a wheel-like yoke. In the simplest case the motions of these inceptors are linked directly to external control surfaces, often by cables, pulleys, and bell cranks.

Again speaking of the simplest case, there is generally one external control surface, or ganged set of control surfaces acting as one surface, associated with each of the motions of the inceptors. Aerodynamic loads on the surfaces are transmitted through the mechanical linkage to the control inceptor in what is called a *reversible* control system. The length of the control stick provides mechanical advantage to counter the air loads on the surfaces. Some reversible control systems may be boosted, almost always hydraulically, to relieve some of the aerodynamic forces transmitted to the control inceptor. (Early A-4 Skyhawk airplanes had a control stick that could be extended to help in the case of boost failure.)

The relationship of inceptor force and displacement required to produce a given airplane response is of great importance in determining the ease with which the pilot can perform flying tasks: the airplane's flying qualities. For example, the longitudinal inceptor force required to generate a given load factor (stick-force per g) has been the focus of extensive research into the longitudinal flying qualities of an airplane. Control of this particular characteristic may be accomplished by modification of the longitudinal stick gearing, or in some cases by modification of the forces through bobweights attached to the stick.

More complex aircraft have fully boosted control surfaces, called *irreversible* control systems. A basic irreversible control system still has a one-to-one correspondence between inceptor motion and that of the corresponding control surface. None of the aerodynamic loads present at the control surfaces is transmitted to the control inceptor, only to the structures to which the actuators (typically hydraulic rams) are mounted. In such cases artificial feel must be provided to the inceptors. This can be as simple as springs and dampers, or more complicated with active elements (such as variable-torque electric motors) that give the inceptors the desired dynamic response.

Irreversible flight control systems may have a control- or stability-augmentation system. A thorough treatment of stability- and control-augmentation systems is found in Stevens and Lewis (2015) and many other excellent books. Pilots flying unaugmented control systems are generating forces and moments that generate the accelerations that shape the response of the airplane. Augmentation systems operate on top of the pilot's inputs, changing the deflections of the surfaces to modify those responses to improve the flying qualities of the airplane. The augmentation system senses some parameter, for example pitch rate or angle of attack, and changes the appropriate surface deflection to achieve the desired response.

The next stage in the evolution of the relationship of control inceptors to the control surfaces is fly-by-wire. The control inceptors are not physically connected to the surfaces, but rather the forces and movements of the inceptors are provided as input to a flight control computer. This computer translates the inputs into a desired response, and drives the control surfaces in a manner designed to achieve that response. The response is often tailored to satisfy a flying-qualities goal.

2.2.3.2   Side-stick

With the advent of fly-by-wire systems, the need for a large center stick was removed. The stick was connected only to a computer, and a smaller, lighter controller like those used for video games was a possibility. Abandoning the center stick took a great amount of confidence in the flight control computer and surface actuation systems, since many irreversible systems had fallback modes in the event of a failure that required a center stick.

Side-stick controllers are becoming more common. They are present in modern tactical airplanes, such as the F-35, and even commercial passenger airplanes, such as the Airbus. Interestingly there is a good amount of variation in the mechanization and artificial feel of the different implementations of side-stick controllers. The F-16 flight control system is cued by side-stick forces primarily. Interestingly, pilots seem to have little trouble adapting to flying with a left-handed side-stick.

2.2.3.3   Rudder Pedals

The rudder pedals are rigged such that application of the left pedal forward, right pedal aft, moves the rudder TEL, which typically yields a nose-left yawing moment, and vice versa for the right pedal. Application of rudder also generates a rolling moment. In many tactical airplanes that lack sophisticated flight control systems, rudder is the primary roll control at high angles of attack, where application of aileron or spoiler could cause a departure from controlled flight.

Aside from up-and-away flight, the rudder pedals have other applications. In ground handling the rudder pedals may be linked to a steerable nose wheel (or tail wheel in at least one instance) to enable turning while taxiing. In hovering flight, as in the X-35, the rudder pedals may be employed to command a yaw rate about the earth-fixed vertical axis, with neutral pedals stopping the yaw in the current orientation.

2.3   Afterword

This chapter has been a very compressed treatment of material that usually requires at least a semester in university to present coherently. There are numerous textbooks that cover the subject matter thoroughly, as a trip to the library of any engineering school will reveal. Everyone has his favorites, the ones that simply fall open to the page sought after years of hard usage. Our favorites for flight dynamics are Etkin (1972), Etkin and Reid (1995), and of course Durham (2013).

Stevens and Lewis (2015) has more to do with automatic flight control than flight dynamics, but is good for both subjects. Stevens and Lewis also provide a simulation code (in FORTRAN) that we found useful for lashing up desktop simulations to test some control-allocation code or other. Their simulation code was also implemented in our manned flight simulator (see Scalera and Durham (1998)) and found to be quite realistic. Their simplex optimization code found its way into our control allocation code, as will be seen in Section 6.6.

A discussion of flight controls, more accurately control effectors, is available in many books, in particular books on general aviation aimed at starting pilots. Pratt (2000), or his contributors, have a much more in-depth treatment, not only of the actuation systems themselves but of flight dynamics, flying qualities, and implementation issues.

One of the chief ideas to be carried forward from this chapter is that of the equations of motion, in Section 2.1.1, and in particular the body-axis moment equations in Section 2.1.1.3. These will feature prominently in the subsequent discussion of dynamic-inversion control laws in Section 3.2, which have as their output the input to the control allocation problem: the desired body-axis moments.

Also of subsequent importance is the control-effectiveness matrix, called B. This is the matrix that relates the moments to the control effectors themselves, defined in Eq. (2.20). The fact that an airplane may have more control effectors than is needed to generate three moments—that the B matrix is wider than it is tall—is what makes control allocation a non-trivial problem.

The rest of the material in this chapter provided the background and context for the posing of the control allocation problem, which we will do in Chapter 4. Before getting to that, however, we will discuss the control laws that were the primary motivation for the authors to learn about control allocation.

References

Durham, W 2013 *Aircraft Flight Dynamics and Control* 1st edn. John Wiley & Sons.

Durham, WC, Lutze, FH, and Mason, W 1994 'Kinematics and aerodynamics of the velocity-vector roll,' *AIAA J. Guidance, Control, and Dynamics*, **17**(6), 1228–1233.

Etkin, B 1972 *Dynamics of Atmospheric Flight*, 1st edn. John Wiley & Sons; Republished 2005 by Dover Publications.

Etkin, B and Reid, LD 1995 *Dynamics of Flight: Stability and Control*, 3rd edn. John Wiley & Sons.

Pratt, RW (ed.) 2000 *Flight Control Systems: Practical Issues in Design and Implementation* Institution of Engineering and Technology Control Engineering Series 57.

Scalera, KR and Durham, W 1998 'Modification of a Surplus Navy 2F122A A-6E OFT for Flight Dynamics Research and Instruction,' AIAA-98-4180 in *AIAA Modeling And Simulation Technologies Conference and Exhibit*.

Shaw, RL 1985 *Fighter Combat: Tactics and Maneuvering*. Naval Institute Press.

Stevens, BL and Lewis, FL 2015 *Aircraft Control and Simulation,* 3rd edn. Wiley-Blackwell.

3

Control Laws

In discussing control allocation it is convenient to separate the allocation from the control law. By control law we mean that which translates the pilot's control inceptor inputs to desired moments, or accelerations, of the airplane.

At one extreme the control 'law' could be trivial: simple cables and pulleys connected to the control stick and rudder pedal at one end, to the control surfaces at the other end. There is no explicit demand of acceleration, but the pilot expects an acceleration (arguably the rate that follows that acceleration, in the case of lateral control) to result from his control application. If there are more control effectors than cockpit inputs, then a somewhat more complicated system that mechanically interconnects the effectors (ganging) could be employed.

In many cases some form of automatic flight control is used. This can range from systems that modify the airplane's dynamic response (stability augmentation), to control-augmentation systems that translate pilot inputs into specific responses, such as making a given longitudinal stick input command a fixed pitch rate, independent of the airplane's state. At the other extreme one may have a full-authority fly-by-wire system in which the pilot's inputs go directly to a computer that interprets the inputs and drives the control effectors and forms the airplane's response.

In each form of flight control there is a point at which a desired control effector force or moment is present or may be inferred. It is at this point that control allocation begins. Some control law designs have such desired effects as their immediate outputs, and are clearly distinguished from the function of control allocation. Other control laws are designed around the control effectiveness, making the distinction between control law and control allocation less obvious.

3.1 Flying Qualities

The design of flight control systems for airplanes is not so much concerned with stability per se, but rather with ensuring that a human pilot can perform the airplane's mission with minimal compensation on his part. The amount of pilot compensation required to perform a task is a measure of an airplane's *flying qualities*. Alternatively this measure is called *handling qualities*, although there are some distinctions between the two terms.

Aircraft Control Allocation, First Edition. Wayne Durham, Kenneth A. Bordignon and Roger Beck.
© 2017 John Wiley & Sons, Ltd. Published 2017 by John Wiley & Sons, Ltd.
Companion website: www.wiley.com/go/durham/aircraft_control_allocation

3.1.1 Requirements

Flying qualities guidance (Department of Defense 1990) is determined by evaluating a great number of airplanes that have different response characteristics, and by using airplanes and simulators whose response characteristics can be changed. The variable-stability airplanes used for these evaluations are highly modified to permit uncoupled control of most if not all linear and angular accelerations. The control laws that modify these airplanes' handling characteristics alter the eigenvalues and the dynamic relationships between the states (eigenvectors) of the response.

Based on large numbers of trials with various tasks, and with a large number of pilots who rate the different responses according to the ease with which they can perform the task, flying qualities guidance and specifications are determined. Military flying qualities requirements are quite extensive; for instance, the guidance for flying qualities associated with an airplane's short-period pitching response runs to around one hundred pages (Department of Defense 1990).

3.1.2 Control Law Design to Satisfy Flying Qualities Requirements

In the past, airplane designers typically selected designs that afforded bare-airframe responses that yielded desired characteristics based on flying qualities requirements. Such designs often required trade-offs with other requirements, such as the airplane's performance. Thus one could improve fuel economy with an aft center of gravity, but in so doing the longitudinal static stability is reduced to a point where the bare-airframe would be difficult to control.

In cases where an airplane's intrinsic flying qualities need improvement, the flight control system may incorporate stability- and control-augmentation systems.

With the advent of full-authority fly-by-wire flight control systems the bare-airframe response characteristics of an airplane are less important. Longitudinal static stability may be relaxed to improve certain performance characteristics, with the flight control computer continually compensating for the reduced stability.

3.2 Dynamic-inversion Control Laws

3.2.1 Basics

Many modern aircraft utilize a very different form of automatic flight control called *dynamic inversion*. It was demonstrated in the F-15 ACTIVE (Advanced Control Technology for Integrated Vehicles) program and was the basis for the flight control system design of the X-35 (Walker and Allen 2002), which began in the Joint Strike Fighter program and became the F-35 fighter. Dynamic inversion fits very well into a study of control allocation because it makes the distribution of control effectors a separate problem from the determination of the dynamic response of the airplane.

The basic idea behind dynamic inversion is quite simple. Consider a physical phenomenon that can be represented by a scalar equation

$$\dot{x} = f(x(t)) + bu(t) \tag{3.1}$$

The key things to note about Eq. (3.1) are that the function $f(x(t))$ may be nonlinear, but that the control effectiveness $bu(t)$ is a linear relationship. The form of Eq. (3.1) is similar to what we were suggesting in Section 2.1.2. The airplane equations of motion with respect to the states are left in their nonlinear form, but the control effectiveness is assumed to be a linear relationship.

Now say we want to find a control law that calculates control inputs that will generate some desired (subscript *des*) dynamic response $\dot{x}_{des}(t)$. Our study of the physics of the phenomenon leads to approximations $\hat{f}(\cdot) \approx f(\cdot)$ and $\hat{b} \approx b$, and of our measurement $\hat{x}(t) \approx x(t)$. Consider the following control law:

$$u^*(t) = \frac{1}{\hat{b}}\left(\dot{x}_{des} - \hat{f}(\hat{x}(t))\right) \tag{3.2}$$

We substitute the $u^*(t)$ from Eq. (3.2) into Eq. (3.1) to find that

$$\dot{x} = f(x(t)) + bu^*(t) = f(x(t)) + \frac{b}{\hat{b}}\left(\dot{x}_{des} - \hat{f}(\hat{x}(t))\right) \tag{3.3}$$

In the subsequent treatment we will drop the $(\hat{\ })$ notation and assume that our approximations and measurements are perfect. Our main purpose here is to set the table for the control allocation problem, so we leave the question of errors in sensing and parameter identification to another day. Then, taking $\hat{x}(t) \equiv x(t), \hat{f}(\hat{x}(t)) \equiv f(x(t)), \hat{b} \equiv b$,

$$u^*(t) = \frac{1}{b}\left(\dot{x}_{des} - f(x(t))\right) \tag{3.4}$$

With the result that

$$\dot{x} = \dot{x}_{des} \tag{3.5}$$

In other words, if we can reasonably well describe the dynamics of the process, and measure the states accurately, then we may solve for the controls and then subtract the natural dynamics and add the dynamics we want ($\dot{x}_{des}$).

A reader new to dynamic inversion may think that he is the victim of a shell game. Nonetheless, the concept works (modulo the correct modeling and state measurement implied when we took the hats off) and has been implemented in airplanes flying today.

The signal $\dot{x}_{des}$ may be determined in various ways. By assuming perfect inversion, conventional feedback-control design methods may be applied. More detail on this process may be found at Section 8.1.

Alternatively, the pilot's inputs may be fed to a computer that applies them to a mathematical model of an airplane with desired response characteristics and which calculates the resulting time histories of $\dot{x}_{des}$. This is akin to the control systems on variable-stability airplanes, used to present pilots with a range of dynamics to evaluate pilot opinion for flying qualities formulation.

In either event, the computer applies $\dot{x}_{des}$ to a model of the inverted dynamics of the actual aircraft (as in Eq. (3.4)) to calculate the control effector deflections to make the airplane respond in the desired manner (as in Eq. (3.5)).

3.2.2 Types of Equations

To begin, consider that the equations of motion of an airplane are of two distinct types: the *kinematic equations* used to describe the position and orientation of the aircraft, and the force and moment equations. The kinematic equations, aside from the navigation equations, describe the relationships between angles and rates, and no forces or moments are present. We will use the relationships for $\dot{\phi}$, $\dot{\theta}$, and $\dot{\psi}$ given in Section 2.1.

The force and moment equations are those that describe the accelerations $\dot{u}$, $\dot{v}$, $\dot{w}$, $\dot{p}$, $\dot{q}$, and $\dot{r}$ as given in Eqs. (2.7) and (2.12). From our discussion of the control effectors it is seen that they are primarily moment generators and they are the primary inputs into the moment equations for $\dot{p}$, $\dot{q}$, and $\dot{r}$. We call these equations *controlled equations*.

Because moments result from generating forces at a distance and no single effector generates a pure couple, the control effectors have an effect on the force equations $\dot{u}$, $\dot{v}$, and $\dot{w}$, but not in a purposeful way. Any control surface deflection will change the overall drag on the airplane and affect at least $\dot{u}$, but speed is not controlled by control effector deflections. The rudder is primarily used to change the yawing moment, but while it does generate a side force it is not used intentionally to control $\dot{v}$. The force equations are called *complementary equations*.

We have not included engine thrust as a control effector, although its use would certainly seem to make $\dot{u}$ a controlled equation. Thrust acts on a completely different timescale than the moment-generating control effectors, and flight control system design frequently considers airspeed control to be a separate, outer loop. In air-combat maneuvering, stick and rudder are very busy indeed, but thrust is often applied as one of three settings: full afterburner, full non-afterburning, and idle power. The only time we will include thrust in our discussions is in the context of thrust vectoring, primarily a moment-generating device.

The question of which equations should be considered controlled depends on the problem. One could even formulate two problems, one involving the moment equations and another the force equations, and solve them separately. In this book we will concentrate on the three moment, or controlled equations.

3.2.3   The Controlled Equations

3.2.3.1   The Inverse Function

Dynamic inversion relies on there being some way to invert the controlled equations and effectively 'solve' for the control vector **u**. We represent this inversion as some functional **g**. That is, given controlled equations of motion

$$\dot{\mathbf{x}}^{Cont} = \mathbf{f}\,(\mathbf{x}, \mathbf{u}) \tag{3.6}$$

We require an expression for the control vector **u** such that

$$\mathbf{u}^* = \mathbf{g}\,(\dot{\mathbf{x}}_{des}^{Cont}, \mathbf{x}) \tag{3.7}$$

The inverse function is represented as **g** in Eq. (3.7). It is the link between the control effectors and their influence on the dynamics of the airplane, or the *control effectiveness*. Generally, however, our understanding of control effectiveness is found only in tables that yield forces and moments as functions of the airplane state and the control deflections. These tables result from theory and approximations, wind-tunnel tests, and flight tests. A table is not a function, and cannot be inverted unless it is extremely simple.

We therefore turn to linearization, as described in Section 2.20, which produced the control effectiveness matrix, B. The inverse function **g** is some kind of inverse of the B matrix. If B is square and non-singular, $\mathbf{g} = B^{-1}$.

$$\mathbf{u}^* = B^{-1}\left(\dot{\mathbf{x}}_{des}^{Cont}, \mathbf{x}\right) \tag{3.8}$$

If B is not square and invertible, then we seek to determine some function or method that accomplishes the inversion. When we say *method* we mean some procedure $\mathbf{g}$, based on the linear B matrix, that does the inverse mapping. This is control allocation, and the functions and methods that accomplish the inversion are the subject of Chapter 6.

$$\mathbf{u}^* = \mathbf{g}(B)\left(\dot{\mathbf{x}}_{des}^{Cont}, \mathbf{x}\right) \tag{3.9}$$

For example, the rolling moment equation, in principal axes and assuming no thrust effects, is

$$\dot{p} = \frac{1}{I_{xp}}[L - (I_{zp} - I_{yp})qr] \tag{3.10}$$

If we assume a conventional airplane with aileron δ_a as the sole generator of rolling moments, then it is only through that moment, L, that the control appears. There is no analytical function for $L(\delta_a)$, so we linearize the function.

$$L(\delta_a) = L(\delta_a = 0) + \frac{\partial L}{\partial \delta_a}\Delta\delta_a \tag{3.11}$$

Implicit in Eq. (3.11) is that the derivative is evaluated at a reference condition for which $\delta_a = 0$, and this reference condition includes all of the rolling moment's other dependencies (β, p, r, and so on). With the usual assumptions, this becomes $L(\delta_a) = L_{\delta_a}\delta_a$. Then, Eq. (3.7) becomes

$$\delta_a^* = \frac{1}{L_{\delta_a}}\{I_{xp}\dot{p}_{des} - [L' - (I_{zp} - I_{yp})qr]\} \tag{3.12}$$

In Eq. (3.12) L' is the rolling moment evaluated at all of its dependencies *except* δ_a. Notice that there is no need to linearize the entire equation, just enough of it to isolate the control. If the flight control computer, sensors, and on-board aerodynamic model can evaluate the right-hand side of Eq. (3.12), then the nonlinear form is the better choice.

If we had considered both roll and yaw (in principal axes, still ignoring thrust), we would have arrived at

$$\begin{Bmatrix} \delta_a^* \\ \delta_r^* \end{Bmatrix} = \begin{bmatrix} L_{\delta_a} & 0 \\ 0 & N_{\delta_r} \end{bmatrix}^{-1} \begin{bmatrix} I_{xp}\dot{p}_{des} - [L' - (I_{zp} - I_{yp})qr] \\ I_{zp}\dot{r}_{des} - [N' - (I_{yp} - I_{xp})pq] \end{bmatrix} \tag{3.13}$$

In Eq. (3.13)

$$\mathbf{g}(B) = \begin{bmatrix} L_{\delta_a} & 0 \\ 0 & N_{\delta_r} \end{bmatrix}^{-1} \tag{3.14}$$

$$\dot{\mathbf{x}}_{des}^{Cont} = \begin{Bmatrix} \dot{p}_{des} \\ \dot{r}_{des} \end{Bmatrix} \tag{3.15}$$

The term $\dot{\mathbf{x}}_{des}^{Cont}$ indicates the desired dynamics. As with our scalar example, it is easy to show that this control yields

$$\dot{\mathbf{x}}^{Cont} = \dot{\mathbf{x}}_{des}^{Cont} \tag{3.16}$$

From Eq. (3.7) the control laws are obtained by replacing $\dot{\mathbf{x}}^{Cont}$ with $\dot{\mathbf{x}}_{des}^{Cont}$, the desired dynamics.

3.2.3.2   The Desired Dynamics, $\dot{\mathbf{x}}^{Cont}_{des}$

In the simplest form of dynamic inversion the controlled states are given desired dynamics $\dot{\mathbf{x}}^{Cont}_{des}$ in response to commanded values of the states $\mathbf{x}^{Cont}_c$ according to the control law

$$\dot{\mathbf{x}}^{Cont}_{des} = \mathbf{\Omega}\left(\mathbf{x}^{Cont}_{cmd} - \mathbf{x}^{Cont}\right) \tag{3.17}$$

In Eq. (3.17) the matrix $\mathbf{\Omega}$ is normally diagonal, and its diagonal entries determine the dynamics of the controlled states in response to commanded values. As scalar equations we have

$$\begin{aligned}
\dot{p}_{des} &= \omega_p(p_{cmd} - p) \\
\dot{q}_{des} &= \omega_q(q_{cmd} - q) \\
\dot{r}_{des} &= \omega_r(r_{cmd} - r)
\end{aligned} \tag{3.18}$$

The expression for δ_a^* in Eq. (3.12) becomes

$$\delta_a^* = \frac{1}{L_{\delta_a}}\{I_{xp}\omega_p(p_{cmd} - p) - [L' - (I_{zp} - I_{yp})qr]\} \tag{3.19}$$

The pilot's stick deflections are interpreted as p_{cmd}, the airplane sensors measure the angle rates p, q, and r as well as the dependencies in the rolling moment, the on-board aerodynamic model calculates that rolling moment, and the aileron is deflected the right amount.

The inversion of the controlled equations is sufficient to set the table for the rest of this book. We have established that the control of the three moment equations is the principle goal of dynamic inversion, and that the determination of the inverse function—the control allocation problem—is required to set the dynamic response of the airplane. Thus, explicitly, the immediate problem before us is to determine the function, method, or procedure that yields $\mathbf{g}(B)$ in the three moment equations.

3.2.4   The Kinematic and Complementary Equations

While the determination of $\mathbf{g}(B)$ in the three moment equations is the primary focus of this book, it is enlightening to describe what happens with these three moments and their associated rates, p, q, and r in the context of the overall control of the airplane. We therefore proceed to see how the ability to specify the dynamics of the three moment equations leads to control of the airplane.

With the pilot controlling the airplane, inversion of the controlled equations is minimally sufficient. Lateral stick is interpreted as commanded roll rate, longitudinal stick as commanded pitch rate, and pedal application as commanded yaw rate, all with specified dynamics. A benefit of dynamic inversion is that the three responses are uncoupled. In particular, roll is accomplished without generating yaw, and vice-versa.

We still have the kinematic relationships to consider, as well as the force, or complementary, equations. The kinematic equations are easiest to deal with, while the complementary equations are more problematic.

3.2.4.1   Control of the Euler Angles

In dynamic inversion, control of the kinematic equations is accomplished by treating the controllable states as controls. With respect to the kinematic equations, it is easy to see that Eq. (2.1) may be easily solved for p, q, and r. Then, with full control over p, q, and r, we may proceed as we did above. That is, we use desired values of $\dot{\phi}$, $\dot{\theta}$, and $\dot{\psi}$ to generate p_c, q_c, and r_c to apply to the controlled equations to determine the control effector deflections.

As an example, consider the bank angle ϕ. From Eq. (3.18) we may select p_{cmd} and the airplane will attain that roll rate according to the first-order dynamics specified through ω_p.

$$\dot{p}_{des} = \omega_p(p_{cmd} - p) \tag{3.20}$$

The bank angle equation is

$$\dot{\phi} = p + (q\sin\phi + r\cos\phi)\tan\theta \tag{3.21}$$

Treating p as a control, and interpreting it as p_{cmd} for Eq. (3.20), we easily invert the dynamics,

$$p_{cmd} = \dot{\phi}_{des} - (q\sin\phi + r\cos\phi)\tan\theta \tag{3.22}$$

The desired dynamics for ϕ are then selected:

$$\dot{\phi}_{des} = \omega_\phi(\phi_{cmd} - \phi) \tag{3.23}$$

Now working back to Eq. (3.20)

$$
\begin{aligned}
\dot{p}_{des} &= \omega_p(p_{cmd} - p) \\
&= \omega_p[\dot{\phi}_{des} - (q\sin\phi + r\cos\phi)\tan\theta - p] \\
&= \omega_p[\omega_\phi(\phi_{cmd} - \phi) - (q\sin\phi + r\cos\phi)\tan\theta - p]
\end{aligned} \tag{3.24}
$$

The resulting bank angle command system is

$$
\begin{aligned}
\delta_a^* = \frac{1}{L_{\delta_a}}\{I_{xp}\omega_p[\omega_\phi(\phi_{cmd} - \phi) - (q\sin\phi + r\cos\phi)\tan\theta - p] \\
-[L' - (I_{zp} - I_{yp})qr]\}
\end{aligned} \tag{3.25}
$$

This control over the Euler angles is certainly possible, and is useful in autopilot design, but it does not find much application in piloted flight. Lateral stick deflection is used to command roll rates p, not bank angle ϕ. A pilot would find a system wherein lateral stick deflection commands bank angles to be unusual. A similar argument applies to pitch control, with the proviso that often the pilot expects his control inputs to command a load factor, or g-force. And with respect to control of ψ, the heading angle is changed in flight, not with direct application of rudder pedals, but by banking and turning the airplane until the desired heading angle is reached.

The preceding discussion is untrue in at least one special case that may be encountered: vertical takeoff and landing. In those flight phases the required bank and pitch angles are small and must be obtained with some precision. The yaw inceptor generally will not command heading angle, but rather the yaw rate, leaving it to the pilot to close the loop on stopping at the desired heading angle.

3.2.4.2 Complementary Equations

Control of the complementary equations by inversion is possible, but is not seen much in practice. More often the controlled equations are inverted, and the complementary equations are treated in a more conventional control system design using the controlled states as inputs.

The approach wherein the complementary equations are inverted using this technique or some other has been performed, at least on a theoretical level. Control of complementary variables may be mixed in the inversion process, permitting various combinations of interpretations of the commanded variables. Using the three moment equations (and assuming independent control of velocity), Azam and Singh (1994) develop controllers for ϕ, α, and β; and another for p, α, and β. Snell *et al.* (1992) develop a controller for α, β, and ϕ_w. In Snell *et al.*, the assignment of desired dynamics for wind-axis bank angle does not exactly follow the development above in that it incorporates filtering to reduce sensitivity to pilot inputs.

3.3 Model-following Control Laws

We briefly mention model-following control because of its similarity to dynamic inversion. In fact, in some of its forms dynamic inversion is a special case of model-following control. For our present purposes the similarity of interest is that they both require an expression or algorithm that 'solves' for the controls.

The primary distinction between model-following and dynamic-inversion control is as follows. The former is used to follow a trajectory (involving all the states, as determined by a model with specified dynamics), and correct back to the trajectory in the presence of errors. Meanwhile, the latter is used to give the airplane certain response characteristics. Model-following control is the 'Blue Angel' in the slot—he must match the leader's trajectory precisely. Dynamic-inversion control is one of the solo Blue Angels—he just wants to look good.

For more on the relationships between dynamic inversion and model-following control see Durham (1996) and Kocurek and Durham (1997).

3.4 'Conventional' Control Laws

Historically flight control system design has been based on feedback control, typically using transfer functions of control effectors to system outputs to find feedback gains that give desired responses. That is, for an output $y_i(t)$ and input $u_j(t)$, a transfer function $g_{ij}(s)$ is determined.

$$g_{i_j}(s) = \frac{y_i(s)}{u_j(s)} \qquad (3.26)$$

The transfer function is then used in any of the many tools available, such as the root locus, to determine feedback gain k, $u_j = ky_i$, that gives the desired response.

If u_j is to be used in conjunction with another effector, say u_k, and the two effectors move in relation to one another so that a deflection a of u_j is accompanied by a deflection b of u_k

$$g_{i_{jk}}(s) = a\frac{y_i(s)}{u_j(s)} + b\frac{y_i(s)}{u_k(s)} \qquad (3.27)$$

Thus the left and right ailerons may be constrained to move in equal and opposite amounts (as conventional ailerons are) so that $a = 1$ and $b = -1$. When two effectors are thus ganged to each other they may be treated as a single control effector.

If the effectors are free to move independently, with no regard for the command to any other effector, then a and b are unknown, or at least not known *a priori*. Thus there will be no single transfer function, so that $g_{i_{jk}}(s)$ will be unknown and potentially time-varying. One cannot use root-locus methods to determine feedback gains if the control effectors are free to move independently of one another.

A simple example will suffice to illustrate this point. Consider an airplane with both aft-mounted horizontal tails and forward-mounted canards. If only the horizontal tail is used to generate pitching moments then the initial response to a nose-up command will generate less lift at the tail to increase the pitching moment. Thus there will be an initial loss of total lift, and momentary downward acceleration, before the pitch rate generates an increase in angle of attack. With a canard only, the opposite is true: there is an initial increase in lift. Between these two extremes there is a combination of horizontal tail and canard that will produce a pure couple in pitch about the center of gravity, with no transmission zeros.

Airplanes may have more than three control effectors, but historically they are ganged so that there are effectively three. Left and right ailerons are connected by cables and pulleys, and left and right horizontal tails are mounted on the same rigid shaft, so that each pair becomes a single control effector. By freeing the left–right pairs to move independently, classical feedback control design becomes problematic.

While the use of transfer functions of control effectors to system outputs will not work, the use of transfer functions of commanded angle rates does and is used often. Dynamic inversion will yield uncoupled control of roll rate, pitch rate, and yaw rate (p, q, and r). These can then be used as controls (p_{cmd}, q_{cmd}, and r_{cmd}) to form transfer functions for conventional analysis. In Durham (2013) we demonstrate control of sideslip, a complementary equation. First dynamic inversion is used to command yaw rate with the rudder pedals. Then the commanded yaw rate is used to form the transfer function $\beta(s)/r_{cmd}(s)$ and then a root-locus is used to select the feedback gains.

3.5 Afterword

This chapter has introduced the control laws that require the designer to 'solve' for the control vector. It is a requirement for both dynamic-inversion and model-following control. When 'conventional' control systems are considered we pose the problem as one of angle rates that are treated as control inputs to the problem, and these angle-rate commands are generated by dynamic inversion or model following.

The ADMIRE flight simulation, described in Appendix B and available on the companion web site to this book, provides an implementation of a dynamic inversion control law. All the modules are laid out clearly, and the interested reader may learn more about dynamic inversion by studying the code.

If the inverse function that relates controls to desired moments is a simple scalar or matrix inverse, then the problem is easy. If not, then the control allocation problem must be addressed. We will pose this problem in the next chapter.

References

Azam, M and Singh, SN 1994 'Invertibility and trajectory control for nonlinear maneuvers of aircraft,' *J. Guidance, Control, and Dynamics,* **17** (1), 192–200.

Department of Defense 1990 MIL-STD-1797A 'Flying Qualities of Piloted Aircraft'.

Durham, W 1996 'Dynamic inversion and model-following control' AIAA 96-3690 in *AIAA, Guidance, Navigation and Control Conference.*

Durham, W 2013 *Aircraft Flight Dynamics and Control,* 1st edn. John Wiley & Sons, pp. 221–224.

Kocurek, N and Durham, W 1997 'Dynamic inversion and model-following flight control: A comparison of performance robustness,' in *22nd AIAA Atmospheric Flight Mechanics Conference.*

Snell, SA, Enns, DF, and Garrard, WL Jr. 1992 'Nonlinear inversion flight control for a supermaneuverable aircraft,' *J. Guidance, Control, and Dynamics,* **15** (4), 976–984.

Walker, GP and Allen, DA 2002 'X-35B STOVL flight control law design and flying qualities,' in *AIAA Biennial International Powered Lift Conference.*

4

The Problem

4.1 Control Effectiveness

The purely aerodynamic control effectors that we consider modify the airflow over an airplane so as to change the local aerodynamic force and, by virtue of being positioned some distance from the center of mass, the aerodynamic moment about that point. We take their primary purpose to be the generation of an aerodynamic moment. The significance of this decision is to declare that the three moment equations are controlled equations in the sense of Section 3.2.3.

These aerodynamic control effectors are generally of the flapping sort, meaning a surface is deflected about a hinge line. They may be hinged to the front edge (such as leading-edge flaps) or, more often, the rear of a larger surface (such as elevators on the trailing edge of a horizontal tail, or ailerons on the trailing edge of a wing), or they may be the primary surface (such as canards or unit horizontal tails).

Leading- and trailing-edge aerodynamic control effectors generally act by changing the camber of their parent surface, which increases or decreases the lift locally. All-moving aerodynamic control surfaces act by changing their incidence relative to the airflow, likewise increasing or decreasing the lift locally.

The only other type of control effector in common use is thrust vectoring. Some thrust vectoring systems are one-dimensional, usually acting only to create pitching moments. Two one-dimensional thrust vectoring nozzles may be actuated differentially to produce rolling moments. Other systems are axisymmetric, and can generate moments in pitch and yaw. Dual-engine axisymmetric thrust vectoring can generate rolling moments as well.

Through analysis and experimentation, the force- and moment-generating characteristics of these effectors are determined over the expected range of flight conditions. This information is usually presented in tabular form as functions of several independent variables that adequately describe the flight condition. All of these data are incorporated in the on-board model (OBM).

Thus, at any instant in flight, the flight control computer may query the OBM to evaluate the forces and moments that will be generated by a particular control effector when deflected. From this information a graph of control effectiveness is implied, and from this the slope can be extracted to satisfy our definition of the elements of the B matrix (Eq. (2.20)).

Aircraft Control Allocation, First Edition. Wayne Durham, Kenneth A. Bordignon and Roger Beck.
© 2017 John Wiley & Sons, Ltd. Published 2017 by John Wiley & Sons, Ltd.
Companion website: www.wiley.com/go/durham/aircraft_control_allocation

4.2   Constraints

All control effectors have certain position limits, usually specified as positive and negative deflections from a specified position that is taken as zero deflection. The zero position is strictly a matter of convention. The zero position does not necessarily insure zero control-generated moment. Our use of the equation $\mathbf{m} = B\mathbf{u}$ seems to imply the opposite, but recall that in the process of linearization, the partial derivatives in the control effectiveness matrix were evaluated at a reference condition. The reference condition included the control effector positions at the reference condition, thus effectively defining zero deflections.

Control effectors also have limits on the rate they can travel. In fact, hydraulically actuated flight control effectors constitute dynamic systems of at least second order (Pratt, 2000). Normally the actuator dynamics are very fast compared to the responses of the airplane and if they are modeled at all they are usually modeled as first-order systems with a relatively small time constant; of the order of 0.05 s as modeled in Stevens and Lewis (2015).

Actuator dynamics may be fast relative to the responses of the airplane, but they may not be fast within the scale of the computational frame size of the flight control computer. If the speed of response of the actuators vis-a-vis the speed of the computer becomes an issue, then the dynamics of the actuators should be included in the state formulation of the control law design.

In our initial discussion we will assume only position limits for the control effectors. Each problem will be posed as static, to determine the control effector positions that will attain a given $\mathbf{m}_{des}$ irrespective of how they might have arrived at those positions. Later we will examine control allocation assuming fixed position and rate limits.

4.3   Control Allocation

The nub of the control allocation problem is easily expressed: solve $B\mathbf{u} = \mathbf{m}_{des}$ for $\mathbf{u}$.

$\mathbf{m}_{des}$ is a vector of desired moments, the output of some control law. The vector is not always defined as actual moments; refer to Table 2.1. The default manner of speaking of this vector will be as moments unless a specific problem is being addressed. The choice does not affect the geometry of the problem or methods of its solution.

The dimension of $\mathbf{m}_{des}$ is nominally three, but from time to time we will discuss two moments, because problems involving roll and yaw alone are relevant problems, but especially because the pictures are easier to draw. And sometimes we will peer into a fourth dimension, such as that which arises when we consider the drag effects of the control effectors.

We will generically denote the dimension of $\mathbf{m}_{des}$ as n,

$$\mathbf{m}_{des} \in \Re^n \tag{4.1}$$

$\mathbf{u}$ is a vector of control effectors of dimension m where $m > n$.

$$\mathbf{u} \in \Re^m \tag{4.2}$$

B is the control effectiveness matrix,

$$B \in \Re^{n \times m} \tag{4.3}$$

The limits on travel of the control effectors is expressed by two vectors, $\mathbf{u}_{Min}$ and $\mathbf{u}_{Max}$ that contain the lower and upper limits of each effector's travel. When we say the control vector $\mathbf{u}$ is greater or less than one or the other of these limit vectors, we mean on an element-by-element comparison.

$$\mathbf{u}_{Min} \leq \mathbf{u} \leq \mathbf{u}_{Max} \Rightarrow u_{i_{Min}} \leq u_i \leq u_{i_{Max}}, i = 1 \cdots m \qquad (4.4)$$

Control effector displacements that are within their limits are said to be *admissible*. If any one of a suite of control effectors is inadmissible, then that vector of controls is inadmissible. The set of all admissible control deflections is given the symbol Ω.

4.3.1   The Control Allocation Problem

The control allocation problem is as follows. Given B, $\mathbf{m}_{des}$, $\mathbf{u}_{Min}$, and $\mathbf{u}_{Max}$, find $\mathbf{u}$ such that:

$$B\mathbf{u} = \mathbf{m}_{des} \qquad (4.5)$$

and

$$\mathbf{u}_{Min} \leq \mathbf{u} \leq \mathbf{u}_{Max} \qquad (4.6)$$

The set of equations $B\mathbf{u} = \mathbf{m}_{des}$ is under-determined (fewer equations than unknowns) and mathematically has an infinite number of solutions. When control limitations are introduced the equations may have no solutions. This statement is easily demonstrated by considering a maximum-effort, wings-level pullup in which every control effector that generates a nose-up pitching moment is hard at its limit. If $\mathbf{m}_{des}$ requests more pitching moment, there is no solution.

If there is an admissible control vector for a particular moment, then that moment is said to be *attainable*. The set of all attainable moments is given the symbol Φ. It is also called the AMS, which stands for the 'attainable moment subset': a subset of all moments, not a subset of attainable moments.

4.4   Afterword

Our original formulation of the control allocation problem (Durham, 1993) had a narrower view than the above statement of the problem. There we wanted the admissible controls that yielded the maximum attainable moment vector in the direction of the desired moment.

The reasoning was that if the desired moment was less than maximum, then the admissible controls at that maximum could be uniformly scaled to yield the desired moment, solving the problem. If the desired moment was equal to the maximum, then that solution was unique, as we will see when we examine the geometry of the problem in Chapter 5. If the desired moment was greater than the maximum, we suggested that the control law design was lacking but that, in any event, taking the maximum for the solution had at least the virtue of preserving the direction of the desired moment.

Within the interior of the AMS Φ there are an infinite number of solutions; we have, after all, an under-determined system of equations. As one approaches the boundary of Φ this infinitude gets 'smaller' until at the boundary the solution is unique. But while considering the infinite number, the idea arose that some of those solutions were better than others, in particular better than our scaled-down solution from the maximum. Thus the statement is relaxed to find any admissible solution $\mathbf{u}$, including what came to be called *preferred solutions*. We will address preferred solutions throughout the book, such as at Section 6.2.3 and throughout Appendix A.

References

Durham, W 1993 'Constrained control allocation' *AIAA J. Guidance, Control, and Dynamics*, **16** (4), 717–725.

Pratt, RW (ed) 2000 *Flight Control Systems: Practical Issues in Design and Implementation*. Institution of Engineering and Technology, Control Engineering Series 57, Chapter 3.

Stevens, BL and Lewis, FL 2015 *Aircraft Control and Simulation,* 3rd edn. Wiley-Blackwell.

5

The Geometry of Control Allocation

There is a fair amount of new notation introduced in this chapter. New notation is always hard to digest. All flight engineers immediately know what the Greek symbols α and β mean, but would have trouble reading a text that used Cyrillic characters instead. It would take some getting used to. With that in mind we will tend to be repetitive, not only using the notation, but in the same breath spelling out what that notation means.

5.1 Admissible Controls

5.1.1 General

We consider each control effector to be independent of all the others. Thus in a graph of their travels their representations are orthogonal one to the other. Control effector u_1 may move through its range independent of the position of u_2, and vice versa. We may then draw a figure that represents the ranges of each control effector relative to the others, as in Figure 5.1, which is the two-dimensional figure of effectors u_1 and u_2.

Within and on the boundary of Figure 5.1 lies every admissible combination of u_1 and u_2, thus the figure is the set of admissible controls, or what we name Ω. We will occasionally refer to the *control space* of dimension m, the number of controls. Figure 5.1 represents two-dimensional control space with a two-dimensional Ω.

Figure 5.1 generalizes to higher dimensions as more controls are added to the problem. For m controls the set is an m-orthotope, also called a hyper-rectangle or rectangular polytope, the generalization of a plane rectangle for higher dimensions.

5.1.2 Objects

At this point we introduce a notation devised to describe relationships between the various features of diagrams of admissible controls. It is a trinary system with certain rules involving intersections and unions that will be defined later. Simply put, if a particular control effector

Aircraft Control Allocation, First Edition. Wayne Durham, Kenneth A. Bordignon and Roger Beck.
© 2017 John Wiley & Sons, Ltd. Published 2017 by John Wiley & Sons, Ltd.
Companion website: www.wiley.com/go/durham/aircraft_control_allocation

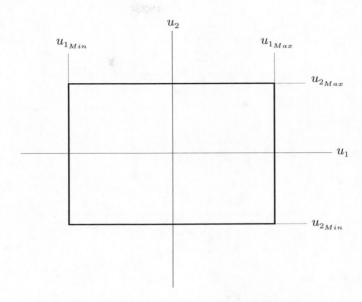

Figure 5.1   Two-dimensional set of admissible controls, Ω_2

Table 5.1   Object notation

Deflection	Value
Min	0
Max	1
Between	2

is at its minimum deflection it is assigned the number 0; if at its maximum deflection it is assigned the number 1; and if it is an arbitrary value or free to vary between minimum and maximum deflection it is assigned the number 2. The numbers do not have value, they are just symbols.[1] The numbering is summarized in Table 5.1.

We will also use this object notation in subscripts to vectors $\mathbf{u}$ and $\mathbf{m}$. When applied to a vector $\mathbf{u}$ it will describe the state of the control effectors in that vector. Thus, $\mathbf{u}_{012}$ means a control vector wherein u_1 is at its lower limit (0), u_2 is at its upper limit (1), and u_3 is arbitrary but admissible (2).

When used as a subscript to $\mathbf{m}$ vectors, it means the moment that results from B times the control vector with the same subscript. Thus $\mathbf{m}_{012} = B\mathbf{u}_{012}$. We will have more to say about this later.

From time to time we will need to designate higher-dimensional structures using object notation. Rather than make up names for these hyper-rectangles, we will call them all *objects*

---

[1] Comparing two integers for equality is more reliable than comparing two floating point numbers; that is, 0 or 1 versus the values for $u_{\min}$ or $u_{\max}$.

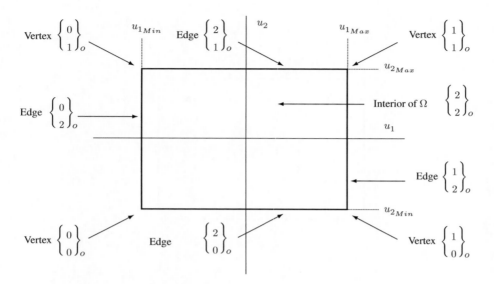

Figure 5.2 Ω_2, with object notation

and designate them with a subscript **o**, such as $\mathbf{o}_{0012201}$. $\mathbf{o}_{0012201}$ is the object in a seven-control problem that results from the first, second, and sixth controls being at their lower limits, the third and seventh at their upper limit, and the fourth and fifth at arbitrary but admissible positions.

With object notation we can add labels to Figure 5.1 that identify the various geometric objects, resulting in Figure 5.2. An object is represented as a vector in { } with a subscript o on the right bracket. For in-line equations, we will represent the objects as row vectors with curly brackets and subscript o.

For clarity, although it will often be unnecessary, we will sometimes also subscript the curly brackets of moment vectors with an m, and of a control vector with a u.

With the 0-1-2 object notation, the dimension of an object is equal to the number of 2s in it, for example, the vertices have no 2s and are of dimension zero, and the edges have one 2 and are of dimension 1. This relationship of the number of 2s is true for any dimension, so that $\mathbf{o}_{0012201}$ is a two-dimensional object within a seven-dimensional object. $\mathbf{o}_{2222221}$ is a six-dimensional object within a seven-dimensional object and is, in fact, one of its bounding polytopes, being of dimension $m - 1$.

An object may be related to the actual control deflections easily; for instance

$$\begin{Bmatrix} 1 \\ 0 \end{Bmatrix}_o \iff \begin{Bmatrix} u_{1_{Max}} \\ u_{2_{Min}} \end{Bmatrix}_u, \begin{Bmatrix} 2 \\ 1 \end{Bmatrix}_o \iff \begin{Bmatrix} u_1 \\ u_{2_{Max}} \end{Bmatrix}_u \tag{5.1}$$

In the second of these two expressions, u_1 is taken to mean an arbitrary admissible value of u_1.

As we consider more control effectors the dimension of control space, and of the set of admissible controls, increases. Figure 5.3 represents three control effectors with 0-dimensional vertices, one-dimensional edges, and now two-dimensional faces.

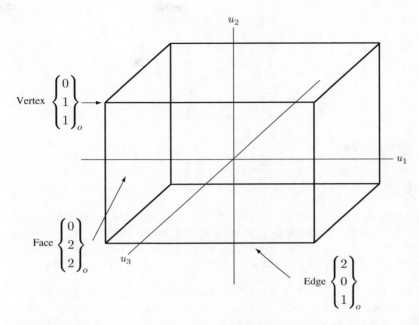

Figure 5.3 Three-dimensional set of admissible controls, Ω_3. Positive deflections are: u_1 to the right, u_2 upwards. u_3 is positive according to the right-hand rule. Three objects are identified

With respect to the objects identified in Figure 5.3:

Vertex $\{0\ 1\ 1\}_o$ Here $u_1 = u_{1_{Min}}$, $u_2 = u_{2_{Max}}$, and $u_3 = u_{3_{Max}}$.

Edge $\{2\ 0\ 1\}_o$ Here $u_2 = u_{2_{Min}}$, $u_3 = u_{3_{Max}}$, and u_1 is free to vary from $u_{1_{Min}}$ to $u_{1_{Max}}$. We can identify the two vertices at the ends of this edge by replacing the 2 in its object notation by 0 and by 1.

Face $\{0\ 2\ 2\}_o$ Here $u_1 = u_{1_{Min}}$, and u_2 and u_3 are free to vary from *Min* to *Max*. We can identify the four edges surrounding this face by replacing one and then the other of the two 2s by 0 and then by 1. We can identify the four vertices by replacing the two 2s in its object notation by all four combinations of 0 and 1, or equivalently by treating each edge as above.

When we consider even more control effectors, the principal problems are with wrapping our minds around the geometry. Four control effectors result in a set of admissible controls of dimension four; the four-dimensional rectangular polytope is bounded by three-dimensional rectangular polytopes within which every point is on the boundary of the Figure Ω_4. The figures get harder and harder to draw because we are actually drawing the projection of a higher-degree figure onto a two-dimensional page.

However, every generalization that we can make about two-dimensional and three-dimensional control spaces extends naturally to higher-order control spaces. In other words, the mathematics will take care of everything if we just believe.

5.1.3 Intersection and Union

There are two simple operations that we can perform on two objects, called *intersection* and *union*.

Table 5.2 Intersection

$\dfrac{obj-1\rightarrow}{\downarrow obj-2}$	0	1	2
0	0	×	0
1	×	1	1
2	0	1	2

5.1.3.1   Intersection

The intersection of two objects is the object that is wholly contained in each of the two. Thus, the top and right edges of Figure 5.1 intersect in a one-dimensional vertex. The top and bottom edges have no intersection.

To determine the intersection of two objects, they are compared on an entry-by-entry basis. Table 5.2 gives the corresponding entry in their intersection based on the entries in $obj-1$ across the top and $obj-2$ (object 1 and 2) in the left column. The two entries × means that if any comparison falls, there the two objects do not intersect, no matter what the other comparisons yield.

With the objects identified in Figure 5.3, the intersection of edge $\{2\ 0\ 1\}_o$ and face $\{0\ 2\ 2\}_o$ is $\{0\ 0\ 1\}_o$, a vertex. Edge $\{2\ 0\ 1\}_o$ and vertex $\{0\ 1\ 1\}_o$ fail in the comparison of u_2 with a 0 in the edge and a 1 in the vertex notation.

The intersection is useful for determining the ordering of objects, as for example the vertices in Figure 5.2.

5.1.3.2   Union

The union is the smallest object of which the two given objects are both members. In Figure 5.1 the union of the upper-left and upper-right vertices is the upper edge; the union of the upper-right vertex and the right edge is the right edge; the union of the upper-left and lower-right vertices is the whole figure; and the union of any two different edges is also the whole figure.

The union is easy to determine. Again, two objects are compared element-by-element. If two corresponding entries are different, that entry in their union is 2. Otherwise it is the number the two objects share. This is summarized in Table 5.3.

With the objects identified in Figure 5.3, the union of edge $\{2\ 0\ 1\}_o$ and face $\{0\ 2\ 2\}_o$ is $\{2\ 2\ 2\}_o$, the entire figure. The union of edge $\{2\ 0\ 1\}_o$ and vertex $\{0\ 1\ 1\}_o$ is $\{2\ 2\ 1\}_o$, a face of the figure.

We will have use of the union in Section 6.10.

Table 5.3 Union

$\dfrac{obj-1\rightarrow}{\downarrow obj-2}$	0	1	2
0	0	2	2
1	2	1	2
2	2	2	2

5.1.4   Convex Hull

Considering an arbitrary set of points in m-dimensional space, the smallest convex figure that totally encloses all the points is called its *convex hull*, denoted by $\partial(\cdot)$. Algorithms are available to determine convex hulls, such as the command `convhull` in MATLAB®.

Since the admissible controls are more than an arbitrary set of points, its boundary is known: it is the totality of its objects of dimension $m - 1$. Thus the hull of a one-dimensional Ω is made up of the two zero-dimensional vertices, the hull of a two-dimensional Ω is made up of the four one-dimensional edges, the hull of a three-dimensional Ω is made up of the six two-dimensional faces, and so on.

The significance of the convex hull of the admissible controls is that points on that hull represent the maximum deflections of the collective control effectors in arbitrary combinations; that is, directions in control space.

5.2   Attainable Moments

For a given control effectiveness matrix B, the set of all attainable moments (Φ) is B times every admissible control. However, not every control vector on the boundary of Ω will result in a moment on the boundary of Φ. We can easily demonstrate this fact using an example of three controls and two moments.

For our three controls consider Figure 5.3, and let the control limits be ± 0.5 for each effector:

$$-0.5 \le u_i \le 0.5, i = 1 \cdots 3 \tag{5.2}$$

Now consider a specially chosen control effectiveness matrix,

$$B = \begin{bmatrix} 1 & 0 & -0.5 \\ 0 & 1 & -0.5 \end{bmatrix} \tag{5.3}$$

The attainable moments that result from this combination of control effectors and their effectiveness are shown, perhaps not to scale but qualitatively correct, in Figure 5.4.

The relationship of Figure 5.4 to Figure 5.3 is obvious, and we have intentionally left vestiges of three edges and two vertices of Ω that mapped to the interior of Φ to emphasize the relationship.

To return to the point about controls on $\partial(\Omega)$, note that

$$B \begin{Bmatrix} 0.5 \\ 0.5 \\ 0.5 \end{Bmatrix}_u = \begin{Bmatrix} 0.25 \\ 0.25 \end{Bmatrix}_m \tag{5.4}$$

But also note that

$$B \begin{Bmatrix} 0.5 \\ 0.5 \\ -0.5 \end{Bmatrix}_u = \begin{Bmatrix} 0.75 \\ 0.75 \end{Bmatrix}_m \tag{5.5}$$

Clearly both control vectors are on the boundary of admissible controls, and they both map to the same *direction* in moment space, but the image of the former is only one third the

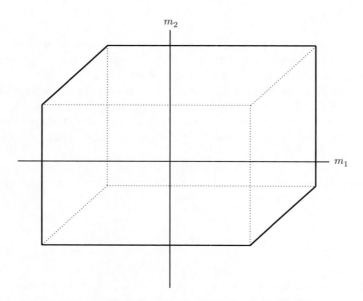

Figure 5.4   Two-dimensional set of attainable moments, Φ

magnitude of that of the latter. Thus although the former is clearly on the boundary of Ω, it maps to a point well within Φ.

At this point we can make some observations about the B matrix, and the mapping $B : \Omega \rightarrow \Phi$. We will continue with our simple example of two moments and three controls.

Rank of B: B is of dimension $n \times m$, where $n < m$. The number of $n \times n$ matrices that may be formed from the columns of B is the number of combinations of m columns taken n at a time. The *rank* of our B matrix is the greatest rank of these $n \times n$ matrices. For our purposes we assume that *every* $n \times n$ matrix thus formed is of rank n. Since we will generally take $n = 3$, we assume that *every* 3×3 matrix thus formed is of rank 3. When we take $n = 2$, we assume that *every* 2×2 matrix thus formed is of rank 2. We will subsequently refer to this condition by saying that B is *robustly* of full rank.

Practically speaking this is not an unreasonable assumption. As we will see, the columns of B correspond to directions in moment space that result from the action of a single control effector. For three columns to be linearly dependent means that the action of a given control effector may be duplicated by that of one or two others. Since we are dealing with aerodynamic phenomena, this seems unlikely.

Consider a single control effector, say the right horizontal tail. Its action will generate moments in all three axes; primarily pitch, but also roll and yaw. The specific combinations of these three moments are given by the column of B that corresponds to its effectiveness. For three columns of B, one of which corresponds to the right horizontal tail, to be linearly dependent means that there must be one or two others that can *exactly* reproduce this combination of rolling, pitching, and yawing moments.

Moreover, the numbers in the B matrix represent aerodynamic data that has limited precision. They are based on wind-tunnel experiments or parameter identification of flight test results. Thus, even if some singularity is observed in one of the sub-matrices of B, one is justified in making small, aerodynamically insignificant additions to remedy the situation. This is not to say that some of these sub-matrices may not become ill-conditioned, and their determinants become very small. A good bit of computer code has been generated by the authors to check for such conditions and to deal with the problem.

Convexity is preserved from $\Omega \rightarrow \Phi$: This statement is almost axiomatic given a convex set of control effector ranges of motion and a linear transformation. A proof may be found in Bordignon (1996).

Columns of B correspond to edges in Φ: The edges in Ω map to edges in Φ in the directions of the corresponding columns of B. Calling these line segments in Φ 'edges' is misleading, since one thinks of edges as being on the bounding surface. We don't have names for the edge-like features that map to the interior of Φ (the thin dotted lines in Figure 5.4).

This means that edges associated with the action of u_i are mapped to directions corresponding to the ith column of B. For example, consider an edge corresponding to u_2:

$$\left\{ \begin{matrix} 0 \\ 2 \\ 1 \end{matrix} \right\}_o \tag{5.6}$$

To generate this edge u_1 is kept at $u_{1_{Min}}$, u_3 is kept at $u_{3_{Max}}$, and u_2 is varied from $u_{2_{Min}}$ to $u_{2_{Max}}$. A vector from the first vertex (at $u_{2_{Min}}$) to the second vertex (at $u_{2_{Max}}$), generically called **v**, is

$$\mathbf{v} = \left\{ \begin{matrix} u_{1_{Min}} \\ u_{2_{Max}} \\ u_{3_{Max}} \end{matrix} \right\}_u - \left\{ \begin{matrix} u_{1_{Min}} \\ u_{2_{Min}} \\ u_{3_{Max}} \end{matrix} \right\}_u = \left\{ \begin{matrix} 0 \\ u_{2_{Max}} - u_{2_{Min}} \\ 0 \end{matrix} \right\}_u \tag{5.7}$$

In the mapping $\mathbf{m} = B\mathbf{u}$, the first and third columns of B multiply zeros, while the second column multiplies a vector that describes the action of u_2.

In going from Figure 5.3 to Figure 5.4, edges generated by u_1 become edges in the direction of the first column of B, all parallel to m_1 in this example. Likewise edges of u_2 went to the direction of m_2. Those of u_3 mapped to directions down and to the left at $45°$.

Points on $\partial(\Phi)$: We assert that not all of the points on $\partial(\Omega)$ map to points on $\partial(\Phi)$, but all points on $\partial(\Phi)$ are maps of points on $\partial(\Omega)$. The first part has been demonstrated, and the second part is a consequence of the convexity being preserved in the mapping.

In words, while it is true that every moment on the convex hull of Φ—the attainable moments—is the image of a set of controls on the convex hull of Ω—the admissible controls—the converse is not true. Moreover, under our robust rank assumption, such moments are the images of *unique* sets of controls.

Null space of B and the boundary of Φ: Some attainable moments are generated by an infinite number of admissible controls. One may visualize the null space of our 2×3 problem by reconsidering Figure 5.4 with some extra detail, as Figure 5.5.

We have identified one moment **m**, plus two edges from Ω, $\{2 \; 0 \; 0\}_o$ and $\{1 \; 2 \; 1\}_o$. Clearly there is some value of u_1, along with $u_{2_{Min}}$ and $u_{3_{Min}}$, that generates **m**. There is also some value of u_2, along with $u_{1_{Max}}$ and $u_{3_{Max}}$, that generates **m**. Moreover, every admissible combination of controls along a line connecting these two points in Ω also generates **m**.

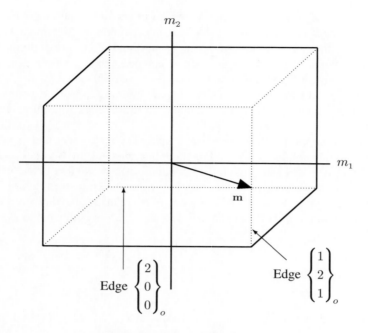

Figure 5.5   Null space of B in Φ

A demonstration of flight in the null space is shown in Figure 5.6. All three airplanes are in steady formation flight; in other words, none is turning and none is generating a net moment. The two wingmen are employing ailerons, rudder, and sideslip (sideslip viewed here as a control) to balance the moments. Different combinations of the three controls generate the same net moment.[2] The two wingmen are performing *steady-heading sideslip* maneuvers. Figure 5.6 also suggests a problem with the utilization of the null space of B in flight control applications. In general, the difficulty is that we have only considered the moments produced by the control effectors in formulating the control allocation problem. The controls affect more than the moments applied to the airplane; they also create increments of drag, for example. The two wingmen in Figure 5.6 are undoubtedly at a higher power setting than the lead F2F because of the unnecessary use of aerodynamic controls.

In addition to increased drag, unnecessary deflections of control effectors also places a higher demand on the mechanism that powers the control system: the pilot's muscles in Figure 5.6, hydraulic pumps in modern applications.

One last observation regarding Figure 5.6: there are structural issues. The application of ailerons and rudders at cross-purposes generates internal torques on the wings, empenage, and fuselage of the airplane. Steady-heading sideslips are not unusual in normal flight—they are often used to bleed off energy in an approach to landing wherein the pilot finds himself high and fast. But the routine use of applications of the control effectors that pointlessly stress the airframe is undesirable.

---

[2] The bank angle is also being used as a control of sorts, but only to balance the side force due to sideslip, using a component of gravity.

Figure 5.6 Null space in flight. VF-2 Grumman F2F-1s doing the Flying Chiefs' famous 'razzle dazzle' section formation over the San Diego area. Assigned to USS Lexington (CV-2) 1937–38. Credit: USN, Bill Swisher collection

5.3 The Two-moment Problem

The method for the two-moment problem is shown by example.

Example 5.1

Determination of Φ_2 Consider again Figure 5.4. We can clearly see the six edges of Φ, but now we want to determine the coordinates of the vertices that define those six edges. For example, the left and right vertical edges clearly mapped from the action of u_2 because the second column of B points in the vertical direction in moment space. The upper and lower horizontal edges are mapped from the action of u_1 for a similar reason.

In fact, we may generalize that a zero in a row of B means that there is an edge orthogonal to the axis (row of B) with the zero, and that edge is generated by the action of the control corresponding to that column. There can't be more than one zero, or our assumption of robust full rank will be violated.

Again with reference to Figure 5.4, we see there are four candidates for the two positions of left and right edge. In object notation they are

$$\left\{\begin{matrix}0\\2\\0\end{matrix}\right\}_o \quad \left\{\begin{matrix}1\\2\\0\end{matrix}\right\}_o \quad \left\{\begin{matrix}0\\2\\1\end{matrix}\right\}_o \quad \left\{\begin{matrix}1\\2\\1\end{matrix}\right\}_o \tag{5.8}$$

Two of these edges in Ω map to the boundary of Φ, while the other two become thin dashed lines in the interior. The determination of which goes where is a trivial problem in optimization. The right edge maximizes the value of m_1, and the left edge minimizes it. Considering the B matrix assumed for this example,

$$B = \begin{bmatrix} 1 & 0 & -0.5 \\ 0 & 1 & -0.5 \end{bmatrix} \tag{5.9}$$

If we wish to maximize the value of m_1 (on the edge farthest to the right), corresponding to the first row of B, then $B(1,1)$ (a positive number) should multiply the greatest value of u_1 (which is $u_{1_{Max}}$); and $B(1,3)$ (a negative number) should multiply the smallest value of u_3 (which is $u_{3_{Min}}$).

In other words, by observing the signs of the entries in a row of B that has a zero in the ith column, we may determine which of the edges associated with u_i are extrema in the directions (in moment space) along the axis of that row. That is, with a zero in $B(1,2)$, a positive entry in $B(1,1)$, and a negative entry in $B(1,3)$, we may immediately conclude that the right-most vertical edge of Φ is

$$\mathbf{u}_{120} = \left\{\begin{matrix}1\\2\\0\end{matrix}\right\}_o \tag{5.10}$$

This corresponds to the physical vectors

$$\mathbf{u}_{120} = \left\{\begin{matrix}u_{1_{Max}}\\u_2\\u_{3_{Min}}\end{matrix}\right\}_u \tag{5.11}$$

Because of symmetry, the left vertical edge of Φ is

$$\mathbf{u}_{021} = \left\{\begin{matrix}0\\2\\1\end{matrix}\right\}_o \tag{5.12}$$

This corresponds to the physical vectors

$$\mathbf{u}_{021} = \left\{\begin{matrix}u_{1_{Min}}\\u_2\\u_{3_{Max}}\end{matrix}\right\}_u \tag{5.13}$$

If we are going to calculate the coordinates of the vertices of $\partial(\Phi)$, we now have four of them:

$$B\left\{\begin{matrix} u_{1_{Max}} \\ u_{2_{Min}} \\ u_{3_{Min}} \end{matrix}\right\}_u \quad B\left\{\begin{matrix} u_{1_{Max}} \\ u_{2_{Max}} \\ u_{3_{Min}} \end{matrix}\right\}_u \quad B\left\{\begin{matrix} u_{1_{Min}} \\ u_{2_{Min}} \\ u_{3_{Max}} \end{matrix}\right\}_u \quad B\left\{\begin{matrix} u_{1_{Min}} \\ u_{2_{Max}} \\ u_{3_{Max}} \end{matrix}\right\}_u \tag{5.14}$$

Using our assumed B matrix and control limits, the coordinates of these four vertices in moment space are, respectively,

$$\left\{\begin{matrix} 0.75 \\ -0.25 \end{matrix}\right\}_m \quad \left\{\begin{matrix} 0.75 \\ 0.75 \end{matrix}\right\}_m \quad \left\{\begin{matrix} -0.75 \\ -0.75 \end{matrix}\right\}_m \quad \left\{\begin{matrix} -0.75 \\ 0.25 \end{matrix}\right\}_m \tag{5.15}$$

The question naturally arises, what if there are no zeroes in a particular column? The answer is, we rotate the axes in moment space to put one there. We could do this by orienting either row to be perpendicular to the edge of interest; in fact, the second row already has a zero corresponding to u_2. To generalize the algorithm somewhat, we will pick the first row and stick with it. Thus to orient the m_1 axis with the edges associated with u_2, we need to rotate the axes 90° counter-clockwise. However, we will not generally know the required rotation angle, so we approach the problem from another direction.

To begin, observe that we do not need a pure rotation matrix. In fact, we don't even care what happens to the second row, as there is no useful information there. We need only the first row of our transformation matrix, applied to the column of interest from B to make the first entry there equal to zero.

Designate the first row of the transformation matrix $\mathbf{t}$ and the ith column of B as $\mathbf{b}_i$,

$$\mathbf{t} \equiv \begin{bmatrix} t_{11} & t_{12} \end{bmatrix} \quad \mathbf{b}_i \equiv \begin{bmatrix} b_{1i} \\ b_{2i} \end{bmatrix} \tag{5.16}$$

If $b_{1i} = 0$ then the operation is unnecessary. Otherwise, we want $\mathbf{t}$ such that

$$t_{11}b_{1i} + t_{12}b_{2i} = 0 \tag{5.17}$$

We may arbitrarily choose a non-zero value for either t_{11} or t_{12} and solve for the other. We pick $t_{12} = 1$ and solve for t_{11}:

$$t_{11} = -\frac{b_{2i}}{b_{1i}} \tag{5.18}$$

Therefore,

$$\mathbf{t} = \begin{bmatrix} -\frac{b_{2i}}{b_{1i}} & 1 \end{bmatrix} \tag{5.19}$$

Now, multiplying $\mathbf{t}B$ will result in a zero in $B(1, i)$ and the signs of the other entries in the first row of B tell us which of the other controls should be set at their minimum and which at their maximum values to define the two edges.

Applying this method to our presumed B matrix,

$$B = \begin{bmatrix} 1 & 0 & -0.5 \\ 0 & 1 & -0.5 \end{bmatrix} \tag{5.20}$$

For u_1, $\mathbf{t} = \begin{bmatrix} 0 & 1 \end{bmatrix}$ and $\mathbf{t}B = \begin{bmatrix} 0 & 1 & -1 \end{bmatrix}$. Thus the top horizontal edge is

$$\mathbf{u}_{210} = \left\{ \begin{array}{c} 2 \\ 1 \\ 0 \end{array} \right\}_o \tag{5.21}$$

The bottom horizontal edge is

$$\mathbf{u}_{201} = \left\{ \begin{array}{c} 2 \\ 0 \\ 1 \end{array} \right\}_o \tag{5.22}$$

The edges for u_2 were previously determined. For u_3, $\mathbf{t} = \begin{bmatrix} -1 & 1 \end{bmatrix}$ and $\mathbf{t}B = \begin{bmatrix} -1 & 1 & 0 \end{bmatrix}$. The two diagonal edges are then

$$\mathbf{u}_{012} = \left\{ \begin{array}{c} 0 \\ 1 \\ 2 \end{array} \right\}_o \tag{5.23}$$

and

$$\mathbf{u}_{102} = \left\{ \begin{array}{c} 1 \\ 0 \\ 2 \end{array} \right\}_o \tag{5.24}$$

This completes the determination of the geometry of the attainable moments. The order of the controls that generate the vertices, beginning with the upper-right and proceeding clockwise, is

$$\left\{ \begin{array}{c} 1 \\ 1 \\ 0 \end{array} \right\}_o \quad \left\{ \begin{array}{c} 1 \\ 0 \\ 0 \end{array} \right\}_o \quad \left\{ \begin{array}{c} 1 \\ 0 \\ 1 \end{array} \right\}_o \quad \left\{ \begin{array}{c} 0 \\ 0 \\ 1 \end{array} \right\}_o \quad \left\{ \begin{array}{c} 0 \\ 1 \\ 1 \end{array} \right\}_o \quad \left\{ \begin{array}{c} 0 \\ 1 \\ 0 \end{array} \right\}_o \tag{5.25}$$

In the same order, the actual moments associated with those vertices are:

$$\left\{ \begin{array}{c} 0.75 \\ 0.75 \end{array} \right\}_m \quad \left\{ \begin{array}{c} 0.75 \\ -0.25 \end{array} \right\}_m \quad \left\{ \begin{array}{c} 0.25 \\ -0.75 \end{array} \right\}_m \quad \left\{ \begin{array}{c} -0.75 \\ -0.75 \end{array} \right\}_m \quad \left\{ \begin{array}{c} -0.75 \\ 0.25 \end{array} \right\}_m \quad \left\{ \begin{array}{c} -0.25 \\ 0.75 \end{array} \right\}_m \tag{5.26}$$

It is clear that, given our robust-rank assumption, each of the vertices of Φ corresponds one-to-one with a unique control vector in Ω. When we do not care about the numerical values of the vertices in moment space we will give them a notation that reflects the controls that generated that vertex. Thus, the notation $\mathbf{m}_{110}$ refers to the moment generated by the control vector whose object notation is

$$\left\{ \begin{array}{c} 1 \\ 1 \\ 0 \end{array} \right\}_o \tag{5.27}$$

In other words,

$$\mathbf{m}_{110} = \left\{ \begin{array}{c} 0.75 \\ 0.75 \end{array} \right\}_m \tag{5.28}$$

We will also have occasion to refer to vectors in moment space according to the vertices that define it, so $\mathbf{m}_{102}$ is a vector from $\mathbf{m}_{100}$ to $\mathbf{m}_{101}$, always from the minimum to the maximum of the varied control effector.

$$\mathbf{m}_{102} = \mathbf{m}_{101} - \mathbf{m}_{100} = \left\{ \begin{array}{c} 0.25 \\ -0.75 \end{array} \right\}_m - \left\{ \begin{array}{c} 0.75 \\ -0.25 \end{array} \right\}_m = \left\{ \begin{array}{c} -0.50 \\ -0.50 \end{array} \right\}_m \tag{5.29}$$

The current example may seem trivial, but the numbers of edges of Ω do add up quickly as more controls are added to the problem. Observe Figure 5.7, which represents four control effectors and two moments. The bounding edges of that rather complicated-looking figure may be calculated with only one more step than was used above.

5.3.0.1 Summary of the Determination of Φ_2

Given a $2 \times m$ control effectiveness matrix B and control effector limits $\mathbf{u}_{Min} \leq \mathbf{u} \leq \mathbf{u}_{Max}$:

1. Repeat the following for $i = 1 \cdots m$.
2. Take the ith column of B, $\mathbf{b}_i$.
3. If the first entry of $\mathbf{b}_i$ is zero, skip ahead to step 6.

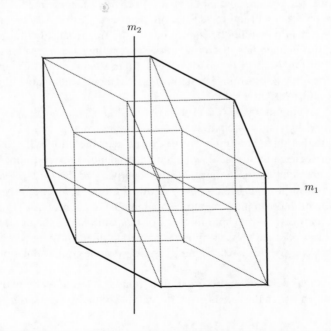

Figure 5.7   A four-dimensional Ω mapped to to a two-dimensional Φ

4. Calculate **t**:

$$\mathbf{t} = \left[-\frac{b_{2i}}{b_{1i}} \; 1 \right]$$

(5.30)

5. Calculate **t**B.
6. Evaluate the maximum edge in object notation according to the sign of the entries in the first row of **t**B (or the first row of B if steps 4 and 5 were skipped). If an entry is positive, assign the number 1. If it is negative, assign the number 0. If it is zero, assign the number 2. If there is more than one zero in the first row of **t**B, the problem is degenerate.
7. Evaluate the minimum edge in object notation according to the sign of the entries in the first row of **t**B. If an entry is positive, assign the number 0. If it is negative, assign the number 1. If it is zero, assign the number 2.
8. Evaluate the two vertices associated with each of the two edges found in steps 6 and 7 in object notation. One vertex of each is found by replacing the sole number 2 with a 0, another by replacing it with a 1, repeated for each of the maximum and minimum edges.
9. Convert the four vertices found in step 8 to control notation. For each entry that is 0, assign the lower limit of the corresponding control effector. For each entry that is 1, assign the upper limit of the corresponding control effector.
10. Calculate the four moment vectors that comprise four vertices of the convex hull of the attainable moment set, $\partial(\Phi_2)$ by multiplying each of the vectors found in step 9 by B.

5.3.1 Area Calculations

We propose as a measure for comparison of different control allocation schemes, the area of the two-dimensional set of attainable moments. As we will see, some methods of solving the control allocation problem will fail for desired moments, $\mathbf{m}_{des}$, that are demonstrably attainable; in other words, that lie within the convex hull of Φ. This means that the set of all points within Φ that can be obtained lies strictly within $\partial(\Phi)$, and thus has a smaller area.

The argument for this basis of comparison is that one should not be leaving out some of the theoretical capabilities of a control system. If the moment is attainable, the flight control system should be able to attain it.

There are many arguments against the criteria of maximum area, most having to do with the price to be paid for achieving it. We will explore this in more depth later.

One could argue that there are regions of Φ that no sensible pilot will ever visit. Most of these regions are associated with nose-down pitching moments, since restraint systems in most tactical airplanes are deficient in the presence of negative g forces, and pilots are averse to having their head pressed against the canopy. However, one need find only one maneuver that requires all the nose-down pitching moment attainable to counter this argument, and that maneuver is spin recovery. During spin recovery of many airplanes, the procedure places the stick full forward as well as full left or right. If there are regions of Φ that are seldom visited, those rare occasions may be those when all the capability of the control effectors is needed for survival.

The area calculations of Φ are quite simple: If a triangle is specified by two vectors $\mathbf{v}_1$ and $\mathbf{v}_2$ originating at a point, then the area is given by half that of the corresponding parallelogram:

$$A_\Delta = \frac{1}{2} |\mathbf{v}_1 \times \mathbf{v}_2|$$

(5.31)

Given a list of the vertices of $\partial(\Phi)$ (as moments), we can treat each edge as the base of a triangle that has the origin of moment space as its vertex, and sum them to get the total area of Φ.

We have an ordered list of the vertices of $\partial(\Phi)$ in our continuing example in Eq. (5.26). Therefore, the area of this particular Φ (Figure 5.4), A_Φ, is

$$
\begin{aligned}
A_\Phi = {} & \tfrac{1}{2}|\mathbf{m}_{110} \times \mathbf{m}_{100}| + \tfrac{1}{2}|\mathbf{m}_{100} \times \mathbf{m}_{101}| + \tfrac{1}{2}|\mathbf{m}_{101} \times \mathbf{m}_{001}| \\
& + \tfrac{1}{2}|\mathbf{m}_{001} \times \mathbf{m}_{011}| + \tfrac{1}{2}|\mathbf{m}_{011} \times \mathbf{m}_{010}| + \tfrac{1}{2}|\mathbf{m}_{010} \times \mathbf{m}_{110}|
\end{aligned}
\tag{5.32}
$$

To get proper cross products using MATLAB$^\circledR$, the vectors need to be three-dimensional, so we augment the two-dimensions of moment space with a third dimension that has coordinate zero:

$$
A_\Phi = \frac{1}{2} \left| \left\{ \begin{matrix} 0.75 \\ 0.75 \\ 0 \end{matrix} \right\}_m \times \left\{ \begin{matrix} 0.75 \\ -0.25 \\ 0 \end{matrix} \right\}_m \right| + \frac{1}{2} \left| \left\{ \begin{matrix} 0.75 \\ -0.25 \\ 0 \end{matrix} \right\}_m \times \left\{ \begin{matrix} 0.25 \\ -0.75 \\ 0 \end{matrix} \right\}_m \right| + \cdots
\tag{5.33}
$$

For our example, this results in

$$
A_\Phi = 1.625
\tag{5.34}
$$

The units of A_Φ are the square of whatever units of moment are being used.

5.4   The Three-moment Problem

The three-moment problem is a natural extension of the two-moment problem. The biggest problem will be our ability to draw pictures of an n-dimensional hyper-rectangle mapped into three-dimensional space, and then drawn on a two-dimensional surface (a page of this book).

5.4.1   Determination of Φ_3

With respect to our previous discussion of the two-moment problem, the primary difference is that the set of admissible controls (Ω) is now mapped to a three-dimensional moment space. The three-dimensional set of attainable moments (Φ) is bounded by two-dimensional facets. Whereas edges in moment space are defined by varying a single control effector, facets are defined by varying two control effectors while all the others are at one or the other of their limits.

When seeking the convex hull of Φ we must align an axis (assumed hereinafter to be the first row, L, C_ℓ, or $\dot{p}$) of the control effectiveness matrix B such that it is perpendicular to a facet. That is, we seek a transformation matrix T that results in not one but *two* zeroes in the first row of TB. Say we want the facet that corresponds to the actions of u_i and u_j. We designate the first row of the transformation matrix $\mathbf{t}$, and the ith and jth columns of B as $\mathbf{b}_i$ and $\mathbf{b}_j$. For the three-moment problem, $\mathbf{t}$, $\mathbf{b}_i$, and $\mathbf{b}_j$ each have three components.

$$
\mathbf{t} \equiv \begin{bmatrix} t_{11} & t_{12} & t_{13} \end{bmatrix} \qquad
\mathbf{b}_i \equiv \begin{bmatrix} b_{1i} \\ b_{2i} \\ b_{3i} \end{bmatrix} \qquad
\mathbf{b}_j \equiv \begin{bmatrix} b_{1j} \\ b_{2j} \\ b_{3j} \end{bmatrix}
\tag{5.35}
$$

If $b_{1i} = 0$ and $b_{1j} = 0$ then the operation is unnecessary. Otherwise, we want $\mathbf{t}$ such that

$$\mathbf{t}[\mathbf{b}_i \quad \mathbf{b}_j] = [0 \quad 0] \tag{5.36}$$

$$t_{11}b_{1i} + t_{12}b_{2i} + t_{13}b_{3i} = 0$$

$$t_{11}b_{1j} + t_{12}b_{2j} + t_{13}b_{3j} = 0 \tag{5.37}$$

We arbitrarily choose a non-zero value for t_{13} and solve for the other two elements of $\mathbf{t}$. We pick $t_{13} = 1$ so that

$$t_{11}b_{1i} + t_{12}b_{2i} = -b_{3i}$$

$$t_{11}b_{1j} + t_{12}b_{2j} = -b_{3j} \tag{5.38}$$

As a matrix equation,

$$\begin{bmatrix} b_{1i} & b_{2i} \\ b_{1j} & b_{2j} \end{bmatrix} \begin{Bmatrix} t_{11} \\ t_{12} \end{Bmatrix} = - \begin{Bmatrix} b_{3i} \\ b_{3j} \end{Bmatrix} \tag{5.39}$$

If

$$det \begin{bmatrix} b_{1i} & b_{2i} \\ b_{1j} & b_{2j} \end{bmatrix} \neq 0 \tag{5.40}$$

then

$$\begin{Bmatrix} t_{11} \\ t_{12} \end{Bmatrix} = - \begin{bmatrix} b_{1i} & b_{2i} \\ b_{1j} & b_{2j} \end{bmatrix}^{-1} \begin{Bmatrix} b_{3i} \\ b_{3j} \end{Bmatrix} \tag{5.41}$$

If the determinant is zero, this means that $t_{13} = 0$. In that case we apply t_{11} and t_{12} to the first two entries of $\mathbf{b}_i$ and $\mathbf{b}_j$, similar to what was done in the two-moment problem. In any event, our robust-rank requirement guarantees us a solution.

Now, multiplying $\mathbf{t}B$ will result in a zero in $B(1, i)$ and $B(1, j)$, and the signs of the other entries in the first row of B tell us which of the other controls should be set at their minimum and which at their maximum values to define the two u_i–u_j facets. That is, setting all the controls that have positive entries in the first row of B to their maximum limits, and all those with negative entries to their minimum limits, will yield one facet that bounds Φ. Then by applying the opposite rule, positive entries to minimum limits and negative entries to maximum will yield another.

To determine the entire set of facets that define $\partial(\Phi)$ requires the calculations described above to be performed for every combination of two control effectors. Given m control effectors and three moments, the number of required calculations n_{Calcs} is

$$n_{Calcs} = \frac{m!}{2!(m-2)!} = \frac{1}{2}m(m-1) \tag{5.42}$$

As before, m is the number of control effectors. The number 2 appears in this calculation because we are considering a three-moment problem, and the set of attainable moments is a three-dimensional polytope (not rectangular) bounded by two-dimensional facets. Each such facet is generated by the action of $m - 2$ control effectors being set at their upper or lower limits, and two control effectors being allowed to vary.

Since each calculation yields two facets, there are $m(m - 1)$ facets on $\partial(\Phi)$.

Example 5.2

Determination of Φ_3 for the ADMIRE Simulation Appendix B will detail the flight simulation used for examples throughout most of this book. Using representative low, slow data, the control effectiveness matrix (Eq. (2.20) in terms of the angular accelerations $\dot{p}$, $\dot{q}$, and $\dot{r}$) and control limits were determined. These are shown in Eqs. (5.43) and (5.44). The units of the control effectors are radians, and the B-matrix entries are $(rad/s^2)/rad$.

$$B = \begin{bmatrix} 0.7073 & -0.7073 & -3.4956 & -3.0013 & 3.0013 & 3.4956 & 2.1103 \\ 1.1204 & 1.1204 & -0.7919 & -1.2614 & -1.2614 & -0.7919 & 0.0035 \\ -0.3309 & 0.3309 & -0.1507 & -0.3088 & 0.3088 & 0.1507 & -1.2680 \end{bmatrix} \quad (5.43)$$

$$\mathbf{u}_{Min} = \left\{ \begin{matrix} -0.9599 \\ -0.9599 \\ -0.5236 \\ -0.5236 \\ -0.5236 \\ -0.5236 \\ -0.5236 \end{matrix} \right\}_u \quad \mathbf{u}_{Max} = \left\{ \begin{matrix} 0.4363 \\ 0.4363 \\ 0.5236 \\ 0.5236 \\ 0.5236 \\ 0.5236 \\ 0.5236 \end{matrix} \right\}_u \quad (5.44)$$

Table 5.4 Control effectors by name

u_i	Name
u_1	Right canard
u_2	Left canard
u_3	Right outboard elevon
u_4	Right inboard elevon
u_5	Left inboard elevon
u_6	Left outboard elevon
u_7	Rudder

Using the methods described in this section, the bounding facets of the attainable moments were determined. We illustrate the procedure using u_1 and u_3.

From the first (u_1) and third (u_3) columns of B we obtain

$$\begin{bmatrix} b_{11} & b_{21} \\ b_{13} & b_{23} \end{bmatrix} = \begin{bmatrix} 0.7073 & 1.1204 \\ -3.4956 & -0.7919 \end{bmatrix} \quad (5.45)$$

$$\left\{ \begin{matrix} b_{31} \\ b_{33} \end{matrix} \right\} = \left\{ \begin{matrix} -0.3309 \\ -0.1507 \end{matrix} \right\} \quad (5.46)$$

Performing the indicated operation yields

$$\mathbf{t} = [-0.1284 \quad 0.3764 \quad 1] \quad (5.47)$$

$$\mathbf{t}B = [\, 0 \quad 0.8434 \quad 0 \quad -0.3983 \quad -0.5513 \quad -0.5961 \quad -1.5376 \,] \quad (5.48)$$

$\mathbf{t}B$ is the first row of B after the m_1 ($\dot{p}$) axis has been rotated to be perpendicular to the facets formed by u_1 and u_3. The signs of the non-zero entries tell us that a maximum facet in that direction is given in object notation by $\mathbf{o}_{2120000}$, and the actual vectors are found by setting

$$u_2 = u_{2_{Max}}$$
$$u_4 = u_{4_{Min}}$$
$$u_5 = u_{5_{Min}}$$
$$u_6 = u_{6_{Min}}$$
$$u_7 = u_{7_{Min}}$$

The four vertices of this facet are found by setting u_1 and u_3 to the four combinations of $u_{1_{Min}}$, $u_{1_{Max}}$, $u_{3_{Min}}$, and $u_{3_{Max}}$. The minimum facet is found by swapping the subscripts *Min* and *Max* in the previous discussion.

The control deflections that determine the vertices of the maximum facet are therefore

$$\mathbf{u}_{0100000} = \left\{ \begin{array}{c} -0.9599 \\ 0.4363 \\ -0.5236 \\ -0.5236 \\ -0.5236 \\ -0.5236 \\ -0.5236 \end{array} \right\}_u \qquad \mathbf{u}_{0110000} = \left\{ \begin{array}{c} -0.9599 \\ 0.4363 \\ 0.5236 \\ -0.5236 \\ -0.5236 \\ -0.5236 \\ -0.5236 \end{array} \right\}_u$$

$$\mathbf{u}_{1100000} = \left\{ \begin{array}{c} 0.4363 \\ 0.4363 \\ -0.5236 \\ -0.5236 \\ -0.5236 \\ -0.5236 \\ -0.5236 \end{array} \right\}_u \qquad \mathbf{u}_{1110000} = \left\{ \begin{array}{c} 0.4363 \\ 0.4363 \\ 0.5236 \\ -0.5236 \\ -0.5236 \\ -0.5236 \\ -0.5236 \end{array} \right\}_u \qquad (5.49)$$

The coordinates of the vertices in moment space are found by multiplying each of these four vectors by B.

$$\mathbf{m}_{0100000} = \left\{ \begin{array}{c} -2.0925 \\ 1.5617 \\ 1.1259 \end{array} \right\}_m \qquad \mathbf{m}_{0110000} = \left\{ \begin{array}{c} -5.7531 \\ 0.7325 \\ 0.9681 \end{array} \right\}_m$$

$$\mathbf{m}_{1100000} = \left\{ \begin{array}{c} -1.1050 \\ 3.1260 \\ 0.6639 \end{array} \right\}_m \qquad \mathbf{m}_{1110000} = \left\{ \begin{array}{c} -4.7655 \\ 2.2968 \\ 0.5061 \end{array} \right\}_m \qquad (5.50)$$

The corresponding results for the minimum facet are:

$$\mathbf{u}_{0001111} = \left\{ \begin{array}{c} -0.9599 \\ -0.9599 \\ -0.5236 \\ 0.5236 \\ 0.5236 \\ 0.5236 \\ 0.5236 \end{array} \right\}_u \qquad \mathbf{u}_{0011111} = \left\{ \begin{array}{c} -0.9599 \\ -0.9599 \\ 0.5236 \\ 0.5236 \\ 0.5236 \\ 0.5236 \\ 0.5236 \end{array} \right\}_u$$

$$
\mathbf{u}_{1001111} = \left\{ \begin{array}{c} 0.4363 \\ -0.9599 \\ -0.5236 \\ 0.5236 \\ 0.5236 \\ 0.5236 \\ 0.5236 \end{array} \right\}_u
\quad
\mathbf{u}_{1011111} = \left\{ \begin{array}{c} 0.4363 \\ -0.9599 \\ 0.5236 \\ 0.5236 \\ 0.5236 \\ 0.5236 \\ 0.5236 \end{array} \right\}_u
\tag{5.51}
$$

$$
\mathbf{m}_{0001111} = \left\{ \begin{array}{c} 4.7655 \\ -3.4700 \\ -0.5061 \end{array} \right\}_m
\quad
\mathbf{m}_{0011111} = \left\{ \begin{array}{c} 1.1050 \\ -4.2993 \\ -0.6639 \end{array} \right\}_m
$$

$$
\mathbf{m}_{1001111} = \left\{ \begin{array}{c} 6.1234 \\ -1.3191 \\ -1.1414 \end{array} \right\}_m
\quad
\mathbf{m}_{1011111} = \left\{ \begin{array}{c} 2.4628 \\ -2.1484 \\ -1.2992 \end{array} \right\}_m
\tag{5.52}
$$

Proceeding with other pairs of controls, all $m(m-1) = 42$ facets were found. They are plotted, with shading to show the front-side, in Figure 5.8.

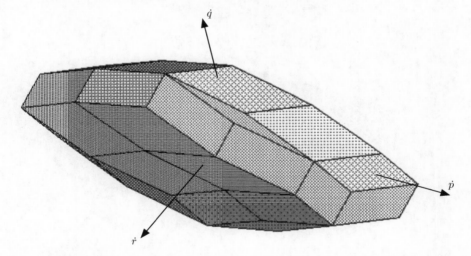

Figure 5.8 Attainable moments for the ADMIRE simulation. Trimmed at Mach 0.22 at 20 m of altitude. View is from $(\dot{p}\ \dot{q}\ \dot{r}) = (1\ 1\ 1)$

As to the efficiency of the algorithm presented, note that there are many other facets of Ω that mapped to the interior of Φ. In the present case, each of the five controls u_2, u_4–u_7, may be set on an upper or lower limit to define the two u_1–u_3 facets in Ω. This results in 2^5 possible combinations. The general formula is 2^{m-2}.

Figures such as 5.8 are useful for visualizing the maximum capabilities of a suite of control effectors. Later we will see how to combine the three-dimensional Π, (to be defined in Sections 5.5.1 and 6.10) with the figure to further visualize what one sacrifices in capability by using different control allocation schemes.

We may also observe Φ from viewpoints along each of the three axes, permitting us to see the capabilities of the suite of control effectors in the planes $\dot{p}$–$\dot{q}$, $\dot{q}$–$\dot{r}$, and $\dot{p}$–$\dot{r}$. These are shown in Figures 5.9, 5.10, and 5.11, respectively. To help give an idea of the scale of the figures, the maximum capabilities are given in Table 5.5. Note that the maximum values shown are those obtained about the origin; that is, with the other two accelerations at zero.

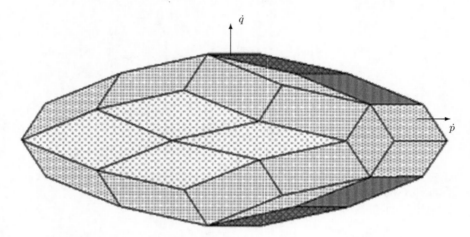

Figure 5.9   Plane of $\dot{p}$–$\dot{q}$

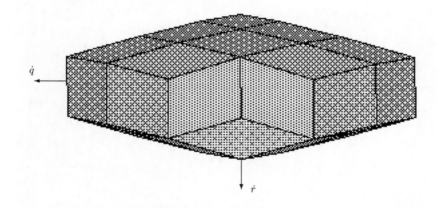

Figure 5.10   Plane of $\dot{q}$–$\dot{r}$

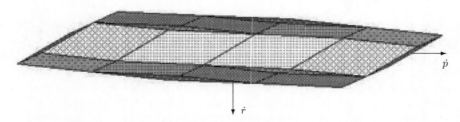

Figure 5.11   Plane of $\dot{p}$–$\dot{r}$

Table 5.5 Minimum and maximum accelerations about the origin (rad/s^2)

Coefficient	Minimum	Maximum	Positive
$\dot{p}$	−7.74	7.74	Right wing down
$\dot{q}$	−4.30	3.13	Nose up
$\dot{r}$	−1.40	1.40	Nose right

5.4.1.1 Summary of the Determination of Φ_3

Given a $3 \times m$ control effectiveness matrix B and control effector limits $\mathbf{u}_{Min} \le \mathbf{u} \le \mathbf{u}_{Max}$:

1. Repeat the following for $i = 1 \cdots m - 1$ and $j = i + 1 \cdots m$. This yields every combination of m controls taken two at a time.
2. Take the ith and jth columns of B $\mathbf{b}_i$ and $\mathbf{b}_j$ to make the 3×2 matrix B_{ij}.

$$B_{ij} \equiv \begin{bmatrix} \mathbf{b}_i & \mathbf{b}_j \end{bmatrix} \tag{5.53}$$

3. If the first row of B_{ij} are zeros, skip ahead to step 6.
4. Calculate $\mathbf{t} = \begin{bmatrix} t_{11} & t_{12} & t_{13} \end{bmatrix}$ from

$$\mathbf{t}B_{ij} = \begin{bmatrix} 0 & 0 \end{bmatrix} \tag{5.54}$$

   There are a few different possibilities that arise in the solving of Eq. (5.54). In the most common case the equation may be solved by setting $t_{13} = 1$, rearranging the terms, and solving for the rest of $\mathbf{t}$. In any event, the problem is guaranteed to have a solution if the robust-rank requirement we have imposed on the control effectiveness matrix B is honored.
5. Calculate $\mathbf{t}B$.
6. Evaluate the maximum facet in object notation according to the sign of the entries in the first row of $\mathbf{t}B$ (or the first row of B if steps 4 and 5 were skipped). If an entry is positive, assign the number 1. If it is negative, assign the number 0. If it is zero, assign the number 2. If there are more than two zeros in the first row of $\mathbf{t}B$, the problem is degenerate.
7. Evaluate the minimum facet in object notation according to the sign of the entries in the first row of $\mathbf{t}B$. If an entry is positive, assign the number 0. If it is negative, assign the number 1. If it is zero, assign the number 2.
8. Evaluate the eight vertices associated with the two facets found in steps 6 and 7 in object notation. The four vertices of each facet are found by setting the two 2s to the four combinations of 0 and 1; that is, (0 0), (0 1), (1 1), and (1 0).
9. Convert the eight vertices found in step 8 to control notation. For each entry that is 0, assign the lower limit of the corresponding control effector. For each entry that is 1, assign the upper limit of the corresponding control effector.
10. Calculate the eight moment vectors that comprise eight vertices of the convex hull of the attainable moment set, $\partial(\Phi_3)$ by multiplying each of the vectors found in step 9 by B.

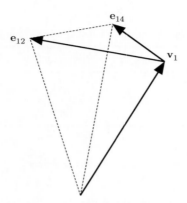

Figure 5.12   Tetrahedron defined by three vectors

5.4.2   *Volume Calculations*

It is easy to see that all our facets are parallelograms that, together with the origin, form the bases of pyramids. There are several ways to calculate the volume of a pyramid, but since we have expressed the vertices as vectors from the origin, we will use the vector triple product. Later when we seek the volume of the moments attainable by other methods we will generalize this method.

Consider Figure 5.12. It shows a vector $\mathbf{v}_1$ from the origin and two edges, $\mathbf{e}_{12}$ and $\mathbf{e}_{14}$, emanating from the tip of that vector. These three vectors define a tetrahedron, completed by the three dashed lines.

The volume of this tetrahedron is given by geometers (Hausner 1965) as one sixth the magnitude of the vector triple product (in any order) of $\mathbf{v}_1$, $\mathbf{e}_{12}$, and $\mathbf{e}_{14}$. We choose the order shown:

$$V = \frac{1}{6}|\mathbf{v}_1 \cdot (\mathbf{e}_{12} \times \mathbf{e}_{14})| \tag{5.55}$$

The geometers also tell us that the triple product may be evaluated as the determinant of a 3×3 matrix whose columns are $\mathbf{v}_1$, $\mathbf{e}_{12}$, and $\mathbf{e}_{14}$:

$$V = \frac{1}{6}|[\ \mathbf{v}_1 \quad \mathbf{e}_{12} \quad \mathbf{e}_{14}\ ]| \tag{5.56}$$

The symbols $|\cdot|$ have two meanings here. They may be used to denote the determinant of a matrix, or the absolute value of the enclosed quantity. In Eq. (5.56) we mean both: evaluate the determinant, and take its absolute value.

Our pyramids each consist of two tetrahedrons that share the origin and two vertices. Due to the symmetry of the facets, the volume of the two tetrahedrons will be the same, so we just double the volume of one or the other of the tetrahedrons.

Example 5.3

Volume of Φ_3 for the ADMIRE Simulation   To relate these volume calculations to the facets defined above in Section 5.4.1, we know the coordinates of the four vertices that define a facet. These vertices define the four edges. We need an *ordered* list of the edges so that we can associate two of them with a particular vertex in order to evaluate the triple product. To get this ordered list we may use object notation: Beginning with any vertex the two edges of the facet that have this vertex in common are found by changing one of its defining controls to a 2.

The result of the operations in Section 5.4.1 is information as to which controls should be constrained to their minimum value (object notation 0), or their maximum value (object notation 1), or left free to vary across the facet (object notation 2). Thus in Figure 5.8 we consider the facet defined by the two controls u_1 and u_2, determined to be $o_{2200110}$. This leads to Figure 5.13.

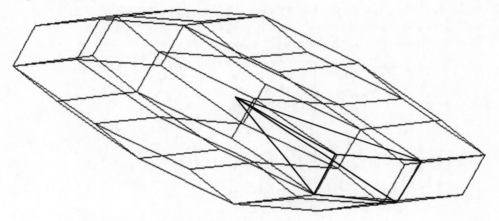

Figure 5.13   The facet $\mathbf{o}_{2200110}$ (from $u_1 - u_2$) and its vertices *in situ*

In Figure 5.13 the AMS is shown in wire-frame, and one of the two facets associated with u_1 and u_2 and its four vertices is highlighted. This same facet is removed for clarity, as shown in Figure 5.14.

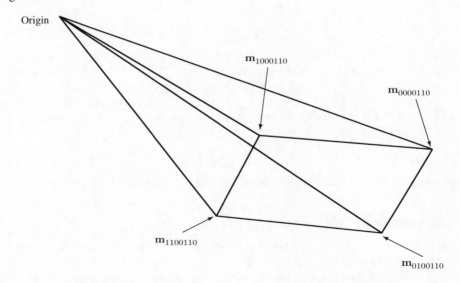

Figure 5.14   The facet $\mathbf{o}_{2200110}$ and its vertices, removed for examination

In Figure 5.14 the edges may be defined as described above. Take as the starting vertex $\mathbf{m}_{0000110}$. The two edges that share it are $\mathbf{m}_{0200110}$ and $\mathbf{m}_{2000110}$. A vector in moment space along the first of these is given by $\mathbf{m}_{0000110} - \mathbf{m}_{0100110}$, and along the second by $\mathbf{m}_{0000110} - \mathbf{m}_{1000110}$. With respect to Figure 5.14, this is simple vector addition.

Taking $\mathbf{m}_{0000110}$ as $\mathbf{v}_1$ and the two edges as $\mathbf{e}_{12}$ and $\mathbf{e}_{13}$, Eq. (5.56) becomes

$$V_{1/2} = \frac{1}{6} |[\ \mathbf{m}_{0000110} \quad (\mathbf{m}_{0000110} - \mathbf{m}_{0100110}) \quad (\mathbf{m}_{0000110} - \mathbf{m}_{1000110})\ ]| \tag{5.57}$$

Note that this is only half the volume of the pyramid in Figure 5.14 ($V_{1/2}$), since our pyramid is made up of two tetrahedrons.

The moments corresponding to the defining controls are calculated, yielding

$$\mathbf{m}_{0000110} = \begin{Bmatrix} 5.6986 \\ -2.1528 \\ 1.1451 \end{Bmatrix}_m \tag{5.58}$$

$$\mathbf{m}_{0000110} - \mathbf{m}_{0100110} = \begin{Bmatrix} 5.6986 \\ -2.1528 \\ 1.1451 \end{Bmatrix}_m - \begin{Bmatrix} 4.7111 \\ -0.5885 \\ 1.6071 \end{Bmatrix}_m = \begin{Bmatrix} 0.9875 \\ -1.5638 \\ -0.4620 \end{Bmatrix}_m \tag{5.59}$$

$$\mathbf{m}_{0000110} - \mathbf{m}_{1000110} = \begin{Bmatrix} 5.6986 \\ -2.1528 \\ 1.1451 \end{Bmatrix}_m - \begin{Bmatrix} 6.6861 \\ -0.5885 \\ 0.6831 \end{Bmatrix}_m = \begin{Bmatrix} -0.9875 \\ -1.5638 \\ 0.4620 \end{Bmatrix}_m \tag{5.60}$$

$$V_{1/2} = \frac{1}{6}\ det \begin{bmatrix} 5.6986 & 0.9875 & -0.9875 \\ -2.1528 & -1.5638 & -1.5638 \\ 1.1451 & -0.4620 & 0.4620 \end{bmatrix} = -1.9619 \tag{5.61}$$

The value of the determinant actually is negative, a consequence of the order in which we placed the two edges. We could either reverse the order of these edges (swap the second and third columns of the matrix) and get a positive result, or just ignore the sign.

The volume of half of Figure 5.14 is 1.9619 $(\text{rad}/\text{s}^2)^3$. When we apply this procedure to each of the facets of Φ, Figure 5.8, and double each to get the total volume of each pyramid, we obtain the volume of Φ, 177.1 $(\text{rad}/\text{s}^2)^3$.

$$V_\Phi = 177.1\ (\text{rad}/\text{s}^2)^3 \tag{5.62}$$

5.5   Significance of the Maximum Set

The *maximum set* is the set of all attainable moments. We showed in Sections 5.1 and 5.4.1 how to determine the geometry of the maximum set, and in Sections 5.3.0.1 and 5.4.2 ways to calculate measures of the size of it. We will next examine some of the uses to be made of this information.

5.5.1 As a Standard of Comparison of Different Methods

In Chapter 6 we will examine various methods of control allocation; that is, that find vectors **u** that satisfy $B\mathbf{u} = \mathbf{m}_{des}$ for given B and $\mathbf{m}_{des}$. We will see that many of these methods when presented with some $\mathbf{m}_{des}$ that is clearly attainable will return a solution **u** that is not admissible. In other words, the moments for which the particular algorithm returns admissible controls will be a proper subset of Φ.

We name this subset Π and we illustrate it notionally in the two-dimensional case by the polygonal shape in Figure 5.15.

In Figure 5.15 three desired moments are shown. If the hypothetical control allocation algorithm is given $\mathbf{m}_{d_1}$ then the solution **u** that it returns will be admissible. If given $\mathbf{m}_{d_2}$, however, one or more of the control effectors in the calculated vector **u** will be commanded to an inadmissible position. If any control allocation scheme is required to produce $\mathbf{m}_{d_3}$, then no admissible control effector vector **u** will be found.

For some control allocation methods we may be able to determine all the moments in the subset Π analytically, and possibly to calculate the area or volume of moment space that they occupy. In Figure 5.15 it is easy to visualize the triangles formed by the edges of the polyhedron $\partial(\Pi)$ and anticipate calculating the areas of these triangles for comparison with V_Φ of the two-dimensional moments.

In other cases, however, analytical solutions will not be available. In those cases we will, for given control effectiveness matrices B and control effector limits, present the algorithm with $\mathbf{m}_{des}$ in a variety of directions in moment space and for each we will vary the length of the vector by successively smaller amounts about attainability, to the point at which one or more control effectors in the solution is just saturated. That is, for some control allocation solution, method, or algorithm, we will methodically generate a large number of directions in moment

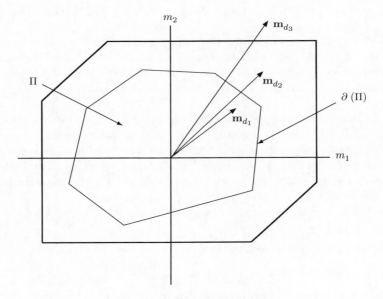

Figure 5.15 Π in Φ

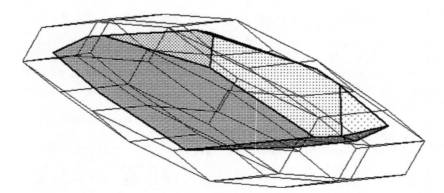

Figure 5.16   Pseudo-inverse Π within the three-dimensional wire-frame of $\partial(\Phi)$. The viewpoint is identical to that in Figure 5.8

space. For each we will begin with a nominal $\mathbf{m}_{des}$ in that direction and ask the method or algorithm for its solution. If that solution is admissible, we will increase the magnitude of $\mathbf{m}_{des}$ by some step, repeating until at least one control effector is commanded to a position beyond its limits: minimum or maximum. Depending on the precision desired in determining Π for this method or algorithm, we may then step back by a smaller step until the solution is once again admissible, then increase by an even smaller step, and so on until some criteria is reached. Useful methods for increasing and decreasing the step sizes are bisection and the golden mean.

Once determined, those points lie on $\partial(\Pi)$, at least approximately. With enough points we may determine its convex hull using any suitable algorithm, such as the aforementioned command `convhull` in MATLAB®. The output of `convhull` is sufficient to approximate the area of Π (approximate, because there is not always a guarantee that the method being analyzed creates convex sets).

Figure 5.16 repeats Figure 5.8 but with only the wire-frame of $\partial(\Phi)$ shown. Within the wire-frame nestles the moments that can be attained *using admissible controls* by the pseudo-inverse solution, Π. The moments attainable by this control allocation method may be determined analytically, so the use of a convex-hull algorithm was not necessary to calculate its Π.

Figure 5.16 clearly illustrates the significance of volume as a comparative measure of the performance of different methods of allocating the control effectors. The data used to create the figure are representative of a real tactical fighter aircraft, and the pseudo-inverse, or something like it, was for many years the default method of allocating redundant control effectors. But as the figure shows, the use of this straightforward solution sacrifices a great amount of the potential moment-generating capabilities of the control effectors.

While the data supporting Figure 5.15 is made-up, the basis of Figure 5.16 decidedly is not.

5.5.2   Maneuver Requirements

5.5.2.1   Specified Maneuver

In something of an extension to the concepts of dynamic inversion and model-following control is the idea of determining the control effector displacements required to perform a particular maneuver. The description of the maneuver may come from mission requirements, such as the many maneuvers performed in air combat maneuvering (Shaw 1985), (Wilson *et al.* 1993).

For example, a velocity–vector roll is often desired for tactical airplanes maneuvering in a high angle-of-attack flight condition (Durham *et al.* 1994). At high angle of attack, a roll around the principal x-body axis will generally generate large amounts of sideslip during the roll. Large angles of sideslip are generally undesirable. The flight control computer is therefore expected to perform the roll about the velocity vector, which means keeping angle of attack constant while regulating sideslip to zero, or something close to zero. This in turn requires very well-coordinated control of moments and accelerations in all three axes.

The process of converting a desired trajectory into control effector requirements usually begins with a time history of the airplane states during the maneuver. These time histories may be determined analytically, or generated by recording actual in-flight or simulator data.

The trajectories are converted into suitable input for a dynamic inversion or model-following control law as time histories of the states (Munro 1992), often involving numerical differentiation. The output of the control law is, as usual, the input to the control allocator.

The time histories of m_1, m_2, and m_3 may then be overlaid on the maximum set, and interpreted as $\mathbf{m}_{des}$. From this we can tell at a glance whether or not our control effector suite has the capability to perform the maneuver and, if so, with what margin of capability. This idea is shown with a time history of desired moments[3] $\mathbf{m}_{des}$ (in this case, the angular accelerations $\dot{p}$, $\dot{q}$, and $\dot{r}$) and the AMS from Figure 5.8, in Figures 5.17 (wire-frame only) and in Figure 5.18 (filled).

It is clear from Figure 5.17, and especially from Figure 5.18, that the aircraft being studied will be incapable of performing the maneuver that resulted in the time history of moments in the figures. This is true for any control allocation method, since the requirement lies beyond the maximum capabilities of the effector suite. If the maneuver is essential to the mission performance of the airplane then the control effectors must be physically re-engineered: there is no software solution to this problem.

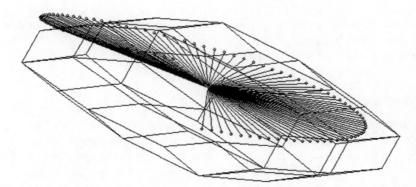

Figure 5.17   Time history of desired moments within the three-dimensional wire-frame of $\partial(\Phi)$. Viewpoint and axes as in Figure 5.8

---

[3] The maneuver that generated the time history in the figures in generically known as an *octafloogaron*. The description of such a maneuver is often preceded by the phrase *eye-watering*, and it is usually performed inadvertently. A lomcovák is an octafloogaron if you did not mean to do it.

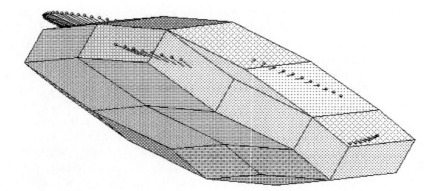

Figure 5.18   Time history of desired moments within the three-dimensional filled picture of $\partial(\Phi)$. Viewpoint and axes as in Figure 5.8

Discussion regarding design considerations based on the maximum set may be found in Chapter 8.

5.5.3   Control Failure Reconfiguration

One of the more attractive features of separating a control law such as dynamic inversion from the control allocation function itself is that control effector failure modes may be accommodated. For example, if a particular control effector becomes completely useless then its column in the B matrix is removed and allocation proceeds with the remaining effectors. Figure 5.19 demonstrates the effect of losing u_3.

The top of Figure 5.19 is the original AMS for the ADMIRE simulation. On the bottom is the AMS that results from removing u_3 from the problem completely, drawn to the same scale and with the same viewpoint. Obviously we may combine the idea of plotting maneuver moment time histories with figures representing control effector failures. This would contribute to the determination of the deterioration in flying qualities that results from such a failure and the assignment of the appropriate flying qualities' level for that failure.

5.6   Afterword

Our experience in introducing newcomers to our approach to the geometry of control allocation is that there is a wide range in people's ability to wrap their minds around the ideas of m-dimensional hyper-rectangles being projected into n-dimensional space to form polytopes. We have therefore made the development from three controls and two moments to the more general case as clear as possible. Some of the early figures may appear all too obvious, but they tend to get complicated quickly.

We have found the object notation to be very useful in implementing algorithms that explore the geometry. It is not used in real-time applications, but primarily in the design phase wherein

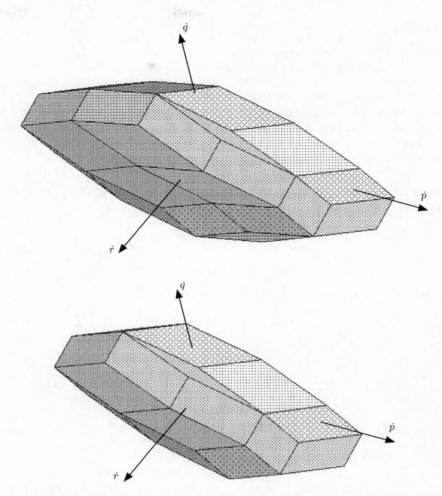

Figure 5.19 Normal AMS (top) and with failed u_3 control effector (bottom), to the same scale and from the same viewpoint. The control u_3 in the simulation is the outboard elevon on the right wing

different allocation methods are compared. Object notation provides a very easy way to write code to examine the relationships between objects, such as the ordering of vertices.

Finally we considered the maximum set (AMS, or Φ). Φ may be used to determine if it contains all the moments or accelerations required to perform a certain maneuver, such as spin recovery. Also, the different Φ that results from control failures may be used to determine how the flight envelope should be restricted. And finally, as we will see in Section 6.10, different allocation methods may not be able to return admissible solutions for all attainable moments. The set of moments for which they do return admissible solutions may be determined and compared to Φ to see what capabilities are being sacrificed.

References

Bordignon, KA 1996 *Constrained Control Allocation for Systems with Redundant Control Effectors,* PhD Thesis, Virginia Polytechnic Institute & State University, Appendix A.

Durham, WC, Lutze, FH, and Mason, W 1993 'Kinematics and aerodynamics of the velocity–vector roll,' *AIAA J. Guidance, Control, and Dynamics*, **17** (6), 1228–1233.

Hausner, M 1965 *A Vector Space Approach to Geometry*. Prentice-Hall Inc.

Munro, BC 1992 *Airplane Trajectory Expansion for Dynamic Inversion,* MS Thesis, Virginia Polytechnic Institute & State University.

Shaw, RL 1985 *Fighter Combat: Tactics and Maneuvering*. Naval Institute Press.

Wilson, DJ, Riley, DR, and Citurs, KD 1993 'Aircraft Maneuvers for the Evaluation of Flying Qualities and Agility, Vol. 2: Maneuver Descriptions And Selection Guide,' Wright Laboratory Technical Report WL-TR-93-3082.

6

Solutions

6.1 On-line vs. Off-line Solutions

6.1.1 On-line Solutions

On-line solutions are those that may be calculated in real time. That is, the control effectiveness matrix B, the control effector limits $\mathbf{u}_{Min}$ and $\mathbf{u}_{Max}$, and some some desired moment $\mathbf{m}_{des}$ are received from the flight control computer and a solution is generated.

There are many advantages to on-line solutions. First and foremost, there are usually a great number of flight conditions that will be encountered in flight. The consequence is that the control effectiveness matrix B may change frequently and over a wide range of values. It would be impractical to store all possible B matrices, but it is practical to query an on-board model (OBM) in real time to construct a current B matrix.

On-line solutions are desirable for dealing with control failure and reconfiguration. Assuming the nature of the control failure can be identified (say, a missing or jammed control effector) the failed control effector can be removed from the control effectiveness matrix, or the B matrix can be modified in some suitable way to accommodate the failure.

6.1.2 Off-line Solutions

Off-line solutions are pre-computed in whole or in part, but in either case, require foreknowledge of, at a minimum, the control effectiveness matrix B. Off-line solutions are usually simple closed-form solutions, typically a matrix or matrices that multiply a new $\mathbf{m}_{des}$ to yield $\mathbf{u}$.

Off-line solutions are invariably faster in execution than on-line solutions. As we will see, most real-time solutions are fairly complex algorithms, but off-line solutions have already done the hard computations in preparing for execution. With the increasingly greater speed of flight control computers, execution speed may not be a great concern.

Control allocation based on simple and known matrix multiplications are desirable for determining stability margins. The classic gain and phase margin approach to stability analysis can easily accommodate a single matrix multiplication in the loop. These methods of analysis are not so easily applied to solution methods that are made up 'on the fly'.

Aircraft Control Allocation, First Edition. Wayne Durham, Kenneth A. Bordignon and Roger Beck.
© 2017 John Wiley & Sons, Ltd. Published 2017 by John Wiley & Sons, Ltd.
Companion website: www.wiley.com/go/durham/aircraft_control_allocation

6.2 Optimal vs. Non-optimal Solutions

There are several meanings of the word 'optimal' in use in control allocation. To be clear, we are talking about the linear, underdetermined problem laid out in Chapter 4. We assume as given the control effectiveness matrix B, control effector lower and upper limits $\mathbf{u}_{Min}$ and $\mathbf{u}_{Max}$, and desired moments $\mathbf{m}_{des}$ such that $\mathbf{m}_{des} = B\mathbf{u}$.

6.2.1 Maximum Capabilities

The solution returns admissible controls for every theoretically attainable moment. This is the set of attainable moments Φ, either two-dimensional (Section 5.1) or three-dimensional (Section 5.4.1).

Control allocation that is optimal in this sense will always be a method or algorithm since, as we will show, there is no closed-form solution (single matrix multiplication of $\mathbf{m}_{des}$) that yields admissible solutions for all moments in Φ.

There are many possible objections to control allocation algorithms that are optimal in the sense of attaining maximal capabilities, mostly concerned with predictability. This concern arises when establishing phase and gain stability margins of a control loop. However, there are regions of flight in which such guarantees are irrelevant, mostly occurring in the corners of the flight envelope. Thus one may design a flight control system that is well-mannered almost everywhere, but that resorts to radical means to accomplish a specific task. These specific tasks are almost always those wherein survival is at issue. Recovery from a flat spin need not satisfy any gain or phase margins. Avoidance of a pilot-induced oscillation due to rate limiting needs all the capability available (Acosta *et al.* 2015). Likewise, defeating a surface-to-air missile needs no such guarantees. This situation is analogous to the provision of 'War Emergency Power', wherein maximum power is demanded of the engine, despite the fact that the engine may never be usable again after the flight.

6.2.2 Maximum Volume

This optimization criteria appears in formulating a method to determine a closed-form solution to the control allocation problem, referred to as a *generalized inverse,* whose own volume of attainable moments (necessarily less than the globally attainable volume) is the greatest of all generalized inverses.

It is possible to apply the notion of maximum volume to other methods and algorithms (not generalized inverses) that are not optimal in the sense of maximum capabilities, but the authors are not aware of any efforts that have been made to do so.

6.2.3 Nearest to Preferred

This optimality criterion applies whenever the calculated solution to the allocation problem is not unique. Lack of uniqueness occurs within the interior of the AMS where there are an infinite number of control vectors that can solve the problem (through the null space of B).

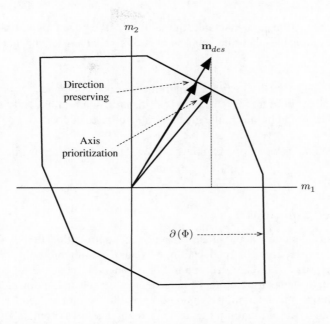

Figure 6.1   An unattainable moment in two-dimensional moment space. Two possible options are shown: direction-preserving, and axis-prioritization

In such a case control allocation is a two-step process. First, find any admissible solution to the problem. Second, move the effectors through the null space of B toward a control effector configuration that is preferred in some sense over all others.

6.2.4   Unattainable Moments

A special case arises if the desired moment $\mathbf{m}_{des}$ is unattainable. This 'optimality' condition describes a strategy to deal with moment demands that exceed those attainable. The authors' position is that no self-respecting control law should ask for such moments, but such things may happen, especially in the presence of hardware and software failures, or in the corners of the flight envelope.

There are several strategies to deal with unattainable moments; two common approaches are shown in Figure 6.1.

Direction preserving

Direction preserving finds the solution at the point where $\mathbf{m}_{des}$ intersects the convex hull of the attainable moments, $\partial(\Phi)$ (or $\partial(\Pi)$ for allocations methods that are not optimal in the sense of maximum volume). As we will see, some control allocation methods first consider just the direction in moment space of $\mathbf{m}_{des}$, and locate the intersection of a half-line in this direction with the convex hull, $\partial(\Phi)$. That is, they find the greatest attainable moment in that direction. Then if $\mathbf{m}_{des}$ is attainable, the solution at the boundary is scaled to achieve it. If $\mathbf{m}_{des}$ is not attainable, then the solution at the boundary is taken; that is, direction is preserved.

Axis prioritization

Axis prioritization seeks to satisfy the requirement of one component of $\mathbf{m}_{des}$ to the greatest extent possible. In Figure 6.1 the axis-prioritization solution has fully satisfied the m_1 component of $\mathbf{m}_{des}$ and has taken the greatest value of m_2 that does so.

Lacking a specific application, in this book we elect to preserve direction over axis prioritization.

6.3 Preferred Solutions

The idea of preferred solutions first arose when certain control allocation algorithms (particularly frame-wise, or discrete-time allocation, Chapter 7) exhibited a phenomenon referred to as 'windup'. It was noted that after a period of (simulated) maneuvering and returning to steady level flight, the control effectors, while generating the desired moment, were at various asymmetric positions: the solution was in an odd place in the null space of B. The condition was much like that of the airplanes in Figure 5.6, except that it was completely unintentional.

The algorithms in question are those in which the control allocation problem is divided into discrete time, as in the successive frames of a flight control computer. It turns out that many discrete-time solutions are path dependent: the total control configuration at any particular instant depends on the history of allocation problems that preceded the current time.

An obvious solution is to add an extra step to the allocation algorithm that drove the effectors through the null space of B to the one that minimized the two-norm of the control vector. In effect, this established zero deflection as a preferred solution. This process was referred to as *restoring*, since the control effectors were being restored to a preferred solution. Alternatively, the controls could be driven toward their trimmed positions. This is effectively zero deflection with a displaced origin.

Despite the historical origin of restoring, the idea that there are an infinite number of solutions when the moment is attainable is quite appealing in that some solutions among that infinitude might have desirable features. Aside from zero deflection, among these are the control configuration that minimizes the effector-induced drag. In a similar vein, one could possibly minimize radar cross-section if sufficient data were available.

Preferred solutions will be addressed in more detail in Chapter 7 and Appendix A.

6.4 Ganging

In Section 8.5 we will see an example of *control ganging*. Individual control effectors are interconnected so that the movement of one depends on that of another. This is the way conventional airplanes are configured: left and right ailerons are mechanically interconnected so that they move in opposite directions. The left and right elevators (or horizontal tails) are constrained to move as a single unit.

Example 6.1

Ganged Controls for the ADMIRE Simulation The control effectors in the ADMIRE simulation (Appendix B and Section 8.5) may be ganged in this manner:

- roll effector: outboard elevons only commanded asymmetrically
- pitch effector: canards only commanded symmetrically
- yaw effector: rudder
- inboard elevons: not used for moment-generation.

Such an arrangement is sometimes called *pseudo-controls*. There are several other methods of dealing with redundant control effectors that use the same name, and they are often unrelated one to the other. See, for example, Lallman (1985).

With respect to the control effectiveness matrix B, the control effectors are related to the entries in $\mathbf{u}$ in Table 5.4. Thus, to account for the outboard elevons being commanded only asymmetrically, the sixth column is subtracted from the third, and the result replaces the two, adopting the sign convention of u_3. Likewise, to account for the canards, the first and second columns of B are added, the result replacing the two, and the sign convention of either is used. The fourth and fifth columns (inboard elevons) are simply removed. The resulting B matrix and control effectors are:

$$
B_{Ganged} = \begin{bmatrix} 0 & -6.9912 & 2.1103 \\ 2.2408 & 0 & 0.0035 \\ 0 & -0.3014 & -1.2680 \end{bmatrix}
\tag{6.1}
$$

$$
\mathbf{u}_{Min_{Ganged}} = \begin{Bmatrix} -0.9599 \\ -0.5236 \\ -0.5236 \end{Bmatrix}_u \qquad \mathbf{u}_{Max_{Ganged}} = \begin{Bmatrix} 0.4363 \\ 0.5236 \\ 0.5236 \end{Bmatrix}_u
\tag{6.2}
$$

Table 6.1 Ganged control effectors by name

u_i	Name
u_1	Ganged canards
u_2	Ganged outboard elevons
u_3	Rudder

With respect to the limits, Eq. (6.2), had the elevons not had symmetric positive and negative limits ($\pm$ 0.5236 rad), then some thought would have to go into deciding how to define the zero position and upper and lower limits of the ganged elevons.

Ganging is addressed here only because it is a valid means of control allocation, although not one that bears greatly on the principal concerns of this book. As can be seen B_{Ganged} is a square, invertible matrix, and $\mathbf{m} = B_{Ganged}\mathbf{u}$ has a unique solution. $\mathbf{u}$ will be either admissible or it will not, and no mathematics are involved. It is possible to gang some of the control effectors so that the ganged plus the others still results in an under-determined system, in which case the resulting problem is of concern to the present treatment.

The consequences of ganging controls in terms of moment-generating capabilities is described in Section 8.2.3, Example 8.5.

6.5   Generalized Inverses

6.5.1   The General Case, and the Significance of P_2

A *generalized inverse* is taken to be a matrix P (with distinguishing subscript in particular cases) that, without regard to the limits on control effectors, finds

$$\mathbf{u}_P = P\mathbf{m}_{des} \tag{6.3}$$

such that

$$B\mathbf{u}_P = \mathbf{m}_{des} \tag{6.4}$$

For this to hold there is the obvious requirement that

$$BP = I_n \tag{6.5}$$

With respect to dimensions, we have taken B to be $n \times m$, where $n = 2$ or $n = 3$ is the number of moments (or moment coefficients, angular accelerations) being considered, and m is the number of control effectors, $m > n$. To be conformal, P is $m \times n$, and the identity matrix in Eq. (6.5) is $n \times n$.

All such generalized inverses may be represented as

$$P = N[BN]^{-1} \tag{6.6}$$

The matrix N is $m \times n$, and is arbitrary so long as the product BN is not singular. While N has $m \cdot n$ arbitrary elements, P does not; it has to satisfy Eq. (6.6) so there are dependencies among the elements. To see this dependency, consider the following.

First, partition the control effectiveness matrix B into B_1, an $n \times n$ ($n = 2$ or 3) matrix using the first n columns, and B_2 consisting of the remaining columns.

$$B = \begin{bmatrix} B_1 & B_2 \end{bmatrix} \tag{6.7}$$

Partition P conformally.

$$P = \begin{bmatrix} P_1 \\ P_2 \end{bmatrix} \tag{6.8}$$

From Eq. (6.5),

$$BP = B_1P_1 + B_2P_2 = I_n \tag{6.9}$$

Any n columns of B are linearly independent by our robust-rank requirement, so we may solve Eq. (6.9) for P_1,

$$P_1 = B_1^{-1}[I_n - B_2 P_2] \tag{6.10}$$

In Eq. (6.10), all the elements of P_2 are arbitrary. As we will soon demonstrate, we may select the elements of P_2 to create generalized inverses P with special properties.

Equations (6.7) and (6.8) may be used to show why we asserted that no closed-form solution (that is, no matrix P) can yield admissible solutions for all moments in Φ. We denote the solutions that P does attain as $\mathbf{u}_P$. These solutions are $\mathbf{u}_P = P\mathbf{m}$. But $\mathbf{m} = B\mathbf{u}$ for any $\mathbf{u}$, including $\mathbf{u}_P$ so

$$\mathbf{u}_P = P\mathbf{m} \Rightarrow \mathbf{u}_P = PB\mathbf{u}_P \tag{6.11}$$

P has only n columns, so the rank of PB can be no more than n and in fact *is* n because of our robust-rank requirement for B. Thus all vectors $\mathbf{u}_P$ that are included in the solutions $\mathbf{u}_P = P\mathbf{m}_{des}$ satisfy Eq. (6.12):

$$[PB - I_m]\mathbf{u}_P = \mathbf{0} \tag{6.12}$$

That is, all the solutions $\mathbf{u}_P$ that P will return lie in the null space of the matrix $[PB - I_m]$

$$\mathbf{u}_P \in \mathcal{N}\,[PB - I_m] \tag{6.13}$$

We can learn about the rank of $[PB - I_m]$ using our partitions in Eqs 6.7 and 6.8 :

$$
\begin{aligned}
[PB - I_m] &= \begin{bmatrix} P_1 B_1 - I_n & P_1 B_2 \\ P_2 B_1 & P_2 B_2 - I_{m-n} \end{bmatrix} \\
&= \begin{bmatrix} B_1^{-1}[I_n - B_2 P_2]B_1 - I_n & B_1^{-1}[I_n - B_2 P_2]B_2 \\ P_2 B_1 & P_2 B_2 - I_{m-n} \end{bmatrix} \\
&= \begin{bmatrix} B_1^{-1}I_n B_1 - B_1^{-1}B_2 P_2 B_1 - I_n & B_1^{-1}I_n B_2 - B_1^{-1}B_2 P_2 B_2 \\ P_2 B_1 & P_2 B_2 - I_{m-n} \end{bmatrix} \\
&= \begin{bmatrix} -B_1^{-1}B_2[P_2 B_1] & -B_1^{-1}B_2[P_2 B_2 - I_{m-n}] \\ [P_2 B_1] & [P_2 B_2 - I_{m-n}] \end{bmatrix}
\end{aligned}
\tag{6.14}
$$

Thus the first n rows are a matrix multiplication of the last $m - n$ rows. That is, the last $m - n$ rows are

$$\begin{bmatrix} [P_2 B_1] & [P_2 B_2 - I_{m-n}] \end{bmatrix} \tag{6.15}$$

and the first n rows are

$$-B_1^{-1}B_2 \begin{bmatrix} [P_2 B_1] & [P_2 B_2 - I_{m-n}] \end{bmatrix} \tag{6.16}$$

Therefore, for $\mathbf{u}_P$ to be in the null space of $[PB - I_m]$, it is sufficient that

$$\begin{bmatrix} [P_2 B_1] & [P_2 B_2 - I_{m-n}] \end{bmatrix} \mathbf{u}_P = \mathbf{0} \tag{6.17}$$

The conclusion is that the rank of $[PB - I_m]$ is $m - n$, so its null space is n-dimensional. However, the vectors that define the convex hull $\partial(\Phi)$ are in Ω in control space, which is of dimension m. In other words, the solutions $\mathbf{u}_P$ do not form a basis for m-space.

Example 6.2

Dimension of a Generalized Inverse Solution We can more easily see what is going on with the simple example in Eq. (5.3):

$$B = \begin{bmatrix} 1 & 0 & -0.5 \\ 0 & 1 & -0.5 \end{bmatrix} \tag{6.18}$$

We form a generalized inverse using Eq. (6.6) with

$$N = \begin{bmatrix} 1 & 0 \\ 0 & 1 \\ 1 & 2 \end{bmatrix} \tag{6.19}$$

$$P = N[BN]^{-1} = \begin{bmatrix} 0 & -2 \\ -1 & -1 \\ -2 & -4 \end{bmatrix} \tag{6.20}$$

$$[PB - I_3] = \begin{bmatrix} -1 & -2 & 1 \\ -1 & -2 & 1 \\ -2 & -4 & 2 \end{bmatrix} \tag{6.21}$$

It is easy to see that $[PB - I_3]$ has rank 1, so its null space is two-dimensional. That is, all the control vectors that this generalized inverse returns lie in a two-dimensional plane in three-dimensional control space. The intersection of this plane with the admissible controls in Ω is shown in Figure 6.2.

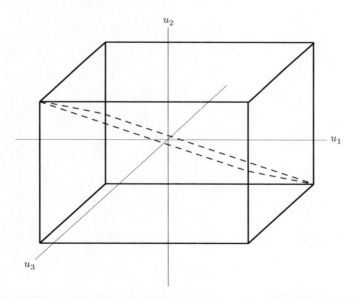

Figure 6.2 Three-dimensional set of admissible controls, from Figure 5.3. The two-dimensional plane of $\mathcal{N}\,[PB - I_3]$ intersects $\partial(\Omega)$ along the dotted lines; the plane is almost parallel to axis u_3

When the intersection of the null space with the admissible controls is mapped via B to moment space, Figure 6.3 results. This particular generalized inverse will yield admissible solutions in only a very small region of all the attainable moments.

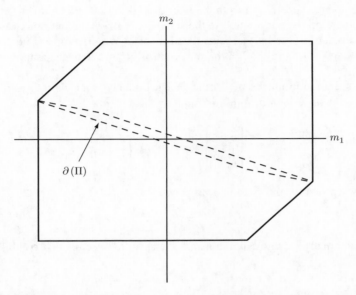

Figure 6.3 Two-dimensional set of attainable moments, Φ, and Π, the subset of moments that P can attain using admissible controls

Figure 5.16 showed the moments for which a generalized inverse returns admissible solutions to the three-moment problem.

The discussion above regarding unattainable moments (Section 6.2.4) deals with the entire attainable moment set. A similar concern exists regarding the use of generalized inverses to solve the control allocation problem. When generalized inverses are used there is usually no explicit information available regarding Π, the moments that may be attained using admissible controls. Often the best one can hope to do in the case of control effector saturation is to scale the solution uniformly so that no effector is saturated, thus preserving the direction of the solution.

6.5.2 Tailored Generalized Inverses

It is possible to 'tailor' a generalized inverse so that it will provide admissible solutions at a number of points in the set of attainable moments. That is, points in the set Π that result from the generalized inverse P may be specified. We will show this using Eqs. (6.7), (6.8), and (6.10). Since P_2 completely determines P, we may use the $n \cdot (m - n)$ elements of P_2 for this task.

The number of points in Π that may be specified is n, either two or three depending on the problem. If the specified point in Π is also in $\partial(\Phi)$ then other points that are also in Π and $\partial(\Phi)$ may result from symmetry or linearity. For each point to be specified, the coordinates of the moment and the control vector that generates it must be known.

It does little good to specify points in the interior of the attainable moment set, since if one specifies a moment on the convex hull in the same direction as the interior point, the interior point will also be attainable by the generalized inverse. We therefore assume the moments specified to be on $\partial(\Phi)$, and that the unique control vectors that generate those points are known.

Denote the specified moments $\mathbf{m}_i^{Spec}$ and their associated controls $\mathbf{u}_i^{Spec}$, $i = 1 \cdots n$. Further, partition $\mathbf{u}_i^{Spec}$ the same as P into $\mathbf{u}_{i_1}^{Spec}$ and $\mathbf{u}_{i_2}^{Spec}$ so that

$$\mathbf{u}_i^{Spec} = \begin{bmatrix} \mathbf{u}_{i_1}^{Spec} \\ \mathbf{u}_{i_2}^{Spec} \end{bmatrix} = P\mathbf{m}_i^{Spec} = \begin{bmatrix} P_1 \\ P_2 \end{bmatrix} \mathbf{m}_i^{Spec} \tag{6.22}$$

From this we extract the information regarding P_2,

$$\mathbf{u}_{i_2}^{Spec} = P_2 \mathbf{m}_i^{Spec} \tag{6.23}$$

Next we make matrices whose columns are the specified vectors and combine the n relationships as

$$\begin{bmatrix} \mathbf{u}_{1_2}^{Spec} & \cdots & \mathbf{u}_{n_2}^{Spec} \end{bmatrix} = P_2 \begin{bmatrix} \mathbf{m}_1^{Spec} & \cdots & \mathbf{m}_n^{Spec} \end{bmatrix} \tag{6.24}$$

In Eq. (6.24) the matrix $\begin{bmatrix} \mathbf{m}_1^{Spec} & \cdots & \mathbf{m}_n^{Spec} \end{bmatrix}$ will be either 2×2 or 3×3 and, if the specified moments were chosen to be linearly independent, will be nonsingular. Therefore,

$$P_2 = \begin{bmatrix} \mathbf{u}_{1_2}^{Spec} & \cdots & \mathbf{u}_{n_2}^{Spec} \end{bmatrix} \begin{bmatrix} \mathbf{m}_1^{Spec} & \cdots & \mathbf{m}_n^{Spec} \end{bmatrix}^{-1} \tag{6.25}$$

Numerical examples of tailored generalized inverses are at Section 6.11.1, Examples 6.4 and 6.5.

If the control effectiveness matrix and control limits are known *a priori*, then tailored generalized inverses may be calculated to fit every facet of the AMS. These solutions would then be stored. Then, employing a relatively simple switching routine, the correct tailored generalized inverse could be called up to yield the same solution as direct allocation.

6.5.3 'Best' Generalized Inverse

'Best' in the present context means maximum volume, as described in Section 6.2.2. The approach is straightforward: we systematically vary the $n \cdot (m - n)$ elements of P_2, calculate P_1, assemble P, calculate the volume of its Π, and repeat until the volume is maximized.

Use of this method clearly requires *a priori* knowledge of the control effectiveness matrix and the control limits, since the computational requirements of its solution are much greater than any other method we have introduced.

No serious research has been conducted into the characteristics of this optimization problem, except to note that the algorithm used (a simplex algorithm by Stevens and Lewis (2003)) always converged, and it always converged to the same generalized inverse for any starting values of P_2. The solution that is returned offers no obvious insight into why that solution generated more volume in moment space than any other generalized inverse.

The calculation of the 'best' generalized inverse for the ADMIRE simulation data is at Section 6.11.1, example 6.6.

6.5.4 Pseudo-inverses

6.5.4.1 The Moore–Penrose Pseudo-inverse

A particular generalized inverse is $P_{|2|}$ that will minimize the 2-norm of our control vector **u** when solving the control allocation problem. The 2-norm of a vector is just its length, the positive square-root of the sum of the squares of the individual controls. We call this generalized inverse the *minimum norm pseudo-inverse*, sometimes without the hyphen, and often the *Moore–Penrose pseudo-inverse*. When one hears reference to *the* pseudo-inverse it is usually this one that is meant. There are many other vector norms that could be used, so there may be some competition for the name 'minimum-norm'.

The optimization to be performed is to minimize $\mathbf{u}^T\mathbf{u}$ (the sum of the squares of the controls) subject to the constraint that $B\mathbf{u} = \mathbf{m}$ for arbitrary **m**. We do not require the square root of $\mathbf{u}^T\mathbf{u}$ since if $\mathbf{u}^T\mathbf{u}$ is minimum, so is its positive square root.

Using LaGrange multipliers we define the scalar function

$$\mathcal{H}(\mathbf{u}, \boldsymbol{\lambda}) = \frac{1}{2}\mathbf{u}^T\mathbf{u} + \boldsymbol{\lambda}^T(\mathbf{m} - B\mathbf{u}) \tag{6.26}$$

Here $\boldsymbol{\lambda}$ is an n-vector of LaGrange multipliers. The factor of $1/2$ anticipates that there will be a 2 to cancel. $\mathcal{H}$ will be a minimum (or maximum) when

$$\frac{\partial \mathcal{H}}{\partial \mathbf{u}} = 0, \ \frac{\partial \mathcal{H}}{\partial \boldsymbol{\lambda}} = 0 \tag{6.27}$$

Performing the operations yields

$$\frac{\partial \mathcal{H}}{\partial \mathbf{u}} = \mathbf{u}^T - \boldsymbol{\lambda}^T B = 0 \tag{6.28}$$

Hence we require that $\mathbf{u}^T = \boldsymbol{\lambda}^T B$, or $\mathbf{u} = B^T\boldsymbol{\lambda}$.

$$\frac{\partial \mathcal{H}}{\partial \boldsymbol{\lambda}} = \mathbf{m} - B\mathbf{u} = 0 \tag{6.29}$$

So that $\mathbf{m} = B\mathbf{u}$. Now combining the two results,

$$\mathbf{m} = B\mathbf{u} = BB^T\boldsymbol{\lambda} \tag{6.30}$$

Since B is full rank, BB^T is too, and since BB^T is square it is invertible. Thus

$$\boldsymbol{\lambda} = [BB^T]^{-1}\,\mathbf{m} \tag{6.31}$$

Since $\mathbf{u} = B^T\boldsymbol{\lambda}$ we have $\mathbf{u} = B^T[BB^T]^{-1}\,\mathbf{m}$ and

$$P = B^T[BB^T]^{-1} \tag{6.32}$$

With respect to Eq. (6.6), $N = B^T$ for this case.

The principal claim made about the minimum-norm pseudo-inverse is that because it minimizes the sum of the squares of the control effector displacements, it thus minimizes some measure of energy. It is useful to ask why this is a good thing. The zero position (that which this formulation seeks) of any control effector is a matter of convention, sometimes involving physics, but oftentimes just a reflection of a datum that was established during the design of the control effectors.

For control effectors such as spoilers, zero may have a physical meaning: when the spoiler is at zero deflection, it is tucked away out of the problem. With flapping control surfaces on the trailing edges of larger airfoils, such as rudders and elevators mounted on vertical and horizontal stabilizers, the zero position is that which makes the section of the larger airfoil containing the effector a natural extension of the rest of the airfoil; in other words, the parts without the control effector.

But some control effectors have no 'natural' zero position. Most modern tactical airplanes do not have an elevator on a horizontal tail, but a single surface that may be varied in incidence. Canards are similar. The position defined as zero displacement for these surfaces is purely a matter of convention. The trim position of a unit horizontal tail or canard is generally not zero.

It follows then that a solution that drives such surfaces as canards to preferred positions that are arbitrarily defined is not a necessarily good thing to do.

6.5.4.2   The Weighted Pseudo-inverses

The Moore–Penrose is just one of an infinity of closed-form inverses that minimize a vector norm. Entire families of these solutions can be obtained from optimization problems that aim to minimize other norms of $\mathbf{u}$. The weighted 2-norm is seen frequently, and the problem is to minimize $\mathbf{u}^T W^T W \mathbf{u}$, where W is a positive diagonal matrix.

The diagonal terms of W may be chosen with the aim to spare activity in some controls at the expense of more demands on the others. Alternatively, W is sometimes chosen in an attempt to introduce the control effectors' deflection limits into the solution. Typically the diagonal terms are of the form

$$W_{ii} = \frac{1}{|u_{i_{Max}} - u_{i_{Min}}|} \tag{6.33}$$

Alternatively,

$$W_{ii} = \frac{1}{(u_{i_{Max}} - u_{i_{Min}})^2} \tag{6.34}$$

In either case, the optimization proceeds as before, and results in

$$P_W = W^{-1}B^T[BW^{-1}B^T]^{-1} \tag{6.35}$$

In Eq. (6.35) we have assumed that W is a diagonal matrix. If it is not, then W should be replaced with W^T.

Note, however, that Eq. (6.35) is valid for any choice of W for which $[BW^{-1}B^T]$ is non-singular. Although there are infinitely many ways to select W, there does not appear to be much physical significance to the off-diagonal terms, or to a diagonal matrix with negative entries.

6.5.5   Methods that Incorporate Generalized Inverses

6.5.5.1   Daisy Chaining

Daisy chaining was introduced at the High-Angle-of-Attack Projects and Technology Conference held at the NASA Dryden Flight Research Facility in April 1992. The main idea was to use conventional controls until one or more is commanded past its limits, and then to bring other controls into the solution. In one application, the non-conventional control effectors used paddles in the exhaust stream to vector the thrust, and it was desired that their use should be reduced relative to that of the conventional controls to lower thermal stresses.

The method works by dividing the controls into two or more groups. Here we will describe only the two-group method, since other groupings are extensions of this grouping. The first group, called $\mathbf{u}_1$, include control effectors that are to be used all the time. They are typically the more conventional control effectors, such as ailerons and rudders. The second grouping, $\mathbf{u}_2$, are to be used only if the first grouping is unable to satisfy the moment demand.

$$\mathbf{u} = \begin{Bmatrix} \mathbf{u}_1 \\ \mathbf{u}_2 \end{Bmatrix} \tag{6.36}$$

The control effectiveness matrix, B, is partitioned into B_1 and B_2, corresponding to the two control groupings.

$$B = \begin{bmatrix} B_1 & B_2 \end{bmatrix} \tag{6.37}$$

Each partition with its associated controls constitutes its own control allocation problem,

$$\begin{aligned} \mathbf{m}_1 &= B_1 \mathbf{u}_1 \\ \mathbf{m}_2 &= B_2 \mathbf{u}_2 \end{aligned} \tag{6.38}$$

While the original system is under-determined (B is wide) the two new systems are not necessarily under-determined. Thus B_1 or B_2 may be square or even over-determined (more equations than unknowns). If a system is under-determined we will use its pseudo-inverse or weighted pseudo-inverse for a solution. If it is square we will simply invert it. Over-determined systems have a solution of sorts, that which minimizes the error in the solution. The derivation proceeds similarly to that for the pseudo-inverse (Section 6.5.4) and results in

$$P_R = [B_{over}^T B_{over}]^{-1} B_{over}^T \tag{6.39}$$

so that

$$P_R B_{over} = I \tag{6.40}$$

For now we will designate the inverse matrix of any of these cases as P_1 and P_2, so that

$$\begin{aligned} \mathbf{u}_1 &= P_1 \mathbf{m}_1 \\ \mathbf{u}_2 &= P_2 \mathbf{m}_2 \end{aligned} \tag{6.41}$$

Now, if $\mathbf{m}_{des}$ is attainable using only $\mathbf{u}_1 = P_1 \mathbf{m}_{des}$, then that solution is used and $\mathbf{u}_2 = \mathbf{0}$. However, if the solution for $\mathbf{u}_1$ is not admissible, then we may proceed in one of two ways:

1. Scale $\mathbf{u}_1$ uniformly so that it is admissible.
2. Set the inadmissible controls in $\mathbf{u}_1$ to their nearest limits, leaving the remaining effectors alone.

Choice 1 preserves the direction while choice 2 does not. There are some interesting consequences to using choice 2, to be mentioned in Section 6.10. In either case it will be true that $\mathbf{m}_1 = B_1\mathbf{u}_1 \neq \mathbf{m}_{des}$. The deficiency in moment is $\mathbf{m}_{des} - \mathbf{m}_1$, and the second group of controls is employed to attempt to satisfy the deficiency.

$$\mathbf{u}_2 = P_2(\mathbf{m}_{des} - \mathbf{m}_1) \tag{6.42}$$

The solution vector is then reassembled.

$$\mathbf{u} = \begin{Bmatrix} \mathbf{u}_1 \\ \mathbf{u}_2 \end{Bmatrix} \tag{6.43}$$

6.5.5.2 Cascaded Generalized Inverses

One generalized-inverse-based control allocation method consistently performs well compared to all others, and it is relatively fast and easy to implement. The method is called *cascaded generalized inverses* (CGI) and it is the method used in the design of the X-35 control laws (Bordignon and Bessolo 2002).

In its simplest form the CGI method uses a single generalized inverse (we will assume the unweighted Moore–Penrose pseudo-inverse) exclusively until one or more of the control effectors is commanded to a position beyond saturation. Then:

1. The saturated controls are left at whatever limit they were when saturated and removed from the problem. That is, if one were commanded to a position beyond the upper limit, it is left at the upper limit, and vice versa for the lower limit. This step is applied to both the physical control effectors and mathematically in the calculations.
2. The columns of the B matrix corresponding to the saturated control effectors are removed from the B matrix.
3. The entries corresponding to the saturated control effectors are removed from $\mathbf{u}$, $\mathbf{u}_{Min}$, and $\mathbf{u}_{Max}$.
4. The moment generated by the saturated controls is subtracted from the desired moment, $\mathbf{m}_{des}$. Denote the number of such saturated controls by n_{Sat}, a vector of saturated control effectors as $\mathbf{u}_{Sat}$, and the columns of the (original) control effectiveness matrix as $\mathbf{b}_i$. The controls in $\mathbf{u}_{Sat}$ are at either their lower or upper limits as described above. Then the moment due to the saturated controls, $\mathbf{m}_{Sat}$ is

$$\mathbf{m}_{Sat} = \sum_{i=1}^{n_{Sat}} \mathbf{b}_i u_{Sat_i} \tag{6.44}$$

In a slightly different form,

$$\mathbf{m}_{Sat} = \begin{bmatrix} \mathbf{b}_1 & \cdots & \mathbf{b}_{n_{Sat}} \end{bmatrix} \mathbf{u}_{Sat} \tag{6.45}$$

5. The saturated moment ($\mathbf{m}_{Sat}$) is removed from the original desired moment $\mathbf{m}_{des}$ to create a new desired moment that remains to be satisfied ($\mathbf{m}'_{des}$).

$$\mathbf{m}'_{des} = \mathbf{m}_{des} - \mathbf{m}_{Sat} \tag{6.46}$$

6. The control allocation process is repeated with the reduced problem. How that problem is dealt with depends on which of three cases obtains with respect to the remaining control effectors:

(a) The remaining system is under-determined. This is the case we have been discussing throughout: more controls than moments. If this is the case, then the pseudo-inverse of this system is calculated and applied to whatever remains of the desired moment with consideration of the moments generated by the saturated controls.

(b) The remaining system is square, with three controls for the three moments. Here the matrix inverse is calculated and applied to the remaining moments.

(c) The remaining system is over-determined, with fewer controls than moments. The solution method of choice is another kind of pseudo-inverse that minimizes the square of the error between what is desired and what may be attained. The solution is similar to the pseudo-inverse for the over-determined case. Call the over-determined B matrix B_{Over} and the pseudo-inverse P_R:

$$P_R = [B_{over}^T B_{over}]^{-1} B_{over}^T \tag{6.47}$$

So that

$$P_R B_{over} = I \tag{6.48}$$

It should be clear that if the process has reached the under-determined stage, the solution is not exact, and an error has been introduced.

7. The entire procedure is repeated until either $\mathbf{m}_{des}$ is satisfied, or no control effectors are left to allocate.

There are two major variations on the method of cascaded generalized inverses. Both cases pertain to the first step in the above procedure, in which saturated controls are removed from the problem.

No Scaling

Every unsaturated control is left at the deflections commanded and the saturated controls simply truncated. For instance, if the solution is

$$\mathbf{u} = \begin{Bmatrix} u_1 \\ \vdots \\ u_i \\ \vdots \\ u_m \end{Bmatrix} \tag{6.49}$$

and only one control u_i is commanded past its maximum limit, say $u_i = a u_{i_{Max}}$, $a > 1$, then the commanded positions after the present step is:

$$\mathbf{u} = \begin{Bmatrix} u_1 \\ \vdots \\ u_{i_{Max}} \\ \vdots \\ u_m \end{Bmatrix} \tag{6.50}$$

Scaling

All the controls in the solution are scaled uniformly such that no effector is commanded past saturation. For instance, if the calculated solution is

$$
\mathbf{u} = \left\{ \begin{array}{c} u_1 \\ \vdots \\ u_i \\ \vdots \\ u_m \end{array} \right\} \tag{6.51}
$$

and the most severely saturated control u_i is commanded past its maximum limit, say $u_i = au_{i_{Max}}$, $a > 1$, then the commanded positions after the first step is $\mathbf{u}/a$:

$$
\mathbf{u}/a = \left\{ \begin{array}{c} u_1/a \\ \vdots \\ au_{i_{Max}}/a \\ \vdots \\ u_m/a \end{array} \right\} = \left\{ \begin{array}{c} u_1/a \\ \vdots \\ u_{i_{Max}} \\ \vdots \\ u_m/a \end{array} \right\} \tag{6.52}
$$

With no scaling, if the algorithm terminates without satisfying $\mathbf{m}_{des}$ then in general the direction will not be preserved nor will an axis be prioritized, as described in Section 6.2.4 (p. 75). The principal advantage to scaling is that it does preserve direction at each step.

6.6   Direct Allocation

What has become known in the literature as *direct allocation* is nothing more than a simple extension of the method for the determination of the AMS, or Φ. For the two-moment problem this was described in Section 5.1, and for the three-moment problem in Section 5.4.1. These procedures were summarized at the ends of their respective sections, pp. 52–53 and 60–62. In both cases the allocation procedure is similar:

1. Given $\mathbf{m}_{des}$ consider a half-line from the origin in the direction of $\mathbf{m}_{des}$.
2. Construct the AMS, testing each edge or facet thus determined to see if the half-line from step 1 intersects it.
3. When the correct edge or facet is found, use the geometry of the intersection to calculate the control vector that generated the moment at the intersection.
4. The moment at the intersection will be in the same direction as $\mathbf{m}_{des}$.
   (a) If $\mathbf{m}_{des}$ is greater in length than the moment at the intersection, $\mathbf{m}_{des}$ is not attainable, and the discussion in Section 6.2.4 applies. The moment at the intersection may be used if direction preservation is desired. Axis prioritization will require other actions.
   (b) If $\mathbf{m}_{des}$ is less than or equal in length to the moment at the intersection, $\mathbf{m}_{des}$ is attainable. If it is equal in length then the control vector calculated in step 3 is unique. If it is less, than the solution from step 3 may be scaled to create a solution, or other methods involving the null space of B may be employed.

The procedure may be terminated after a solution is found, since there will be only one intersection. The only reason to continue would be in a real-time application wherein a worst-case scenario would be assumed for timing purposes, since the correct edge or facet may not be found until all the edges or facets of the attainable moment set have been found.

The testing of each edge or facet suggested in step 2 above is simple vector algebra. Beginning with a vector to any vertex of the edge or facet, and defining the relevant edge or edges common to that vertex, the possible intersection with the edge or facet is hypothesized and the hypothesis is tested.

6.6.1   The Direct Method for the Two-moment Problem

In the determination of the attainable moment set, Φ_2, the last step was to calculate the moments that defined the two vertices of the edge determined. Call either vertex $\mathbf{m}_1$ and the other $\mathbf{m}_2$. The edge, from $\mathbf{m}_1$ to $\mathbf{m}_2$, is $\mathbf{m}_2 - \mathbf{m}_1$. Then the intersection of the half-line in the direction of the desired moment $\mathbf{m}_{des}$ with this edge may be expressed as the vector sum:

$$a\mathbf{m}_{des} = \mathbf{m}_1 + b\,(\mathbf{m}_2 - \mathbf{m}_1) \tag{6.53}$$

On the left, $a\mathbf{m}_{des}$ for $a > 0$ is the half-line. On the right is the vector sum of the first vertex and some distance b along the edge $\mathbf{m}_2 - \mathbf{m}_1$. Clearly b must be between 0 and 1 or the vector sum does not yield a point on the edge, and $a \geq 1$ for $\mathbf{m}_{des}$ to be attainable. Rearranging,

$$\begin{bmatrix} \mathbf{m}_{des} & -(\mathbf{m}_2 - \mathbf{m}_1) \end{bmatrix} \begin{Bmatrix} a \\ b \end{Bmatrix} = \mathbf{m}_1 \tag{6.54}$$

From which,

$$\begin{Bmatrix} a \\ b \end{Bmatrix} = \begin{bmatrix} \mathbf{m}_{des} & -(\mathbf{m}_2 - \mathbf{m}_1) \end{bmatrix}^{-1} \mathbf{m}_1 \tag{6.55}$$

Figure 6.4 illustrates the relationships implied by Eq. (6.53).

Since the controls that generate solutions on the convex hull are unique, the control vector that generates $a\mathbf{m}_{des}$ is unique; denote it $\mathbf{u}^*$, $a\mathbf{m}_{des} = B\mathbf{u}^*$. Denote the control vector that generates $\mathbf{m}_1$ as $\mathbf{u}_1$ ($\mathbf{m}_1 = B\mathbf{u}_1$) and that which generates $\mathbf{m}_2$ as $\mathbf{u}_2$ ($\mathbf{m}_2 = B\mathbf{u}_2$), then

$$B\mathbf{u}^* = B\mathbf{u}_1 + b\,(B\mathbf{u}_2 - B\mathbf{u}_1) \tag{6.56}$$

This is true only if

$$\mathbf{u}^* = \mathbf{u}_1 + b\,(\mathbf{u}_2 - \mathbf{u}_1) \tag{6.57}$$

The control vector $\mathbf{u}^*$ is that which generates $a\mathbf{m}_{des}$ on the convex hull of the AMS, Φ_2. Thus, we generate the edges of Φ_2 and as each is found we evaluate Eq. (6.55). Then we perform the following tests:

1. If $0 \leq b \leq 1$ then we have found the correct edge. Otherwise, keep searching.
2. Calculate $\mathbf{u}^* = \mathbf{u}_1 + b(\mathbf{u}_2 - \mathbf{u}_1)$, the control vector that satisfies $a\mathbf{m}_{des} = B\mathbf{u}^*$.
   (a) If $a = 1$ then the desired moment $\mathbf{m}_{des}$ is on the convex hull of the AMS, and the only solution to the problem is $\mathbf{u}^*$.

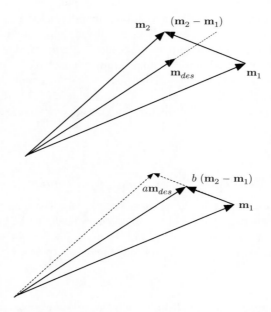

Figure 6.4   An edge, $(\mathbf{m}_2 - \mathbf{m}_1)$, that contains the intersection of a half-line along $\mathbf{m}_{des}$. $a\mathbf{m}_{des} = \mathbf{m}_1 + b(\mathbf{m}_2 - \mathbf{m}_1)$. As drawn here, $a > 1$ and $0 < b < 1$ so the intersection is on the edge itself and $\mathbf{m}_{des}$ is attainable

(b) If $a < 1$ then the desired moment $\mathbf{m}_{des}$ is unattainable using admissible controls. One may use $\mathbf{u}^*$ as the direction-preserving solution, or apply other means to find an acceptable solution.

(c) If $a > 1$ then the desired moment $\mathbf{m}_{des}$ is attainable by an infinity of solutions. One of these solutions is $\mathbf{u}^*/a$ from $a\mathbf{m}_{des} = B\mathbf{u}^*$, $\mathbf{m}_{des} = B\mathbf{u}^*/a$. All other solutions will be of the form

$$\mathbf{u} = \frac{\mathbf{u}^*}{a} + \mathbf{u}^{\perp} \qquad (6.58)$$

Where $\mathbf{u}^{\perp}$ is any vector lying in the null space of B,

$$B\mathbf{u}^{\perp} = \mathbf{0} \qquad (6.59)$$

Thus $\mathbf{u}^{\perp}$ is available to satisfy any manner of secondary objectives in the allocation of the control effectors.

6.6.2   The Direct Method for the Three-moment Problem

The direct method for the three-dimensional problem is completely analogous to that for the two-dimensional problem, except that we now seek the intersection of the half-line from the origin in the direction of $\mathbf{m}_{des}$ with a facet. Figure 6.5 shows the basic geometry.

As each facet is found (Section 5.4.1) one vertex in moment space is selected as $\mathbf{m}_1$. Of the other three vertices, the two that share an edge with $\mathbf{m}_1$ are those whose object notation differs

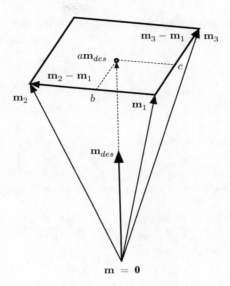

Figure 6.5 A facet of Φ_3

from that of $\mathbf{m}_1$ in exactly one position; that is, whose union with $\mathbf{m}_1$ is an edge. These two are denoted $\mathbf{m}_2$ and $\mathbf{m}_3$.

In Section 5.4.1 the facet we found for illustration was $\mathbf{o}_{2120000}$. The four vertices were

1. $\mathbf{m}_{0100000}$
2. $\mathbf{m}_{1100000}$
3. $\mathbf{m}_{1110000}$
4. $\mathbf{m}_{0110000}$

In the order enumerated, each of these four vertices shares an edge with the adjacent vertex in the list, cyclically, since the union of each pair is an edge. Thus we could choose

$$\mathbf{m}_1 = \mathbf{m}_{0100000} \quad \mathbf{m}_2 = \mathbf{m}_{1100000} \quad \mathbf{m}_3 = \mathbf{m}_{0110000} \tag{6.60}$$

Once the three vertices are identified we calculate a, b, and c from

$$a\mathbf{m}_{des} = \mathbf{m}_1 + b\,(\mathbf{m}_2 - \mathbf{m}_1) + c\,(\mathbf{m}_3 - \mathbf{m}_1) \tag{6.61}$$

$$\begin{Bmatrix} a \\ b \\ c \end{Bmatrix} = \begin{bmatrix} \mathbf{m}_{des} & -(\mathbf{m}_2 - \mathbf{m}_1) & -(\mathbf{m}_3 - \mathbf{m}_1) \end{bmatrix}^{-1} \mathbf{m}_1 \tag{6.62}$$

Again we denote the unique control vector that generates $a\mathbf{m}_{des}$ as $\mathbf{u}^*$, $a\mathbf{m}_{des} = B\mathbf{u}^*$ Denote the control vector that generates $\mathbf{m}_1$, $\mathbf{m}_2$, and $\mathbf{m}_3$ as $\mathbf{u}_1$, $\mathbf{u}_2$, and $\mathbf{u}_3$, then

$$\mathbf{u}^* = \mathbf{u}_1 + b\,(\mathbf{u}_2 - \mathbf{u}_1) + c\,(\mathbf{u}_3 - \mathbf{u}_1) \tag{6.63}$$

The control vector $\mathbf{u}^*$ is that which generates $a\mathbf{m}_{des}$ on the convex hull of the AMS, Φ_3. Thus, we generate the facets of Φ_3 and as each is found we evaluate Eq. (6.62). Then we perform the following tests:

1. If $0 \le b \le 1$ *and* $0 \le c \le 1$ then we have found the correct facet. Otherwise, keep searching.
2. We calculate $\mathbf{u}^* = \mathbf{u}_1 + b(\mathbf{u}_2 - \mathbf{u}_1) + c(\mathbf{u}_3 - \mathbf{u}_1)$, the control vector that satisfies $a\mathbf{m}_{des} = B\mathbf{u}^*$.
   (a) If $a = 1$ then the desired moment $\mathbf{m}_{des}$ is on the convex hull of the AMS, and the only solution to the problem is $\mathbf{u}^*$.
   (b) If $a < 1$ then the desired moment $\mathbf{m}_{des}$ is unattainable using admissible controls. One may use $\mathbf{u}^*$ as the direction-preserving solution, or apply other means to find an acceptable solution.
   (c) If $a > 1$ then the desired moment $\mathbf{m}_{des}$ is attainable by an infinity of solutions. One of these solutions is $\mathbf{u}^*/a$. All other solutions will be of the form

$$\mathbf{u} = \frac{\mathbf{u}^*}{a} + \mathbf{u}^\perp \tag{6.64}$$

This process is illustrated in Section 6.11.2, Example 6.7.

Direct allocation is far from the fastest algorithm that is optimal in the sense of maximum capabilities. Its primary advantage is the transparency of its operation, and its relative robustness to certain problems that can arise in control allocation.

The most common problem is a control effectiveness matrix that is not robustly full rank. The easy answer to this concern is to add small random numbers to the entries of the B matrix. These numbers must be large enough that they are significant in terms of the machine precision of the flight control computer but physically insignificant in terms of the moments or accelerations involved. Moreover, it is not required that new random numbers be generated each time a new B matrix is presented: a single $n \times m$ matrix of suitable numbers may be used every time.

6.7 Edge and Facet Searching

One of the problems with direct allocation as described in Section 6.6 is the possibility that one will have to construct the entire attainable moment set before the correct edge or facet is found. Moreover, there are other algorithms that are less computationally demanding.

Edge- and facet-searching methods seek to remove the requirement to test edges or facets by generating them and checking for an intersection until the correct one is found. This method instead relies on different properties of the geometry, and of the correct edge or facet.

The first property is that for any rotation of the attainable moment set, plane or three-dimensional, the coordinates of the maximum moment in the direction of any axis are easily determined by observing the signs of the entries in a particular row of the rotated B matrix, yielding the vector of admissible controls that generate that point. Then the actual moment at that point is simply B times that control vector.

The second property is that any two vertices in moment space share an edge if their corresponding control vectors differ by a single entry (their union is an edge), either in control or object notation.[1] Given two vertices on the convex hull of the attainable moments, we may determine if they share a common edge by comparison of the control vectors that generated them. Thus, if we have two known vertices of Φ, we may determine if they define an edge by observation.

Extending the previous property, if we know two edges of a three-dimensional Φ, we may determine whether or not they define a facet by comparison of the control vectors that generated them, here testing whether or not the two edges differ by exactly two entries (their union is a facet), either in control or object notation.

6.7.1   Two-dimensional Edge Searching

We begin with Figure 6.6. There we see some two-dimensional set of attainable moments and a particular desired moment, $\mathbf{m}_{des}$. On this figure we put a set of x–y axes such that the x-axis is aligned with $\mathbf{m}_{des}$. Only the x axis is shown. Some of the vertices in the AMS are difficult to distinguish, so as each is determined we have placed a • to show where it is.

In the following there are three different coordinate systems at play.

- One is the original moment-space coordinate system, m_1 and m_2, and it is not shown.
- The consequence of the rotation to x–y, with x aligned with $\mathbf{m}_{des}$, is fixed throughout all six figures. The y-axis is important since an important feature of each vertex identified is whether its y component is positive or negative. Thus if we determine a vertex in control space, and convert it to moment space ($B\mathbf{u}$), we may observe whether its y component is below or above the x-axis to determine the sign of its y component.
- The result of rotating the x-axis through various angles to resulting directions is a new x-axis, labeled x'. This axis is important in the determination of the vertex of Φ that is maximum in its direction in each step. The associated y' is of no interest in this method. Rather than add increasing numbers of $'$s to the superscript, we will call them all x', but they should all be distinguished from the original x.

Figure 6.6a shows the initial orientation. Not shown in the figure are the coordinates of moment space. We are interested only in determining the edge of the two-dimensional Φ that contains the intersection of a half-line in the direction of $\mathbf{m}_{des}$ (shown as a horizontal dashed line in each part of Figure 6.6).

In Figure 6.6a is shown the vertex (•) that is the maximum in the direction of x. The coordinates of this point in control space are determined by the signs of the entries in the first row of the rotated B matrix that accomplished this initial alignment with $\mathbf{m}_{des}$ (as was done in Sections 5.1 and 5.4.1). We observe that the sign of its y component is negative (the vertex is below the dashed line).

We now seek a vertex whose y component is positive in this initial coordinate system. We therefore look upward by rotating the x–y system $\pi/2$ radians anti-clockwise and determining the maximum vertex in this direction. The result is shown in Figure 6.6b. Again, the maximum vertex is highlighted by a •. We note that the sign of its y component is positive (the vertex is above the dashed line).

---

[1] It is generally better to compare the vertices in object notation, as machine precision is not a factor when dealing with integers.

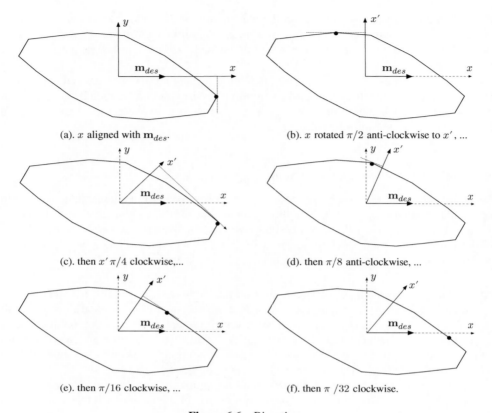

(a). x aligned with $\mathbf{m}_{des}$.

(b). x rotated $\pi/2$ anti-clockwise to x', ...

(c). then x' $\pi/4$ clockwise,...

(d). then $\pi/8$ anti-clockwise, ...

(e). then $\pi/16$ clockwise, ...

(f). then $\pi/32$ clockwise.

Figure 6.6   Bisections

Now we have two vertices, one with a positive and one with a negative y component (in the original x–y coordinate system). We may now test whether these two vertices constitute an edge by comparing the controls that generated them. We would discover that they differ in more than one entry, and therefore do not constitute an edge.

This initial rotation of $\pi/2$ radians was too large, so we rotate x' clockwise at a reduced angle. Here we choose a rotation of $\pi/4$, intending to bisect the angle at each step until we either find the correct edge, or reach some pre-established criterion.

The result of this clockwise rotation is shown in Figure 6.6c, with a • for the maximum vertex in this direction. It is the same vertex that was determined in the first step, Figure 6.6a. We bisect the rotation angle again and rotate $\pi/8$ the other way, anti-clockwise.

By now the pattern should be clear. We are seeking two consecutive vertices, one that is above and the other that is below, the intersection of the original x-axis with the convex hull. This is determined by the signs of the y component of the two vertices. Once we have such a pair, we check to see if they form an edge by comparing their corresponding control vectors. They form an edge if they differ in exactly one entry. The vertices found in Figures 6.6e and 6.6f satisfy both criteria, and the correct edge is identified. This edge is shown in Figure 6.7a.

Shown in Figure 6.7b is a line connecting the vertices found in Figures 6.6c and 6.6d. In some problems a set of vertices, including those defining the correct edge, may be nearly

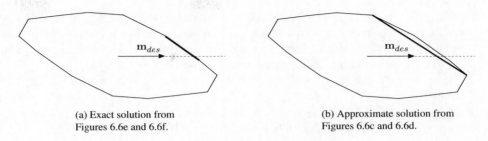

(a) Exact solution from Figures 6.6e and 6.6f.

(b) Approximate solution from Figures 6.6c and 6.6d.

Figure 6.7   Exact and approximate solutions

co-linear. If the algorithm is implemented with a fixed maximum number of bisections the correct edge may not be found. In that case an approximate solution may be obtained by connecting two vertices, one above and one below the x-axis, and finding the intersection of the x-axis with that line. That intersection may be used in the same manner that the intersection with an edge would have been; that is, scaling the vector to the intersection to the size of $\mathbf{m}_{des}$.

With respect to the described method, we have chosen to bisect the angle at each step, but other methods of dividing the angles may prove more efficient. The golden mean may be used, for instance. The rotations of the x'-axis require sines and cosines of the angles, but these may be pre-computed and stored for retrieval as needed.

6.7.1.1   Summary of Two-dimensional Edge Searching

The following summary is written using an initial rotation of $\pi/2$ and subsequent bisections of the rotation angle. A counter i is initialized to $i = 0$. The algorithm will branch back to step 1 with each bisection and i is incremented.

Before beginning, rotate the problem so that m_1 is aligned with $\mathbf{m}_{des}$. Denote the rotated axis x as in Figure 6.6. The simple plane rotation is:

$$\mathbf{m}_{des} = \begin{Bmatrix} m_{d_1} \\ m_{d_2} \end{Bmatrix}, \quad T\mathbf{m}_{des} = \begin{Bmatrix} |\mathbf{m}_{des}| \\ 0 \end{Bmatrix} \tag{6.65}$$

$$T = \begin{bmatrix} \cos\theta & \sin\theta \\ -\sin\theta & \cos\theta \end{bmatrix} \tag{6.66}$$

$$\theta = \arctan\frac{m_{d_2}}{m_{d_1}} \tag{6.67}$$

$$B_0 = TB \tag{6.68}$$

The arctan function should be quadrant-specific.

1. From the first row of B_i determine the control vector, $\mathbf{u}_i$, that generates the maximum vertex in the x direction. If an entry is positive the corresponding control effector is assigned its maximum value in control notation and the number 1 in object notation. If the entry is negative its minimum value is used and the number 0 is assigned. If there are any zeros:

(a) If there is exactly one zero, x is perpendicular to an edge. This edge may be the solution, and it should be tested as in two-dimensional control allocation. If it is not the solution edge, take the vertex that is nearest to the x-axis (magnitude of the y component of $B_i\mathbf{u}_i$ is least, see step 2) for $\mathbf{u}_i$ and proceed with the algorithm.

(b) If two or more entries are zero, the control effectiveness matrix was not robustly full rank.

2. Multiply $B_0\mathbf{u}_i$. At this point we need only the y component of the result, y_i, so only the second row of B_0 is needed to multiply $\mathbf{u}_i$. B_0 is the *initial* matrix from Eq. (6.68).

3. If $y_i = 0$ then $\mathbf{u}_i$ is the solution to the allocation problem.

4. If $i > 0$ and the signs of y_i and y_{i-1} are different, test $\mathbf{u}_i$ and $\mathbf{u}_{i-1}$ to see if they comprise an edge. The test is to compare them element by element. If they differ in exactly one element, that edge is the solution edge. Otherwise:

(a) If y_i is positive, rotate B_i clockwise, $\theta = \pi/2/2^i$ in Eq. (6.66).

(b) If y_i is negative, rotate B_i anti-clockwise, $\theta = -\pi/2/2^i$ in Eq. (6.66).

5. Increment i and go to step 1.

Note that B_i is used to determine a maximum vertex in step 1, but B_0 is used to determine the y coordinate in each visit to step 2.

Allocation proceeds as in Section 6.6 from the point where the correct edge is found.

There are a few variations on this method. First, as mentioned, bisection is not necessarily the best way of dividing the angle. The golden mean (ratio) was used in limited testing with indifferent results. Second, the bisection (or other division of the angle) may be performed only when the direction of rotation is reversed, not on every rotation. Again, limited testing did not show a clear preference for one scheme or the other.

6.7.2   Three-dimensional Facet Searching

Three-dimensional facet searching is illustrated in Figure 6.8. The figure is the AMS, and somewhere within is the desired moment $\mathbf{m}_{des}$. A rotation of the problem has been accomplished to align the x-axis with $\mathbf{m}_{des}$, as was done in Section 6.7.1 for the two-dimensional case. As before we denote the initial rotation of the B matrix as B_0.

Figure 6.8a shows the x-axis and the solution facet. In Figure 6.8b the AMS has been rotated through some negative angle (using the right-hand rule) so that now the solution facet is tilted away from our viewpoint. In Figure 6.8c another rotation has placed the solution facet nearly edge-on to our viewpoint. Any further rotation and the farther edge of the solution facet will recede from view.

Now consider the projection of the AMS into the plane of the page. As was done in the two-dimensional case, we take the initial rotation that resulted in Figure 6.8a as our reference in the following, with y upward and z out of the page. Projecting the AMS into the plane of the page means just removing the z component of each vertex. The plane figure that results is effectively a two-dimensional AMS whose edges are projections of the edges of the three-dimensional AMS. The intersection of the x-axis with an edge may be found exactly as done in Section 6.7.1 through two-dimensional edge searching.

Now consider Figure 6.8c again. An edge search will return one of the four projected edges of the solution facet, the one that is about to recede from view. Another small rotation and that

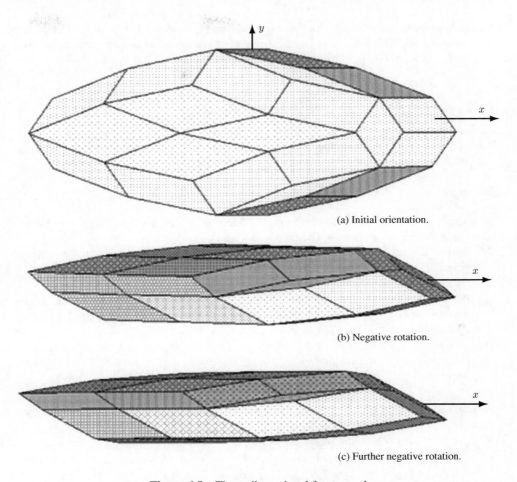

(a) Initial orientation.

(b) Negative rotation.

(c) Further negative rotation.

Figure 6.8   Three-dimensional facet search

edge vanishes from the projection, and another projected edge of the facet takes its place. Two edges define a facet, so the solution facet is determined.

The key to identifying two candidate edges is the sign of the z component of the moment generated by the control vector that was found in the edge search of the projected figure. That is, we treat the AMS as a plane figure to determine the intersection of x with an edge, and as a three-dimensional figure to find its z component.

6.7.2.1   Summary of Three-dimensional Facet Searching

The initial rotation is $\pi/2$ with subsequent bisections of the rotation angle. A counter i is initialized to $i = 0$. The algorithm will branch back to step 1 with each bisection and i is incremented. Two-dimensional edge searches have their own plane rotations independent of the rotations described here.

Before beginning we transform the problem so that m_1 is aligned with $\mathbf{m}_{des}$. Denote the rotated axis x, as in Figure 6.8. There are many ways to accomplish this rotation, since none is unique. Example 6.9 on p. 138 describes one such method.

$$B_0 = T_0 B \tag{6.69}$$

Subsequent transformations will be about the x-axis using

$$T = \begin{bmatrix} 1 & 0 & 0 \\ 0 & \cos\theta & \sin\theta \\ 0 & -\sin\theta & \cos\theta \end{bmatrix} \tag{6.70}$$

Starting with $i = 0$ the steps are:

1. Remove the third row of B_i and perform an edge search with the resulting $2 \times m$ matrix. Denote the two vertices that are returned as $\mathbf{u}_{i,1}$ and $\mathbf{u}_{i,2}$.
2. Multiply $B_0 \mathbf{u}_{i,1}$ and $B_0 \mathbf{u}_{i,2}$. Denote the results $\mathbf{m}_{i,1}$ with z component $z_{i,1}$ and $\mathbf{m}_{i,2}$ with z component $z_{i,2}$. B_0 is the *initial* matrix from Eq. (6.69).
3. If $i = 0$ (the initial orientation):
   (a) If $z_{i,1}$ and $z_{i,2}$ are of different signs then the edge conveys no information, and a new initial orientation is needed. Rotate the problem through $\theta = \pi/2$ using Eq. (6.70). Calculate $B_0 = T_0 B_0$ and repeat step 1, keeping $i = 0$.
   (b) Otherwise $z_{i,1}$ and $z_{i,2}$ are of the same sign. Denote this sign s_i. Increment i to $i = 1$. Rotate the problem through $\theta_1 = \pi/2$ using Eq. (6.70). and calculate $B_1 = T_1 B_0$. Go to step 1.
4. If $z_{i,1}$ and $z_{i,2}$ are of different signs, then set s_i the opposite of s_{i-1}. Bisect the angle and reverse the direction ($\theta_i = -\theta_{i-1}/2$) using Eq. (6.70). Increment i by $+1$, and calculate $B_i = T_i B_{i-1}$. Go to step 1.
5. If $z_{i,1}$ and $z_{i,2}$ are of the same sign denote this sign s_i. Then:
   (a) If $s_i = s_{i-1}$ bisect the angle but do not reverse direction. Increment i, $i = i + 1$, set $\theta_i = \theta_{i-1}/2$, and calculate $B_i = T_i B_{i-1}$. Go to step 1.
   (b) If $s_i \neq s_{i-1}$ test the previous and currently found edges to see if they comprise a facet. If the vectors $\mathbf{u}_{i-,1}$, $\mathbf{u}_{i-,2}$, $\mathbf{u}_{i,1}$, and $\mathbf{u}_{i,2}$ differ in exactly two positions then they are edges of the same facet; exit.
   (c) Otherwise, bisect the angle and reverse direction using Eq. (6.70). Increment i, $i = i + 1$, set $\theta_i = -\theta_{i-1}/2$. and calculate $B_i = T_i B_{i-1}$. Go to step 1.

If a solution is not found after the prescribed number of bisections, the last two distinct edges found may be used to calculate an approximate solution. This approximate solution is calculated by linearly interpolating from the y and z coordinates of the vertices of the edges found to find a point on the x-axis. The controls associated with those vertices are then combined according to the interpolation factors.

6.8 Banks' Method

This control allocation method was proposed by Carl Banks, a graduate student at Virginia Tech. He referred to it as the 'Vertex Jumping Algorithm'. It was devised for the three-moment

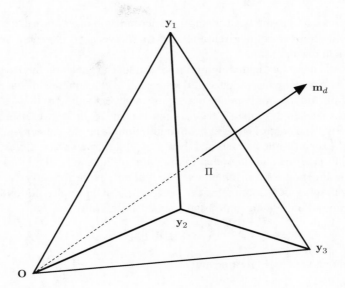

Figure 6.9 Geometry of the vertices. Vertex $\mathbf{y}_2$ is nearest to the viewer

problem, so none of the description that follows deals with the two-moment problem. The following description is based on notes that accompanied a MATLAB® implementation of the algorithm. That implementation is incorporated into the ADMIRE simulation (Appendix B).

Consider Figure 6.9. Three vertices, $\mathbf{y}_1$, $\mathbf{y}_2$, and $\mathbf{y}_3$ with the origin $\mathbf{O}$ form a solid angle bound by the three planes through $(\mathbf{y}_1, \mathbf{y}_2, \mathbf{O})$, $(\mathbf{y}_2, \mathbf{y}_3, \mathbf{O})$, and $(\mathbf{y}_1, \mathbf{y}_3, \mathbf{O})$. Denote by Π the plane triangle whose sides are $(\mathbf{y}_1 - \mathbf{y}_2)$, $(\mathbf{y}_2 - \mathbf{y}_3)$, and $(\mathbf{y}_3 - \mathbf{y}_1)$. Assume these three vertices have the following properties:

1. The vertices are not co-linear.
2. The vertices are ordered such that, when viewed from the side of the plane of Π away from the origin, $(\mathbf{y}_1 - \mathbf{y}_2 - \mathbf{y}_3)$ make a counterclockwise path. This is equivalent to requiring that the determinant of the matrix $\begin{bmatrix} \mathbf{y}_1 & \mathbf{y}_2 & \mathbf{y}_3 \end{bmatrix}$ be positive.
3. The desired moment $\mathbf{m}_{des}$ is inside the solid angle; equivalently, $\mathbf{m}_{des}$ intersects the triangle Π; also equivalently, $\mathbf{m}_{des}$ is always on the left side when traversing $(\mathbf{y}_1 - \mathbf{y}_2 - \mathbf{y}_3)$. The latter is satisfied by the criteria that the signs of the determinants of the three matrices in Eq. (6.71) be positive:

$$\det \begin{bmatrix} \mathbf{m}_{des} & \mathbf{y}_1 & \mathbf{y}_2 \end{bmatrix} > 0$$
$$\det \begin{bmatrix} \mathbf{m}_{des} & \mathbf{y}_2 & \mathbf{y}_3 \end{bmatrix} > 0 \qquad (6.71)$$
$$\det \begin{bmatrix} \mathbf{m}_{des} & \mathbf{y}_3 & \mathbf{y}_1 \end{bmatrix} > 0$$

The algorithm seeks to replace one of the three vertices with another that is farther from the origin than the plane of Π such that the new set of vertices has the same properties as the original set. This process is repeated until no new vertex can be found. The last three vertices will define the facet containing the intersection of $\mathbf{m}_{des}$ with the boundary of the AMS (the solution facet).

A frequently used tool in the algorithm is the determination of a vertex that has the greatest projection onto a vector in some given direction **d**. The vector **d** is generic and is recycled frequently through the algorithm.

Rotations are not necessary in this algorithm so we adopt a simpler method to find maximum vertices. We first find the inner product of **d** with each of the columns of the B matrix, then assign the controls to their minimum or maximum values according to the sign of the result. That is, we first calculate $\mathbf{d}^T B$; then if the ith entry is positive, u_i is set to its maximum limit, and if it is negative, the control is set to is minimum limit. The case of zero entries means an edge of the AMS is perpendicular to **d** and there is no unique maximum vertex.

In the description of the algorithm that follows, use of this tool will be loosely referred to as 'looking in a direction for a maximum vertex'. Slightly more formally, we will just refer to Eq. (6.72) with the understanding that the signs are taken on an element-by-element basis, and then maximum or minimum controls are applied as described.

$$\mathbf{v} = \text{sgn} \, [\mathbf{d}^T B] \tag{6.72}$$

The four main parts to the algorithm are:

1. finding the original three vertices that have the desired properties
2. determining a new vertex
3. determining which of the previous three vertices should be replaced
4. terminating the algorithm.

6.8.1   Finding the Original Three Vertices

There are many ways this step can be accomplished. For example, from the null space of $\mathbf{m}_{des}$ (a plane), three directions may be selected, arbitrarily except for being 120° apart. Then the three vertices that are maximum in these directions, when placed in the right order, may be used to start the algorithm.

Alternatively, the method Banks used was as follows:

1. The first vertex is the maximum in the direction of $\mathbf{m}_{des}$, here denoted $\mathbf{y}_1$. It is found from the signs of the row vector $\mathbf{m}_{des} B$, as described above.
2. A second vertex is found by looking in a direction **d** perpendicular to $\mathbf{y}_1$, and in the plane of **y** and $\mathbf{y}_1$. For this to hold **y** must be a linear combination of a vector in the direction of $\mathbf{y}_1$, say $a\mathbf{y}_1$, and **d**. The factor a is the magnitude of the projection of $\mathbf{m}_{des}$ onto $\mathbf{y}_1$. We can get the angle θ between the two vectors from their dot product,

$$\mathbf{y}^T \mathbf{y}_1 = yy_1 \, \cos \theta \tag{6.73}$$

The factor $a = y \cos \theta = \mathbf{y}^T \mathbf{y}_1 / y_1$, so

$$\mathbf{d} = \mathbf{y} - \frac{\mathbf{y}^T \mathbf{y}_1}{\mathbf{y}_1^T \mathbf{y}_1} \mathbf{y}_1 \tag{6.74}$$

Calculate $\mathbf{y}_2$ from $\mathbf{d}^T B$.

3. A third vertex is found in a direction $\mathbf{d}$ perpendicular to the plane of $\mathbf{y}_1$ and $\mathbf{y}_2$, on the same side of the plane $\mathbf{y}$ is on. The perpendicular is found from the cross product of $\mathbf{y}_1$ and $\mathbf{y}_2$. To determine whether $\mathbf{d}$ points toward the side $\mathbf{y}$ is on, the sign of the inner product of $\mathbf{d}$ and $\mathbf{y}$ is examined. If it is positive the direction is good, otherwise $\mathbf{y}_1$ and $\mathbf{y}_2$ are swapped to keep the orientation correct.

First, calculate $\mathbf{d} = \mathbf{y}_1 \times \mathbf{y}_2$. Then, if $\mathbf{d}^T \mathbf{y} > 0$, calculate $\mathbf{y}_3$ from $\mathbf{d}^T B$.

Otherwise $\mathbf{d}^T \mathbf{y} < 0$, and the swapping goes as follows (with a temporary $\mathbf{y}$, $\mathbf{y}_{Temp}$):

(a) $\mathbf{y}_{Temp} = \mathbf{y}_1$.
(b) $\mathbf{y}_1 = \mathbf{y}_2$.
(c) $\mathbf{y}_2 = \mathbf{y}_{Temp}$.

Now calculate $\mathbf{y}_3$ from $-\mathbf{d}^T B$.

This procedure guarantees that

$$\det \begin{bmatrix} \mathbf{m}_{des} & \mathbf{y}_1 & \mathbf{y}_2 \end{bmatrix} > 0$$

but says nothing about

$$\det \begin{bmatrix} \mathbf{m}_{des} & \mathbf{y}_2 & \mathbf{y}_3 \end{bmatrix}$$

or

$$\det \begin{bmatrix} \mathbf{m}_{des} & \mathbf{y}_3 & \mathbf{y}_1 \end{bmatrix}$$

If, for example, $\det \begin{bmatrix} \mathbf{m}_{des} & \mathbf{y}_2 & \mathbf{y}_3 \end{bmatrix} < 0$ the algorithm replaces $\mathbf{y}_1$ with $\mathbf{y}_3$, and determines a new $\mathbf{y}_3$ by looking in the direction $\mathbf{y}_3 \times \mathbf{y}_2$. Note that replacing $\mathbf{y}_1$ with $\mathbf{y}_3$ means that $\det \begin{bmatrix} \mathbf{m}_{des} & \mathbf{y}_2 & \mathbf{y}_1 \end{bmatrix} < 0$; equivalently, $\det \begin{bmatrix} \mathbf{m}_{des} & \mathbf{y}_1 & \mathbf{y}_2 \end{bmatrix} > 0$. If $\det \begin{bmatrix} \mathbf{m}_{des} & \mathbf{y}_3 & \mathbf{y}_1 \end{bmatrix} < 0$ a similar procedure is employed, replacing $\mathbf{y}_2$ with $\mathbf{y}_3$, and finding a new $\mathbf{y}_2$ by looking in the direction $\mathbf{y}_1 \times \mathbf{y}_3$.

In any event we assume that we have determined three vertices that satisfy Eqs 6.71 and proceed.

6.8.2 Determining a New Vertex

As in the case of finding the intitial three vertices, this step has many possible approaches. Banks' method asks us to look in a direction normal to the current Π and away from the origin and find the maximum vertex. Figure 6.10 shows an edge-on view of the plane of Π, the perpendicular $\mathbf{d}$, the vector in the direction of the desired moment $\mathbf{y}$, and the new vertex.

The ubiquitous $\mathbf{d}$ is determined from:

$$\mathbf{d} = (\mathbf{y}_2 - \mathbf{y}_1) \times (\mathbf{y}_3 - \mathbf{y}_1) \tag{6.75}$$

The next vertex, here denoted $\mathbf{y}_4$, is determined as before from Eq. (6.72).

6.8.3 Replacing an Old Vertex

Figure 6.11 shows the view looking straight down $\mathbf{y}_4$. Four potential points of intersection of $\mathbf{m}_{des}$ with Π are also shown, labeled A, B, C, and D. Considering any one of these points, the

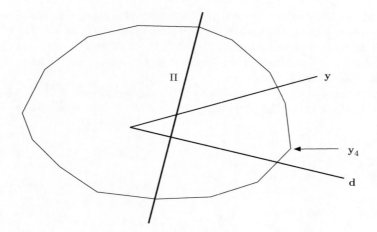

Figure 6.10   Determining the new vertex

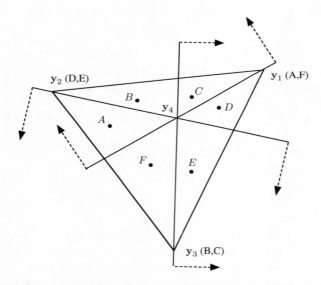

Figure 6.11   Replacing an old vertex. The dashed lines in the figure show the direction of of the normals formed by the cross-products of $\mathbf{y}_4$ with $\mathbf{y}_1$, $\mathbf{y}_2$, and $\mathbf{y}_3$. Six possible intersections of $\mathbf{m}_{des}$ with Π are indicated by the bullet symbol

objective is to retain two of the vertices $\mathbf{y}_1$, $\mathbf{y}_2$, and $\mathbf{y}_3$, and to replace the third with $\mathbf{y}_4$ so that the result still encircles $\mathbf{m}_{des}$ as in Figure 6.9. The letters next to $\mathbf{y}_1$, $\mathbf{y}_2$, and $\mathbf{y}_3$ indicate which will be replaced for the four cases shown.

   We construct three planes from each of $\mathbf{y}_1 - \mathbf{y}_4$, $\mathbf{y}_2 - \mathbf{y}_4$, and $\mathbf{y}_3 - \mathbf{y}_4$ as shown. The dashed lines in the figure show the direction of of the normals formed by the products $\mathbf{y}_4 \times \mathbf{y}_1$, $\mathbf{y}_4 \times \mathbf{y}_2$, and $\mathbf{y}_4 \times \mathbf{y}_3$. By taking the inner product of each of these normals with $\mathbf{m}_{des}$, the side of the plane on which $\mathbf{m}_{des}$ lies is determined. Vertex replacement determination then proceeds from the following logic (with y instead of $\mathbf{m}_{des}$):

```
if det ([y,y4,y1]) > 0 % Case A, B, or C
  if det ([y,y4,y2]) > 0  % Case A
    y1 = y4;
  else % Case B or C
    y3 = y4;
  end
else & Case D, E, or F
  if det ([y,y4,y3]) > 0 % Case D or E
    y2 = y4;
  else % Case F
    y1 = y4;
  end
end
```

For example, for possibility E,

$$\det \begin{bmatrix} \mathbf{m}_{des} & \mathbf{y}_4 & \mathbf{y}_1 \end{bmatrix} < 0$$
$$\det \begin{bmatrix} \mathbf{m}_{des} & \mathbf{y}_4 & \mathbf{y}_2 \end{bmatrix} > 0 \qquad (6.76)$$
$$\det \begin{bmatrix} \mathbf{m}_{des} & \mathbf{y}_4 & \mathbf{y}_3 \end{bmatrix} > 0$$

This combination uniquely identifies $\mathbf{y}_2$ as the vertex to be replaced. Note that the handedness of the three points is not changed by this replacement; that is, $\det \begin{bmatrix} \mathbf{y}_1 & \mathbf{y}_2 & \mathbf{y}_3 \end{bmatrix} > 0$.

6.8.4   Terminating the Algorithm

As each new vertex is identified and replaces an old one, the three vertices are tested to see if they are part of a facet. If so, that is the solution facet and the algorithm terminates. The test is easily performed by determining if their union is a facet; that is, by comparing the three vertices on an element-by-element basis. If they are identical in all but two elements, then they constitute a facet.

Banks' method is demonstrated in Section 6.11, Example 6.0.

6.9   Linear Programming

Linear programming[2] is an extensively studied method of finding optimal solutions to linear sets of equations with constraints on some or all of the variables. Many of the introductory texts couch the problem in terms of the allocation of limited resources to minimize or maximize the outcome of some process involving those resources.

---

[2] Linear programming predates the modern computer age by several decades, and the word *program* originally signified a problem, rather than lines of code to solve the problem. Thus, mixing the two usages, we may speak of a linear program as a problem to be solved by linear programming.

Linear programming has several forms in which the problem is posed. Some are called standard, or canonical. For our purposes we adopt Eq. (6.77):

$$\operatorname*{Min}_{\mathbf{x}} J = \mathbf{c}^T \mathbf{x} \quad A\mathbf{x} = \mathbf{b}, \mathbf{0} \leq \mathbf{x} \leq \mathbf{h} \tag{6.77}$$

The primary variations on Eq. (6.77) are:

- the cost function $\mathbf{c}^T \mathbf{x}$ is maximized
- the constraint is made into an inequality $A\mathbf{x} \geq \mathbf{b}$
- the bounded variable $\mathbf{x}$ has no upper bound, $\mathbf{x} \geq \mathbf{0}$.

All combinations are considered as forms of the linear programming problem.

6.9.1 Casting Control Allocation as a Linear Program

An excellent description of the application of linear programming to the control allocation problem is by Beck (2002), including performance data for implementations of the algorithms in MATLAB®. Most of the formulations incorporate features that we have not yet introduced, primarily that of including preferred solutions. It is not our intention to explore linear programming in great detail in this section, but two examples will serve to give the reader an idea of how the problems are cast. Readers wishing a greatly expanded explanation of linear programming as a control allocation method are referred to Appendix A. Both this section, as a brief introduction, and Appendix A are stand-alone, so there is some repetition between the two sections.

6.9.1.1 Direction Preserving

For this formulation we begin with the control allocation requirement that $\mathbf{m}_{des} = B\mathbf{u}$, and consider a multiplier λ on $\mathbf{m}_{des}$,

$$B\mathbf{u} = \lambda \mathbf{m}_{des} \tag{6.78}$$

If we now take admissible $\mathbf{u}$, constrained to satisfy $B\mathbf{u} = \lambda \mathbf{m}_{des}$, and λ within the range $0 \leq \lambda \leq 1$, then when λ is maximum $\mathbf{u}$ will be the solution on the boundary; what we have called $\mathbf{u}^*$. Then we scale $\mathbf{u}^*$ to achieve the solution.

Because we have posed the linear program as a minimization problem, rather than maximizing λ we minimize $-\lambda$. We also have to deal with the control limits so that zero is the lower limit. This is commonly done by introducing a new variable,

$$\tilde{\mathbf{x}} = \mathbf{u} - \mathbf{u}_{Min} \tag{6.79}$$

The minimization will be performed over $\mathbf{u}$ and λ, equivalently over $\tilde{\mathbf{x}}$ and λ, so we combine them into $\mathbf{x}$,

$$\mathbf{x} \equiv \begin{Bmatrix} \tilde{\mathbf{x}} \\ \lambda \end{Bmatrix} \tag{6.80}$$

The cost function $J = \mathbf{c}^T \mathbf{x}$ therefore requires

$$\mathbf{c}^T \equiv \begin{bmatrix} 0 & \cdots & 0 & -1 \end{bmatrix} \tag{6.81}$$

Then, we construct $A\mathbf{x} = \mathbf{b}$:

$$B\mathbf{u} = B(\tilde{\mathbf{x}} + \mathbf{u}_{Min})$$
$$= B\tilde{\mathbf{x}} + B\mathbf{u}_{Min} = \lambda\mathbf{m}_{des} \tag{6.82}$$

$$B\tilde{\mathbf{x}} - \lambda\mathbf{m}_{des} = -B\mathbf{u}_{Min}$$

$$\begin{bmatrix} B & -\mathbf{m}_{des} \end{bmatrix} \begin{Bmatrix} \tilde{\mathbf{x}} \\ \lambda \end{Bmatrix} = -B\mathbf{u}_{Min} \tag{6.83}$$

$$\begin{bmatrix} B & -\mathbf{m}_{des} \end{bmatrix} \mathbf{x} = -B\mathbf{u}_{Min}$$

$$A \equiv \begin{bmatrix} B & -\mathbf{m}_{des} \end{bmatrix}, \mathbf{b} \equiv -B\mathbf{u}_{Min} \tag{6.84}$$

Finally, the upper limit on $\mathbf{x}$ is given by

$$\mathbf{h} \equiv \begin{Bmatrix} \mathbf{u}_{Max} - \mathbf{u}_{Min} \\ 1 \end{Bmatrix} \tag{6.85}$$

Summarizing, the linear program is

$$\min_{\mathbf{x}} J = \mathbf{c}^T\mathbf{x} | A\mathbf{x} = \mathbf{b}, 0 \le \mathbf{x} \le \mathbf{h} \tag{6.86}$$

where

$$\mathbf{x} = \begin{Bmatrix} \mathbf{u} - \mathbf{u}_{Min} \\ \lambda \end{Bmatrix}$$
$$\mathbf{c}^T = \begin{bmatrix} 0 & \cdots & 0 & -1 \end{bmatrix}$$
$$A = \begin{bmatrix} B & -\mathbf{m}_{des} \end{bmatrix} \tag{6.87}$$
$$\mathbf{b} = -B\mathbf{u}_{Min}$$
$$\mathbf{h} = \begin{Bmatrix} \mathbf{u}_{Max} - \mathbf{u}_{Min} \\ 1 \end{Bmatrix}$$

6.9.1.2 Bodson's Reduced-size, Direction-preserving Formulation

This formulation is somewhat less intuitive than pure direction-preserving, but its nice properties are worth the effort. First we need a suitable cost function. Recall that at the end of the process of direct allocation we arrived at an expression

$$a\mathbf{m}_{des} = B\mathbf{u}^* \tag{6.88}$$

That is, if the desired moment $\mathbf{m}_{des}$ is extended (or contracted, worse) to the boundary of the set of attainable moments, $\partial(\Phi)$, it will intersect at a point that corresponds to $B\mathbf{u}^*$. Note that if we limit our choices only to admissible controls $\mathbf{u}$ that lie in the same direction as $\mathbf{m}_{des}$, when the quantity $\mathbf{m}_{des}^T B\mathbf{u}$ is maximum, $\mathbf{u} = \mathbf{u}^*$.

Since we have posed the problem as one of minimization, we wish to minimize $-\mathbf{m}_{des}^T B\mathbf{u}$. The addition of a constant will not change the $\mathbf{u}$ and moments $\mathbf{m} = B\mathbf{u}$ that minimizes the product, so we take as our cost function

$$J = -\mathbf{m}_{des}^T B\mathbf{u} + const \tag{6.89}$$

Next we deal with the limits on control effector displacements as in Eq. (6.79),

$$\mathbf{x} = \mathbf{u} - \mathbf{u}_{Min} \tag{6.90}$$

The limits on $\mathbf{x}$ become

$$0 \le \mathbf{x} \le \mathbf{h}, \mathbf{h} = \mathbf{u}_{Max} - \mathbf{u}_{Min} \tag{6.91}$$

Our cost function is now

$$J = -\mathbf{m}_{des}^T B\mathbf{u} + const = -\mathbf{m}_{des}^T B(\mathbf{x} - \mathbf{u}_{Min}) + const \tag{6.92}$$

If we take $const = -\mathbf{m}_{des}^T B\mathbf{u}_{Min}$,

$$J = \mathbf{c}^T \mathbf{x}, \mathbf{c}^T = -\mathbf{m}_{des}^T B \tag{6.93}$$

Note that on the boundary $\mathbf{x}^* = \mathbf{u}^* - \mathbf{u}_{Min}$.

Finally we insure that our candidate solutions are in the same direction as $\mathbf{m}_{des}$. We will do this through the constraints $A\mathbf{x} = \mathbf{b}$. We characterize all vectors that are in the same direction as $\mathbf{m}_{des}$ as $\rho\mathbf{m}_{des}$, $\rho \ne 0$, and denote the controls that generate $\rho\mathbf{m}_{des}$ as candidate vectors $\mathbf{u}_c$.

$$\rho\mathbf{m}_{des} = B\mathbf{u}_c \tag{6.94}$$

Denote the elements of $\mathbf{m}_{des}$ as m_{d_i} and the rows of B as B_{r_i}, $i = 1 \cdots n$. Assume that the first element of $\mathbf{m}_{des}$ is non-zero (or that the order of the rows has been rearranged so that it is), $m_{d_1} \ne 0$. Then

$$\rho m_{d_1} = B_{r_1}\mathbf{u}_c \tag{6.95}$$

We solve for ρ and eliminate it from the remaining equations,

$$\rho = \frac{B_{r_1}}{m_{d_1}}\mathbf{u}_c, i = 2 \cdots n \tag{6.96}$$

$$B_{r_i}\mathbf{u}_c = am_{d_i} = \frac{m_{d_i}}{m_{d_1}}B_{r_1}\mathbf{u}_c \tag{6.97}$$

Subtracting the right-hand term from the left in Eq. (6.97)

$$B_{r_i}\mathbf{u}_c - \frac{m_{d_i}}{m_{d_1}}B_{r_1}\mathbf{u}_c = 0 \tag{6.98}$$

$$(B_{r_1}m_{d_i} - B_{r_i}m_{d_1})\mathbf{u}_c = 0 \tag{6.99}$$

The range of i is $i = 2 \cdots n$, so we may arrange those equations into a $(n - 1) \times m$ matrix multiplying $B\mathbf{u}_c$,

$$\begin{bmatrix} m_{d_2} & -m_{d_1} & 0 & \cdots & 0 & 0 \\ m_{d_3} & 0 & -m_{d_1} & \cdots & 0 & 0 \\ \vdots & \vdots & \vdots & \vdots & \vdots & \vdots \\ m_{d_{n-1}} & 0 & 0 & \cdots & -m_{d_1} & 0 \\ m_{d_n} & 0 & 0 & \cdots & 0 & -m_{d_1} \end{bmatrix} B\mathbf{u}_c = \mathbf{0} \tag{6.100}$$

If we designate the matrix of m_{d_i} terms as M,

$$MB\mathbf{u}_c = 0 \tag{6.101}$$

The constraints on the control allocation problem may be expressed as

$$MB\mathbf{x} = MB(\mathbf{u}_c - \mathbf{u}_{Min}) = -MB\mathbf{u}_{Min} \tag{6.102}$$

Now define

$$A \equiv MB, \ \mathbf{b} \equiv -A\mathbf{u}_{Min} \tag{6.103}$$

so that the constraints are

$$A\mathbf{x} = \mathbf{b} \tag{6.104}$$

Equation (6.104) will be satisfied if $\mathbf{x}$ satisfies Eq. (6.102); that is, if $\mathbf{u} = \rho\mathbf{m}_{des}$. Otherwise if $\mathbf{x} = \mathbf{u}_c - \mathbf{u}_{Min} + \delta\mathbf{u}$, then

$$MB\mathbf{x} = MB(\mathbf{u}_c - \mathbf{u}_{Min} + \delta\mathbf{u}) = MB\delta\mathbf{u} - MB\mathbf{u}_{Min} \tag{6.105}$$

That is, $A\mathbf{x} \neq \mathbf{b}$, so Eq. (6.104) is necessary and sufficient to ensure that we only accept $\mathbf{x}$ such that $B\mathbf{u}$ is in the same direction as $\mathbf{m}_{des}$.

In summary, the linear program is:

$$\underset{\mathbf{x}}{Min} \ J = \mathbf{c}^T\mathbf{x} | A\mathbf{x} = \mathbf{b}, \mathbf{0} \leq \mathbf{x} \leq \mathbf{h} \tag{6.106}$$

where

$$
\begin{aligned}
\mathbf{x} &= \mathbf{u} - \mathbf{u}_{Min} \\
\mathbf{c}^T &= -\mathbf{m}_{des}^T B \\
A &= MB \\
M &= \left[\left\{ \begin{matrix} m_{d_2} \\ \vdots \\ m_{d_n} \end{matrix} \right\} \quad -m_{d_1} I_{n-1} \right] \\
\mathbf{b} &= -A\mathbf{u}_{Min} \\
\mathbf{h} &= \mathbf{u}_{Max} - \mathbf{u}_{Min}
\end{aligned}
\tag{6.107}
$$

The term $-m_{d_1} I_{n-1}$ signifies that $-m_{d_1}$ multiplies an identity matrix of dimension $n - 1$. Since we are considering only two-moment and three-moment problems, our M matrices will be either 1×2 or 2×3. Then $A = MB$ will be either a $1 \times m$ or $2 \times m$ matrix, where m is the number of control effectors.

6.9.2 Simplex

Any solution to the linear program that satisfies the constraints $A\mathbf{x} = \mathbf{b}$ and $\mathbf{0} \leq \mathbf{x} \leq \mathbf{h}$ is said to be *feasible*. Obviously the optimal solution is a feasible solution.

A second requirement of the solution is that it be a *basic* solution. Denote the dimension of A as $p \times q$,

$$A \in \mathfrak{R}^{p \times q} \tag{6.108}$$

A basic solution is one for which $q - p$ of the elements of **x** are at their lower limit. When the problem is formulated with the condition $\mathbf{x} \geq \mathbf{0}$ this means that $q - p$ elements are zero. This leaves p columns to satisfy $A\mathbf{x} = \mathbf{b}$, implying at a minimum that A must be of full rank. Also, since there are a finite number of ways to select p columns of A, there are a finite number of basic solutions.

It may be shown that, given a linear program in the form of Eq. (6.77) or its various other forms:

- If there is a feasible solution, there is a basic feasible solution.
- If there is an optimal feasible solution, there is an optimal basic feasible solution.

Simplex algorithms generally require a basic feasible solution to begin. Often this initial solution must be obtained by solving a separate linear program that has an easily determined starting solution. From this initial solution the simplex algorithm systematically replaces the current solution with another that reduces the cost J. When no further improvement is possible, the algorithm terminates.

6.10 Moments Attainable by Various Solution Methods

The volume of the AMS, Φ, was determined in Section 5.4.2. The units of this volume are the cube of whatever objective is being sought, usually moments, moment coefficients, or accelerations. Any control allocation method that cannot return admissible controls for all the moments in Φ is under-utilizing the capabilities of the airplane. There may be sound reasons for accepting this penalty, such as insufficient computation time, or a need for a linear solution. We will continue to use the word 'volume' when talking about the two-dimensional case as an analogous concept.

In this section we describe how the volume of moments attainable by a given allocation scheme may be determined. We will need two new subsets. The first is Θ, which is a subset of Ω (all admissible controls) that a particular allocation algorithm can return when presented with all attainable moments. That is, some allocation schemes may be asked to find controls that correspond to a given attainable moment, but will return a control vector that is inadmissible. Θ is the subset of all solutions that are in fact admissible. Θ is not necessarily a proper subset since, for direct allocation, Θ and Ω are the same.

The other new subset is Π, which is the mapping of Θ to moment space via the control effectiveness matrix B. Π is the set of all moments a given allocation method can solve for admissible controls. It is the volume of Π that we are interested in.

As a figure of merit we will compare the volume of Π to that of Φ as a percentage. This is a little misleading, since a Π with a volume of 75% of that of Φ sounds quite bad, but in fact in a spherical sense the allocation scheme is reaching 90% of the maximum in any one direction. Moreover, the volume does not take into consideration that the allocation scheme may have prioritized an axis and sacrificed capabilities in other directions.

Clearly any volume calculations described here will be performed off-line, primarily in an early design stage. The calculations can be computationally demanding, plus there is no need for the information once a control allocation method has been chosen.

We will discuss the three-moment problem for the general case, since no greater under-standing results from the two-moment problem. For the analysis of generalized inverses the two-moment problem serves as an introduction to the three-moment problem, so we will address both.

6.10.1   General Case (Three-moment Problem)

Many control allocation methods defy analytical calculation of their Θ and Π subsets. In these cases one can create a great number of $\mathbf{m}_{des}$ in different directions in moment space and determine the greatest vector in this direction for which the allocation method does return admissible controls.

Sometimes this determination is straightforward. If it is not then a search along several directions must be performed to find the point at which inadmissible controls are introduced. For the desired moment in a given direction, the moment can be grown or shrunk (scaled) to determine the greatest magnitude for which admissible controls are returned. Each time the returned controls change from admissible to inadmissible the direction is reversed and the step size of the growth is decreased, by bisection or some such scheme, until some tolerance is met. For the solution to be inadmissible, only one control need be commanded beyond its limit.

The control vectors returned by the 'shotgun' method are loosely Θ, and B times each vector is loosely Π. With a suitably large number of points in Π the convex hull $\partial(\Pi)$ may be determined. There are many sources of software to determine the convex hull of a set of points. Most, like the MATLAB® `convhull` command, return the coordinates of a contiguous set of triangles that define the boundary. The command `convhull` has an option that returns the volume (or area if two-dimensional) of the convex hull it finds.

In Section 6.5.5.1 on daisy chaining we mentioned two ways of handling the saturated controls in the first group of the daisy chain. The first required uniform scaling, while the second set the inadmissible controls to their nearest limit. This second method has the consequence that its Π may not be convex. That is, when fitting a convex hull to a collection of barely-admissible solutions from the daisy chain, some of the points will be inside the convex hull. For more details on this phenomenon see Bordignon (1996).

Given the coordinates of a contiguous set of triangles that define the boundary of Π, the volume of moment space enclosed by the convex hull $\partial(\Pi)$ may be determined. Say the convex-hull algorithm has returned a total of n_{tri} triangles that define the boundary. Designate the coordinates of each vertex of the ith triangle, $i = 1 \cdots n_{tri}$ as a row vector from the origin as $\mathbf{r}_{ij}, j = 1 \cdots 3$. Form the matrix M whose rows are these vectors.

$$M_i = \begin{bmatrix} \mathbf{r}_1 \\ \mathbf{r}_2 \\ \mathbf{r}_3 \end{bmatrix}, \ i = 1 \cdots n_{tri} \tag{6.109}$$

Then the volume of the triangular pyramid whose base is the given triangle and whose apex is the origin is given by (Hausner 1965):

$$V_i = \frac{|\det(M_i)|}{3!} \tag{6.110}$$

The total volume of the convex hull, V_Π, is then the sum of all such volumes

$$V_\Pi = \sum_{i=1}^{n_{tri}} V_i \tag{6.111}$$

In Eq. (6.110) the numerator is the absolute value of the determinant of M_i. This is because the determinant may be positive or negative, depending on the order of the three rows. This 'directed volume' is important in other applications, but not here.

Equations (6.109)–(6.111) generalize to other dimensions. For the two-moment problem the convex-hull algorithm would return edges, each of which makes a triangle with the origin. In that case M_i would have two rows, and the denominator of Eq. (6.110) would be 2!.

6.10.2   Generalized Inverses (Two- and Three-moment Problems)

In the case of generalized inverses the volume calculations may be made exact. The key to understanding these calculations is to note that every solution returned by the generalized inverse P will be

$$\mathbf{u}_P = P\mathbf{m}_{des} \qquad (6.112)$$

Denote the columns of P as $\mathbf{c}_i$, $i = 1 \cdots n$. Any control vector $\mathbf{u}_P$ returned by this method will be a linear combination of these columns,

$$\mathbf{u}_P = m_{d_1}\mathbf{c}_1 + \cdots + m_{d_n}\mathbf{c}_n \qquad (6.113)$$

There are only n of these columns, so they can not span the m-dimensional control space. They generate an n-dimensional subspace, which we denote S_P. The intersection of the subspace S_P with the admissible controls in Ω are thus the only admissible controls that can be generated by the generalized inverse, or Θ.

6.10.2.1   Two-moment Problem

This situation is illustrated notionally in Figure 6.12 for a three-control, two-moment problem. The rectangular box is Ω, the admissible controls. The plane shown intersecting Ω represents part of the two-dimensional subspace S_P generated by the columns of the generalized inverse P. The dashed line is part of the boundary of this intersection. [3] The rest of the boundary is on the back side of the figure. Only those control effectors that lie within the intersection will be admissible if returned by $P\mathbf{m}_{des}$ (Θ), and only the moments generated by those controls will be attainable using the generalized inverse (Π).

Figure 6.12 suggests the method for calculating the admissible controls Θ and hence the attainable moments Π that a given generalized inverse can return for all attainable moments Φ. If we calculate the intersection of the two-dimensional subspace that is the span of the columns of P with the admissible controls Ω, then we can map them to moment space via B, yielding Π. Once there, we can calculate the area by summing the areas of all the triangles thus formed.

For two moments ($m = 2$) and more than three controls ($n > 3$) the problem is the same, but with more intersections to be found. The total intersection will still be a two-dimensional figure, but with more edges. The difficult thing to visualize about these higher-dimension problems is that Ω, the subset of admissible controls, will contain objects of every dimension less than n, and that the two-dimensional plane will intersect all objects of dimension $m - n$ in a point.

---

[3] The use of the word *intersection* here is different from the intersection of two objects explained in Section 5.1.3.

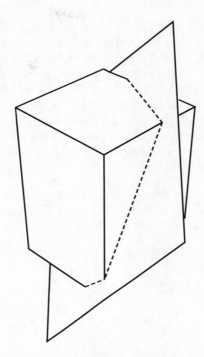

Figure 6.12 Two-dimensional subspace intersecting a three-dimensional set of admissible controls

In other words, with four controls and two moments, the four-dimensional Ω will contain three-, two-, and one-dimensional objects, and the two-dimensional subspace created by the columns of P will intersect the two-dimensional objects at a point.

Generally the problem is to:

1. Find the intersection of the two-dimensional subspace S_P generated by the columns of the generalized inverse P with every object of dimension $m - 2$ of the subset of admissible controls , Ω.
2. If the intersection found is admissible, save it. Otherwise, discard it.
3. Order the list of admissible intersections so that they are contiguous; that is, so that they precede from one to the next along mutual connections. This defines Θ.
4. Map this ordered list of admissible intersections to moment space. This defines Π.
5. Consider each pair of successive intersections in moment space as the base of a triangle with the apex at the origin, and calculate its area.
6. Sum the areas thus found. This is the volume of the attainable moments for the generalized inverse P.

We now expand on each of these steps.

Step 1
Find the intersection of the two-dimensional subspace S_P generated by the columns of the generalized inverse P with every object of dimension $m - 2$ of the subset of admissible controls, Ω.

Consider three controls: $m = 3$. $m - 2 = 1$ so we are looking for points of intersection with one-dimensional edges. These are characterized by having two control effectors fixed and a third one varying along the edge. Now take $m = 4$, and we are looking for points of intersection with two-dimensional faces. Two-dimensional faces are characterized by two fixed controls and two free to vary, so again two will be fixed. This will be true no matter how high the dimension of Ω. We are therefore required to find all combinations of m controls taken two at a time, and for each combination consider all four combinations of minimum and maximum deflections. That is, min–min, min–max, max–min, and max–max. [4]

Group the two controls being considered into $\mathbf{u}_1$ and the remainder into $\mathbf{u}_2$. Partition B and P consistently into B_1, B_2, P_1, and P_2 so that

$$[B_1 \ B_2] \left\{ \begin{matrix} \mathbf{u}_1 \\ \mathbf{u}_2 \end{matrix} \right\} = \mathbf{m}, \qquad \left[\begin{matrix} P_1 \\ P_2 \end{matrix} \right] \mathbf{m} = \left\{ \begin{matrix} \mathbf{u}_1 \\ \mathbf{u}_2 \end{matrix} \right\} \tag{6.114}$$

Since $\mathbf{u}_1$ is known, solve

$$\mathbf{m} = P_1^{-1} \mathbf{u}_1 \tag{6.115}$$

Then solve

$$\mathbf{u}_2 = P_2 \mathbf{m} \tag{6.116}$$

The intersection is at

$$\mathbf{u} = \left\{ \begin{matrix} \mathbf{u}_1 \\ \mathbf{u}_2 \end{matrix} \right\} \tag{6.117}$$

This $\mathbf{u}$ will have to be re-ordered to conform with the original definition of $\mathbf{u}$.

This procedure must be repeated for every combination of two controls and every combination of minimum and maximum deflection.

Note that the same intersection may be found on different evaluations. This will be true if $\mathbf{u}_2$ contains any controls that are at one or the other of its limits. For example, if S_P happens to pass exactly through a vertex of Ω then every control in $\mathbf{u}_2$ will be saturated, and the examination of any and every pair of controls in $\mathbf{u}_1$ that contains that vertex will yield that vertex.

Step 2
If the intersection found is admissible, save it. Otherwise, discard it.

If a single control effector in $\mathbf{u}_2$ is inadmissible, then the intersection is outside Ω and must be discarded.

Step 3
Order the list of admissible intersections so that they are contiguous.

The fast way to accomplish this step is by using a convex-hull algorithm. The intersection of S_P with Ω is indeed convex, as is proven in Bordignon (1996).

The other way to order the list is by using the object numbers of the intersections found in steps 1 and 2. Consider Figure 6.13, which is Figure 6.12 with some labels.

---

[4] The number of combinations of m controls taken n at a time is $m!/(n!(m-n)!)$, and the number of minimum/maximum combinations is 2^n.

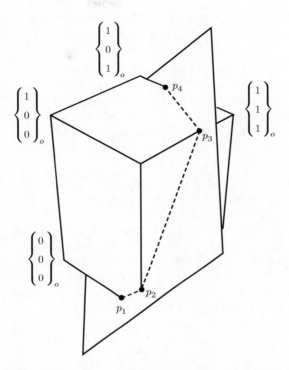

Figure 6.13   Two-dimensional subspace intersecting a three-dimensional set of admissible controls

In Figure 6.13, four points of intersection are labeled, as are four of the vertices of Ω. The following results are consistent with the labeling indicated:

- In determining one of the intersections, in step 1 $\mathbf{u}_1$ was made up of u_1 and u_3, and both were set to their lower limit. $\mathbf{u}_2$ consisted of u_2. In step 2 $\mathbf{u}_2$ was found to be admissible. This intersection was labeled p_1. In object notation, $p_1 = \{0\ 2\ 0\}_o$.
- In determining another of the intersections, in step 1 $\mathbf{u}_1$ was made up of u_2 and u_3, with u_2 at its upper limit and u_3 at its lower. $\mathbf{u}_2$ consisted of u_1. In step 2 $\mathbf{u}_2$ was found to be admissible. This intersection was labeled p_2. In object notation, $p_2 = \{2\ 1\ 0\}_o$.
- In determining another of the intersections, in step 1 $\mathbf{u}_1$ was made up of u_1 and u_2, with both at their upper limits. $\mathbf{u}_2$ consisted of u_3. In step 2 $\mathbf{u}_2$ was found to be admissible. This intersection was labeled p_3. In object notation, $p_3 = \{1\ 1\ 2\}_o$.
- In determining another of the intersections, in step 1 $\mathbf{u}_1$ was made up of u_1 and u_3, with both at their upper limits. $\mathbf{u}_2$ consisted of u_2. In step 2 $\mathbf{u}_2$ was found to be admissible. This intersection was labeled p_4. In object notation, $p_4 = \{1\ 2\ 1\}_o$.

We now utilize the *union* of two objects from Section 5.1.3. Recall that the union of two objects will yield the smallest object of which both are members. To form the union, compare two object numbers entry by entry. If two entries are the same number, the corresponding entry in the union is that number. If they are different, the corresponding entry of the union is 2. Thus:

- the union of the vertex labeled $\{1\ 0\ 0\}_o$ and the vertex labeled $\{1\ 0\ 1\}_o$ is $\{1\ 0\ 2\}_o$, a one-dimensional object (it has one 2), or an edge, which is obvious from Figure 6.13;
- the union of the vertex labeled $\{1\ 0\ 0\}_o$ and the vertex labeled $\{1\ 1\ 1\}_o$ is $\{1\ 2\ 2\}_o$, a two-dimensional object (it has two 2s), or a face, which is also obvious from Figure 6.13;
- the union of the vertex labeled $\{0\ 0\ 0\}_o$ and the vertex labeled $\{1\ 1\ 1\}_o$ is $\{2\ 2\ 2\}_o$, a three-dimensional object (it has three 2s), or a hyper-rectangle, which is Ω.

The union of $p_1 = \{0\ 2\ 0\}_o$ and $p_2 = \{2\ 1\ 0\}_o$ is thus $p_1 \cup p_2 = \{2\ 2\ 0\}_o$, the face on the lower left of the figure. The union of $p_1 = \{0\ 2\ 0\}_o$ and $p_4 = \{1\ 2\ 1\}_o$ is $p_1 \cup p_2 = \{2\ 2\ 2\}_o$, which is the entirety of Ω.

It is clear in Figure 6.13 that two intersections will be connected if and only if their union is of dimension two *or less*. The 'or less' is emphasized because if (in the figure) the two-dimensional subspace S_P had an intersection at a vertex of Ω, this intersection would be found on every search that consisted of any pair of controls at the extrema that define that vertex. The union of any two of these would result in a zero-dimensional object (the vertex).

Still with reference to Figure 6.13, we learn the following:

$$p_1 \cup p_2 = \{2\ 2\ 0\}_o \tag{6.118a}$$

$$p_1 \cup p_3 = \{2\ 2\ 2\}_o \tag{6.118b}$$

$$p_1 \cup p_4 = \{2\ 2\ 2\}_o \tag{6.118c}$$

$$p_2 \cup p_3 = \{2\ 1\ 2\}_o \tag{6.118d}$$

$$p_2 \cup p_4 = \{2\ 2\ 2\}_o \tag{6.118e}$$

$$p_3 \cup p_4 = \{1\ 2\ 2\}_o \tag{6.118f}$$

The two-dimensional connections are thus $p_1 \cup p_2$, $p_2 \cup p_3$, and $p_3 \cup p_4$. We thus order the list of vertices shown as $p_1 - p_2 - p_3 - p_4$, as is clear from Figure 6.13.

For higher order problems, given a list of intersections p_i with m controls and n moments, the intersections p_j and p_k will be connected if and only if the dimension of their union (number of 2s) is

$$\text{dim}\ (p_j \cup p_k) \leq m - n + 1 \tag{6.119}$$

Step 4
Map this ordered list of admissible intersections to moment space. This defines Π.

The ordered list from step 3 will be in object notation. Convert the list to control deflections using the minimum and maximum deflections for $\mathbf{u}_1$ used in step 1 and the admissible controls found in step 2. Multiply each of these by the control effectiveness matrix B.

Step 5
Consider each pair of successive intersections in moment space as the base of a triangle with the apex at the origin, and calculate its area.

This is Eq. (6.111) modified for the two-moment problem.

$$A_i = \frac{|\det(M_i)|}{2!}$$ (6.120)

M_i is a 2×2 matrix whose rows are the two connected intersections determined in step 3.

Step 6
Sum the areas thus found. This is the volume of the attainable moments for the generalized inverse P.

$$A_\Pi = \sum_{i=1}^{n_{tri}} A_i$$ (6.121)

In Eq. (6.121), n_{tri} is the number of triangles thus formed. This number cannot be determined analytically, and will be determined by the results of the preceding steps.

6.10.2.2   Three-moment Problem

The three-moment problem is a natural extension of the two-moment problem, except the figures are harder to draw. There are three such figures that will be shown in Section 6.11.1 (Figures 6.14–6.16), but they show only the finished result, not the steps taken to arrive there.

With three moments we calculate the intersection of the three-dimensional subspace that is the span of the columns of P with the admissible controls Ω. The intersections will consist of plane polygons of varying size, as may be seen from the figures shown later in Section 6.11.1. When mapped to moment space via B they will remain polygons under the linear mapping. There the polygons must be divided into triangles that make the bases of triangular pyramids with the origin as the vertex. Then we apply the volume calculations that we used for convex-hull volume determination, Eq. (6.110).

We are required to find the intersection of the three-dimensional subspace with every $(m - n)$-dimensional object of the m-dimensional subset of admissible controls Ω. Modified from the two-dimensional problem, the problem is to:

1. Find the intersection of the three-dimensional subspace S_P generated by the columns of the generalized inverse P with every object of dimension $m - 3$ of the subset of admissible controls, Ω.
2. If the intersection found is admissible, save it. Otherwise, discard it.
3. Find the pairs of controls that create edges, and find the combinations of edges that form the two-dimensional faces of the three-dimensional Θ.
4. Map these lists of admissible intersections to moment space. This defines Π.
5. Consider each two-dimensional face of the three-dimensional Θ. Take an arbitrary point on the face as a vertex and form triangles with its edges. Treat each triangle as the base of a triangular pyramid with the apex at the origin, and calculate its volume.
6. Sum the volumes thus found. This is the volume of the attainable moments for the generalized inverse P.

We now expand on each of these steps.

Step 1

Find the intersection of the three-dimensional subspace S_P generated by the columns of the generalized inverse P with every object of dimension $m - 3$ of the subset of admissible controls, Ω.

Consider all controls taken three at a time, and each combination of upper and lower limits for them. Group the three controls being considered into $\mathbf{u}_1$ and the remainder into $\mathbf{u}_2$. Partition B and P consistently into B_1, B_2, P_1, and P_2.

Solve

$$\mathbf{m} = P_1^{-1}\mathbf{u}_1 \tag{6.122}$$

Then solve

$$\mathbf{u}_2 = P_2\mathbf{m} \tag{6.123}$$

The intersection is at

$$\mathbf{u} = \left\{ \begin{matrix} \mathbf{u}_1 \\ \mathbf{u}_2 \end{matrix} \right\} \tag{6.124}$$

This $\mathbf{u}$ will have to be re-ordered to conform with the original definition of $\mathbf{u}$.

Step 2

If the intersection found is admissible, save it. Otherwise, discard it.

If a single control effector in $\mathbf{u}_2$ is inadmissible, then the intersection is outside Ω and must be discarded.

Step 3

Find the pairs of controls that create edges, and find the combinations of edges that form the two-dimensional faces of the three-dimensional Θ.

Given a list of intersections p_i with m controls and 3 moments, the intersections of p_j and p_k will be connected by an edge if and only if the dimension of their union (number of 2s) is

$$\dim(p_j \cup p_k) \leq m - 3 + 1 \tag{6.125}$$

Given the same list, the intersections p_j and p_k will be part of the same two-dimensional face if and only if the dimension of their union (number of 2s) is

$$\dim(p_j \cup p_k) \leq m - 3 + 2 \tag{6.126}$$

The first test will tell us which intersections are connected by edges and the second will tell which are on the same face. From this information an ordered list of edges of each face may be derived. To illustrate, consider a three-moment, ten-control problem that yielded the intersections shown in Table 6.2.

The test for an edge connection is $\dim(p_j \cup p_k) \leq (10 - 3 + 1) = 8$ and for a shared face it is $\dim(p_j \cup p_k) \leq (10 - 3 + 2) = 9$. We see that $\dim(p_1 \cup p_2) = 8$ (same edge), $\dim(p_1 \cup p_3) = 8$ (same edge), $\dim(p_1 \cup p_4) = 9$ (same face but not same edge), and so on. Finishing the unions we determine that there is a face with intersections p_1, p_2, p_3, and p_4 in that order (or any cyclic order of those intersections).

From the formation of all the unions we arrive at Tables 6.3 and 6.4.

Table 6.2 Intersections

Intersection	Object number
p_1	$\{2\ 0\ 2\ 2\ 2\ 2\ 0\ 2\ 2\ 0\}_o$
p_2	$\{2\ 0\ 2\ 2\ 2\ 2\ 0\ 2\ 2\ 1\}_o$
p_3	$\{1\ 2\ 2\ 2\ 2\ 2\ 0\ 2\ 2\ 0\}_o$
p_4	$\{1\ 2\ 2\ 2\ 2\ 2\ 0\ 2\ 2\ 1\}_o$
p_5	$\{2\ 1\ 2\ 2\ 2\ 0\ 2\ 2\ 2\ 0\}_o$
p_6	$\{2\ 1\ 2\ 2\ 2\ 0\ 2\ 2\ 2\ 1\}_o$
p_7	$\{0\ 2\ 2\ 2\ 2\ 0\ 2\ 2\ 2\ 0\}_o$
p_8	$\{0\ 2\ 2\ 2\ 2\ 0\ 2\ 2\ 2\ 1\}_o$
p_9	$\{0\ 0\ 2\ 2\ 2\ 2\ 2\ 2\ 2\ 0\}_o$
p_{10}	$\{0\ 0\ 2\ 2\ 2\ 2\ 2\ 2\ 2\ 1\}_o$
p_{11}	$\{1\ 1\ 2\ 2\ 2\ 2\ 2\ 2\ 2\ 0\}_o$
p_{12}	$\{1\ 1\ 2\ 2\ 2\ 2\ 2\ 2\ 2\ 1\}_o$

Table 6.3 Edges

Intersection	Edge with
p_1	p_2, p_3, p_9
p_2	p_1, p_4, p_{10}
p_3	p_1, p_4, p_{11}
p_4	p_2, p_3, p_{12}
p_5	p_6, p_7, p_{11}
p_6	p_5, p_8, p_{12}
p_7	p_5, p_8, p_9
p_8	p_6, p_7, p_{10}
p_9	p_1, p_7, p_{10}
p_{10}	p_2, p_8, p_9
p_{11}	p_3, p_5, p_{12}
p_{12}	p_4, p_6, p_{11}

Step 4

Map these lists of admissible intersections to moment space. This defines Π.

The ordered list from step 3 will be in object notation. Convert the list to control deflections using the minimum and maximum deflections for $\mathbf{u}_1$ and the admissible controls in $\mathbf{u}_2$ found in step 1. Multiply each of these by the control effectiveness matrix B.

Step 5

Consider each two-dimensional face of the three-dimensional Θ or Π. Take an arbitrary point on the face as a vertex and form triangles with its edges. If Π was used map it to moment space. Treat each triangle as the base of a triangular pyramid with the apex at the origin, and calculate its volume.

If a face is triangular, form a triangular pyramid with the origin. Otherwise, take an arbitrary point on the face and not on an edge p_0 and form triangles with each of the edges. One could take a line connecting two intersections not on the same edge and another with the same properties and find their intersection.

Table 6.4   Faces

Face	Intersections in order
1	P_1, P_2, P_4, P_3
2	P_1, P_2, P_{10}, P_9
3	$P_1, P_3, P_{11}, P_5, P_7, P_9$
4	$P_2, P_4, P_{12}, P_6, P_8, P_{10}$
5	P_3, P_4, P_{12}, P_{11}
6	P_5, P_6, P_8, P_7
7	P_5, P_6, P_{12}, P_{11}
8	P_7, P_8, P_{10}, P_9

Once the triangles are found, we may use Eq. (6.111) again.

$$V_i = \frac{|\det(M_i)|}{3!} \tag{6.127}$$

M_i is a 3×3 matrix whose rows are the three intersections determined in finding the triangles.

Step 6
Sum the volumes thus found. This is the volume of the attainable moments for the generalized inverse P.

$$V_\Pi = \sum_{i=1}^{n_{tri}} V_i \tag{6.128}$$

In Eq. (6.121), n_{tri} is the number of triangles formed. This number cannot be determined analytically, and will be determined by the results of the preceding steps.

6.10.2.3   Summary

We have developed the case for the two-moment problem in some detail because it is easier to visualize the steps, especially for the two-moment and three-control problem. Some of the steps there may seem almost intuitive, but one should beware of intuition when extending these results to higher dimensions, where two two-dimensional objects can intersect in a point (or worse).

Most of the results shown are adapted from studies of *n-flats*, almost exclusively drawn from Sommerville (1929). An *n*-flat is like an *n*-dimensional subspace, except it does not contain the origin. Almost every 'object' we have described lives in an *n*-flat (Ω and Φ contain the origin, but the rest generally do not), but is bounded. We took the boundedness into account when we decided whether intersections in Ω were admissible or not.

Finally, the single biggest problem in implementing the algorithms suggested by this section was in determining what was zero. With finite machine precision it was often found that two numbers that should have been the same came up almost so, but not quite. Code was littered with epsilons to be used to see if the two numbers were 'close enough'. At first it seemed this would be a rare event, but experience proved otherwise.

One philosophy that helped eliminate many of these near-zero experiences was to acknowl-edge that one was dealing with aerodynamic data whose accuracy was almost always precise to at most three significant figures. One need only look at the various ways in which the rotary derivatives (like C_{ℓ_p}, the roll-rate damping, and others) are determined. Different approaches will yield different results, often quite at variance with each other (see, for example, Mulkens and Ormerod (1993)). Thus we had no qualms at all about 'dithering' the control effectiveness data by adding small random numbers, aerodynamically insignificant, to the B matrix.

On that note, a caution: If you think some mathematical juxtaposition of numbers in the control allocation problem will never occur, you are wrong.

6.11 Examples

Unless otherwise noted, data for these examples were introduced in Eqs 5.43 and 5.44, and are repeated here for convenience.

$$B = \begin{bmatrix} 0.7073 & -0.7073 & -3.4956 & -3.0013 \\ 1.1204 & 1.1204 & -0.7919 & -1.2614 \\ -0.3309 & 0.3309 & -0.1507 & -0.3088 \end{bmatrix}$$

$$\begin{matrix} 3.0013 & 3.4956 & 2.1103 \\ -1.2614 & -0.7919 & 0.0035 \\ 0.3088 & 0.1507 & -1.2680 \end{matrix}$$

(6.129)

$$\mathbf{u}_{Min} = \begin{Bmatrix} -0.9599 \\ -0.9599 \\ -0.5236 \\ -0.5236 \\ -0.5236 \\ -0.5236 \\ -0.5236 \end{Bmatrix}_u \qquad \mathbf{u}_{Max} = \begin{Bmatrix} 0.4363 \\ 0.4363 \\ 0.5236 \\ 0.5236 \\ 0.5236 \\ 0.5236 \\ 0.5236 \end{Bmatrix}_u$$

(6.130)

6.11.1   Generalized Inverses

Example 6.3

Pseudo-inverse for the ADMIRE Simulation The B matrix in Eq. (6.129) was applied to Eq. (6.32) on p. 84 with the result,

$$P = \begin{bmatrix} 0.0140 & 0.1612 & -0.1585 \\ -0.0140 & 0.1614 & 0.1591 \\ -0.0734 & -0.1140 & -0.0817 \\ -0.0634 & -0.1816 & -0.1573 \\ 0.0634 & -0.1816 & 0.1566 \\ 0.0734 & -0.1141 & 0.0812 \\ 0.0410 & 0.0001 & -0.6100 \end{bmatrix}$$

(6.131)

The moments attainable using $\mathbf{u} = P\mathbf{m}_{des}$ were depicted in Figure 5.16 (p. 67). The volume of those moments were 37% of the total volume.

This solution works well for moderately sized moments (actually, angular accelerations in this case). For instance, given

$$\mathbf{m}_{des} = \begin{Bmatrix} 2.8493 \\ -0.2942 \\ 0.5726 \end{Bmatrix} \tag{6.132}$$

using Eq. (6.131) yields the admissible solution

$$\mathbf{u}_{Pseudo} = \begin{Bmatrix} -0.0984 \\ 0.0037 \\ -0.2223 \\ -0.2173 \\ 0.3238 \\ 0.2892 \\ -0.2324 \end{Bmatrix} \tag{6.133}$$

However, this $\mathbf{m}_{des}$ is attainable by direct allocation,

$$\mathbf{m}_{des} = \begin{Bmatrix} 5.6986 \\ -0.5884 \\ 1.1452 \end{Bmatrix} \tag{6.134}$$

but the pseudo-inverse here yields several inadmissible controls:

$$\mathbf{u}_{Pseudo} = \begin{Bmatrix} -0.1968 \\ 0.0075 \\ -0.4445 \\ -0.4345 \\ 0.6477 \\ 0.5784 \\ -0.4647 \end{Bmatrix} \tag{6.135}$$

That the moment in Eq. (6.134) is actually attainable is easily seen using direct allocation, and the moment is revealed to be on the boundary of the attainable moments.

$$\mathbf{u}_{DA} = \begin{Bmatrix} -0.2620 \\ -0.2617 \\ -0.5236 \\ -0.5236 \\ 0.5236 \\ 0.5236 \\ -0.5236 \end{Bmatrix} \tag{6.136}$$

Example 6.4

A Tailored Generalized Inverse for the ADMIRE Simulation We will tailor a generalized inverse so that it attains, using admissible controls, the maximum attainable positive rolling, pitching, and yawing moments.

The partitions of B are:

$$B_1 = \begin{bmatrix} 0.7073 & -0.7073 & -3.4956 \\ 1.1204 & 1.1204 & -0.7919 \\ -0.3309 & 0.3309 & -0.1507 \end{bmatrix} \tag{6.137}$$

$$B_2 = \begin{bmatrix} -3.0013 & 3.0013 & 3.4956 & 2.1103 \\ -1.2614 & -1.2614 & -0.7919 & 0.0035 \\ -0.3088 & 0.3088 & 0.1507 & -1.2680 \end{bmatrix} \tag{6.138}$$

Using the direct allocation methods of Section 6.6 the three desired moments and their accompanying controls were determined.

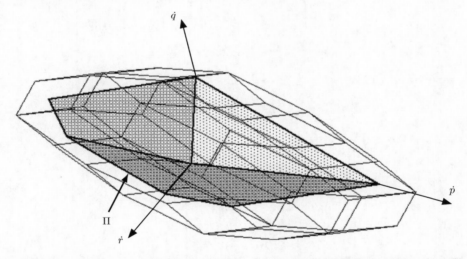

Figure 6.14 The solid figure within the wire-frame AMS for the ADMIRE simulation, from Figure 5.8, is the tailored generalized inverse described in Example 6.4. View is from $(\dot{p}\ \dot{q}\ \dot{r}) = (1\ 1\ 1)$

Maximum $\dot{p}$ with $\dot{q} = \dot{r} = 0$

$$\mathbf{m}_1^{Spec} = \begin{Bmatrix} 7.74 \\ 0 \\ 0 \end{Bmatrix} \quad \mathbf{u}_1^{Spec} = \begin{Bmatrix} 0.436 \\ -0.436 \\ -0.523 \\ -0.523 \\ 0.523 \\ 0.523 \\ 0.151 \end{Bmatrix} \quad \mathbf{u}_{1_2}^{Spec} = \begin{Bmatrix} -0.523 \\ 0.523 \\ 0.523 \\ 0.151 \end{Bmatrix} \tag{6.139}$$

Maximum $\dot{q}$ with $\dot{p} = \dot{r} = 0$

$$\mathbf{m}_2^{Spec} = \left\{ \begin{array}{c} 0 \\ 3.13 \\ 0 \end{array} \right\} \quad \mathbf{u}_2^{Spec} = \left\{ \begin{array}{c} 0.436 \\ 0.436 \\ -0.523 \\ -0.523 \\ -0.523 \\ -0.523 \\ 0.000 \end{array} \right\} \quad \mathbf{u}_{2_2}^{Spec} = \left\{ \begin{array}{c} -0.523 \\ -0.523 \\ -0.523 \\ 0.000 \end{array} \right\} \quad (6.140)$$

Maximum $\dot{r}$ with $\dot{p} = \dot{q} = 0$

$$\mathbf{m}_3^{Spec} = \left\{ \begin{array}{c} 0 \\ 0 \\ 1.40 \end{array} \right\} \quad \mathbf{u}_3^{Spec} = \left\{ \begin{array}{c} -0.959 \\ 0.436 \\ -0.221 \\ -0.523 \\ 0.523 \\ -0.521 \\ -0.523 \end{array} \right\} \quad \mathbf{u}_{3_2}^{Spec} = \left\{ \begin{array}{c} -0.523 \\ 0.523 \\ -0.521 \\ -0.523 \end{array} \right\} \quad (6.141)$$

Now applying Eq. (6.25),

$$
\begin{aligned}
P_2 &= \left[\mathbf{u}_{1_2}^{Spec} \quad \mathbf{u}_{2_2}^{Spec} \quad \mathbf{u}_{3_2}^{Spec} \right] \left[\mathbf{m}_1^{Spec} \quad \mathbf{m}_2^{Spec} \quad \mathbf{m}_3^{Spec} \right]^{-1} \\
&= \begin{bmatrix} -0.523 & -0.523 & -0.523 \\ 0.523 & -0.523 & 0.523 \\ 0.523 & -0.523 & -0.521 \\ 0.151 & 0.000 & -0.523 \end{bmatrix} \begin{bmatrix} 7.74 & 0 & 0 \\ 0 & 3.13 & 0 \\ 0 & 0 & 1.40 \end{bmatrix}^{-1} \\
&= \begin{bmatrix} -0.0676 & -0.1674 & -0.3729 \\ 0.0676 & -0.1674 & 0.3729 \\ 0.0676 & -0.1674 & -0.3715 \\ 0.0196 & 0 & -0.3729 \end{bmatrix}
\end{aligned}
\quad (6.142)
$$

Finally, we use Eq. (6.10) to get P_1 and then assemble P.

$$P_1 = B_1^{-1}[I_n - B_2 P_2]$$

$$
\begin{aligned}
P_1 &= \begin{bmatrix} 0.7073 & -0.7073 & -3.4956 \\ 1.1204 & 1.1204 & -0.7919 \\ -0.3309 & 0.3309 & -0.1507 \end{bmatrix}^{-1} \cdot \\
&\quad \left[I_3 - \begin{bmatrix} -3.0013 & 3.0013 & 3.4956 & 2.1103 \\ -1.2614 & -1.2614 & -0.7919 & 0.0035 \\ -0.3088 & 0.3088 & 0.1507 & -1.2680 \end{bmatrix} \cdot \right. \\
&\quad \left. \begin{bmatrix} -0.0676 & -0.1674 & -0.3729 \\ 0.0676 & -0.1674 & 0.3729 \\ 0.0676 & -0.1674 & -0.3715 \\ 0.0196 & 0 & -0.3729 \end{bmatrix} \right]
\end{aligned}
$$

$$= \begin{bmatrix} 0.0564 & 0.1395 & -0.6837 \\ -0.0564 & 0.1395 & 0.3107 \\ -0.0676 & -0.1674 & -0.1576 \end{bmatrix} \tag{6.143}$$

$$P = \begin{bmatrix} 0.0564 & 0.1395 & -0.6837 \\ -0.0564 & 0.1395 & 0.3107 \\ -0.0676 & -0.1674 & -0.1576 \\ -0.0676 & -0.1674 & -0.3729 \\ 0.0676 & -0.1674 & 0.3729 \\ 0.0676 & -0.1674 & -0.3715 \\ 0.0196 & 0 & -0.3729 \end{bmatrix} \tag{6.144}$$

The subset of attainable moments of this generalized inverse, Π, is shown in Figure 6.14.

The shape of Π in Figure 6.14 is produced by the intersection of the three-dimensional null space of $PB - I$ with the seven-dimensional set of admissible controls Ω. The volume of Π is 25.5% that of Φ.

Example 6.5

A Tailored Generalized Inverse for an Entire Facet of Φ For this example we will tailor a generalized inverse that 'fits' at three of the four vertices defining a facet. We choose a facet of u_1 and u_2 in the ADMIRE simulation. Following the procedure in Section 5.4.1 (see Example 6.7 below) we calculate the controls and moments that define the facets. We will use three of the four vertices from those results for our specified moments. Then, since the fourth vertex is a linear combination of the three specified, the generalized inverse that results will attain admissible solutions for the entire facet.

The facets that result are $\mathbf{o}_{2200110}$ and $\mathbf{o}_{2211001}$. The following results are for $\mathbf{o}_{2200110}$.

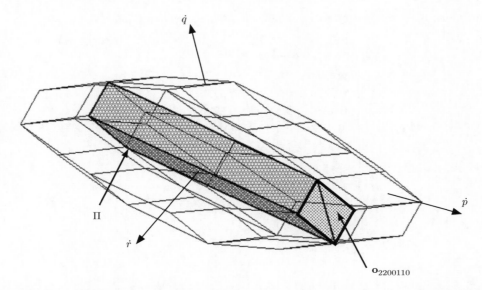

Figure 6.15 The tailored generalized inverse described in Example 6.5. It is tailored to fit the entire facet $\mathbf{o}_{2200110}$. View is from $(\dot{p}\ \dot{q}\ \dot{r}) = (1\ 1\ 1)$

Controls:

$$\mathbf{u}_{0000110} = \left\{\begin{array}{c} -0.9599 \\ -0.9599 \\ -0.5236 \\ -0.5236 \\ 0.5236 \\ 0.5236 \\ -0.5236 \end{array}\right\}_u \quad \mathbf{u}_{0100110} = \left\{\begin{array}{c} -0.9599 \\ 0.4363 \\ -0.5236 \\ -0.5236 \\ 0.5236 \\ 0.5236 \\ -0.5236 \end{array}\right\}_u$$

$$\mathbf{u}_{1000110} = \left\{\begin{array}{c} 0.4363 \\ 0.4363 \\ -0.5236 \\ -0.5236 \\ 0.5236 \\ 0.5236 \\ -0.5236 \end{array}\right\}_u \quad \mathbf{u}_{1100110} = \left\{\begin{array}{c} 0.4363 \\ -0.9599 \\ -0.5236 \\ -0.5236 \\ 0.5236 \\ 0.5236 \\ -0.5236 \end{array}\right\}_u$$

$$(6.145)$$

Moments:

$$\mathbf{m}_{0000110} = \left\{\begin{array}{c} 5.6986 \\ -2.1528 \\ 1.1451 \end{array}\right\}_m \quad \mathbf{m}_{0100110} = \left\{\begin{array}{c} 4.7111 \\ -0.5885 \\ 1.6071 \end{array}\right\}_m$$

$$\mathbf{m}_{1100110} = \left\{\begin{array}{c} 5.6986 \\ 0.9758 \\ 1.1451 \end{array}\right\}_m \quad \mathbf{m}_{1000110} = \left\{\begin{array}{c} 6.6861 \\ -0.5885 \\ 0.6831 \end{array}\right\}_m$$

$$(6.146)$$

We will use the first three controls and moments. Again applying Eq. (6.25),

$$\begin{aligned} P_2 &= \begin{bmatrix} \mathbf{u}_{1_2}^{Spec} & \mathbf{u}_{2_2}^{Spec} & \mathbf{u}_{3_2}^{Spec} \end{bmatrix} \begin{bmatrix} \mathbf{m}_1^{Spec} & \mathbf{m}_2^{Spec} & \mathbf{m}_3^{Spec} \end{bmatrix}^{-1} \\ &= \begin{bmatrix} -0.5236 & -0.5236 & -0.5236 \\ 0.5236 & 0.5236 & 0.5236 \\ 0.5236 & 0.5236 & 0.5236 \\ -0.5236 & -0.5236 & -0.5236 \end{bmatrix} \cdot \\ & \quad \begin{bmatrix} 5.6986 & 4.7111 & 5.6986 \\ -2.1528 & -0.5885 & 0.9758 \\ 1.1451 & 1.6071 & 1.1451 \end{bmatrix}^{-1} \\ &= \begin{bmatrix} -0.0643 & 0.0 & -0.1374 \\ 0.0643 & 0.0 & 0.1374 \\ 0.0643 & 0.0 & 0.1374 \\ -0.0643 & 0.0 & -0.1374 \end{bmatrix} \end{aligned}$$

$$(6.147)$$

The last step uses Eq. (6.10).

$$P_1 = B_1^{-1}[I_n - B_2 P_2]$$

$$P_1 = \begin{bmatrix} 0.7073 & -0.7073 & -3.4956 \\ 1.1204 & 1.1204 & -0.7919 \\ -0.3309 & 0.3309 & -0.1507 \end{bmatrix}^{-1} \cdot$$

$$\left[I_3 - \begin{bmatrix} -3.0013 & 3.0013 & 3.4956 & 2.1103 \\ -1.2614 & -1.2614 & -0.7919 & 0.0035 \\ -0.3088 & 0.3088 & 0.1507 & -1.2680 \end{bmatrix} \cdot \right.$$

$$\left. \begin{bmatrix} -0.0643 & 0.0 & -0.1374 \\ 0.0643 & 0.0 & 0.1374 \\ 0.0643 & 0.0 & 0.1374 \\ -0.0643 & 0.0 & -0.1374 \end{bmatrix} \right]$$ (6.148)

$$= \begin{bmatrix} 0.2125 & 0.4463 & -1.0568 \\ -0.2123 & 0.4463 & 1.0572 \\ -0.0643 & 0.0 & -0.1374 \end{bmatrix}$$

$$P = \begin{bmatrix} 0.2125 & 0.4463 & -1.0568 \\ -0.2123 & 0.4463 & 1.0572 \\ -0.0643 & 0.0 & -0.1374 \\ -0.0643 & 0.0 & -0.1374 \\ 0.0643 & 0.0 & 0.1374 \\ 0.0643 & 0.0 & 0.1374 \\ -0.0643 & 0.0 & -0.1374 \end{bmatrix}$$ (6.149)

Lastly, we check that $\mathbf{u}_{1100110} = P\mathbf{m}_{1100110}$. It does, and this generalized inverse will return admissible solutions for a certain volume of Π, which includes every moment in the pyramid formed by the facet $\mathbf{o}_{2200110}$ and the origin of moment space. The solutions in the pyramid will, in fact, be scaled solutions of those obtained on the facet $\mathbf{o}_{2200110}$. This generalized inverse also fits the opposite facet, $\mathbf{o}_{2211001}$.

The attainable moments for which this generalized inverse returns admissible solutions (Π) is shown in Figure 6.15. The diagonal line across facet $\mathbf{o}_{2200110}$ belongs to Π, and is a consequence of the fact that the four points of that intersection are not precisely planar due to the use of finite precision arithmetic. The total Π is is the convex hull of the two facets $\mathbf{o}_{2200110}$ and $\mathbf{o}_{2211001}$. The volume of this Π is 13.3% of Φ.

We could, of course, calculate a set of generalized inverses, each corresponding to one of the facets of the attainable moment set Φ. By pre-calculating and storing these inverses they could then be searched in real-time to find the inverse that contains a given desired moment, $\mathbf{m}_{des}$. This approach would afford the predictability of generalized inverse solutions yet still be optimal in the sense of maximum capabilities described in Section 6.2.1. A means for determining the correct generalized inverse to use is suggested in Section 6.6.2.

Example 6.6

'Best' Generalized Inverse for the ADMIRE Simulation The simplex algorithm in Stevens and Lewis (2003) was integrated into software of the authors' design that performed all the calculations necessary to assemble a generalized inverse from the sub-matrix P_2 and calculate the volume of the attainable moments for which it returned admissible controls (Π). The starting inverse was the Moore–Penrose pseudo-inverse described in Section 6.5.4.1.

The termination criterion was based on the size (2-norm) of the vector of perturbations in the parameters being varied, in this case the elements of P_2. It was set to 0.01 rad/s^2/rad. The algorithm terminated after 419 iterations when the criterion was satisfied.

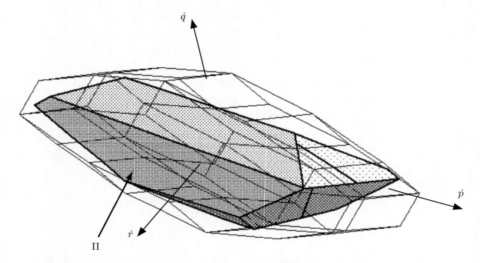

Figure 6.16   The 'best' generalized inverse for the ADMIRE simulation, described in Example 6.6. View is from $(\dot{p}\ \dot{q}\ \dot{r}) = (1\ 1\ 1)$

Figure 6.16 shows the 'best' generalized inverse in the wire-frame of Φ.

The volume of the attainable moment set, Φ, was previously determined to be 177.1 (rad/s^2)^3. The volume of the Moore–Penrose pseudo-inverse is 37.6% of that, or 66.6 (rad/s^2)^3. The volume of the 'best' generalized inverse was 46.7% of the total, or 82.8 (rad/s^2)^3.

$$P_{Best} = \begin{bmatrix} 0.0311 & 0.2399 & -0.3380 \\ -0.0266 & 0.2415 & 0.2649 \\ -0.0668 & -0.1118 & -0.1428 \\ -0.0668 & -0.1120 & -0.1440 \\ 0.0684 & -0.1145 & 0.1141 \\ 0.0708 & -0.1094 & 0.0845 \\ 0.0342 & 0.0001 & -0.5415 \end{bmatrix} \tag{6.150}$$

6.11.2   Direct Allocation

We take for this example a desired moment,

$$\mathbf{m}_{des} = \left\{ \begin{array}{c} 2.8493 \\ -0.2942 \\ 0.5726 \end{array} \right\} \tag{6.151}$$

This $\mathbf{m}_{des}$ is not arbitrary, although we will proceed as if it were. It was selected as that that points directly toward the center of the facet used in Example 6.5, $\mathbf{o}_{2200110}$, with 50% maximum control effector saturation. The moment at the center of the facet was found from averaging two diagonally opposite moments of that facet, for example $\mathbf{m}_{0000110}$ and $\mathbf{m}_{1100110}$. That average was divided by 2 to yield the moment half-way to the facet from the origin.

Example 6.7

Direct Allocation: an Attainable Desired Moment If we were presented the $\mathbf{m}_{des}$ of Eq. (6.151) with no foreknowledge of its origins, we would begin generating facets according to the algorithm in Section 5.4.1.1 (p. 60) by taking $i = 1 \cdots m - 1$ and $j = i + 1 \cdots m$. The first pair generated is $i = 1, j = 2$ (which will reduce our search time greatly since, as was noted, this $\mathbf{m}_{des}$ was not arbitrary).

Next we take the first and second columns of B and form B_{12}.

$$B_{12} = \begin{bmatrix} 0.7073 & -0.7073 \\ 1.1204 & 1.1204 \\ -0.3309 & 0.3309 \end{bmatrix} \tag{6.152}$$

The first row of B_{12} are not zeros, so we calculate $\mathbf{t} = \begin{bmatrix} t_{11} & t_{12} & t_{13} \end{bmatrix}$ from

$$\mathbf{t}B_{12} = \begin{bmatrix} 0 & 0 \end{bmatrix} \tag{6.153}$$

Setting $t_{13} = 1$,

$$\begin{aligned} \mathbf{t}B_{12} &= \begin{bmatrix} t_{11} & t_{12} & 1 \end{bmatrix} \begin{bmatrix} 0.7073 & -0.7073 \\ 1.1204 & 1.1204 \\ -0.3309 & 0.3309 \end{bmatrix} \\ &= \begin{bmatrix} t_{11} & t_{12} \end{bmatrix} \begin{bmatrix} 0.7073 & -0.7073 \\ 1.1204 & 1.1204 \end{bmatrix} + \begin{bmatrix} -0.3309 & 0.3309 \end{bmatrix} \\ &= \begin{bmatrix} 0 & 0 \end{bmatrix} \end{aligned} \tag{6.154}$$

From which,

$$\begin{aligned} \begin{bmatrix} t_{11} & t_{12} \end{bmatrix} &= -\begin{bmatrix} -0.3309 & 0.3309 \end{bmatrix} \begin{bmatrix} 0.7073 & -0.7073 \\ 1.1204 & 1.1204 \end{bmatrix}^{-1} \\ &= \begin{bmatrix} 0.4678 & 0 \end{bmatrix} \end{aligned} \tag{6.155}$$

$$\mathbf{t} = \begin{bmatrix} 0.4678 & 0 & 1 \end{bmatrix} \tag{6.156}$$

$$\mathbf{t}B = \begin{bmatrix} 0 & 0 & -1.7861 & -1.7129 & 1.7129 & 1.7861 & -0.2807 \end{bmatrix} \tag{6.157}$$

This yields facets $\mathbf{o}_{2200110}$ (maximum) and $\mathbf{o}_{2211001}$ (minimum). We will test the positive facet first. The physical controls and moments associated with the facet were given in Example 6.5. The moments at the four vertices are

$$\mathbf{m}_{0000110} = \left\{ \begin{array}{c} 5.6986 \\ -2.1528 \\ 1.1451 \end{array} \right\}_m \quad \mathbf{m}_{0100110} = \left\{ \begin{array}{c} 4.7111 \\ -0.5885 \\ 1.6071 \end{array} \right\}_m$$

$$\tag{6.158}$$

$$\mathbf{m}_{1100110} = \left\{ \begin{array}{c} 5.6986 \\ 0.9758 \\ 1.1451 \end{array} \right\}_m \quad \mathbf{m}_{1000110} = \left\{ \begin{array}{c} 6.6861 \\ -0.5885 \\ 0.6831 \end{array} \right\}_m$$

We pick up the procedure in Section 6.6.2,

$$\mathbf{m}_1 = \mathbf{m}_{0000110} \quad \mathbf{m}_2 = \mathbf{m}_{0100110} \quad \mathbf{m}_3 = \mathbf{m}_{1000110} \tag{6.159}$$

Now we calculate a, b, and c from

$$a\mathbf{m}_{des} = \mathbf{m}_1 + b(\mathbf{m}_2 - \mathbf{m}_1) + c(\mathbf{m}_3 - \mathbf{m}_1) \tag{6.160}$$

$$\left\{ \begin{array}{c} a \\ b \\ c \end{array} \right\} = \begin{bmatrix} \mathbf{m}_{des} & -(\mathbf{m}_2 - \mathbf{m}_1) & -(\mathbf{m}_3 - \mathbf{m}_1) \end{bmatrix}^{-1} \mathbf{m}_1 \tag{6.161}$$

$$\left\{ \begin{array}{c} a \\ b \\ c \end{array} \right\} = \left\{ \begin{array}{c} 2 \\ 0.5 \\ 0.5 \end{array} \right\} \tag{6.162}$$

We could have designated any of the four moments as $\mathbf{m}_1$, and the two others that differ from it by one element of object notation as $\mathbf{m}_2$ and $\mathbf{m}_3$. Note that we could have reversed $\mathbf{m}_2$ and $\mathbf{m}_3$ in Eq. (6.160).

From $0 \le b \le 1$ and $0 \le c \le 1$ we learn that we have found the correct facet, and from $a \ge 1$ we learn that it is attainable.

The controls for this facet are

$$\mathbf{u}_{0000110} = \begin{Bmatrix} -0.9599 \\ -0.9599 \\ -0.5236 \\ -0.5236 \\ 0.5236 \\ 0.5236 \\ -0.5236 \end{Bmatrix}_u \quad \mathbf{u}_{0100110} = \begin{Bmatrix} -0.9599 \\ 0.4363 \\ -0.5236 \\ -0.5236 \\ 0.5236 \\ 0.5236 \\ -0.5236 \end{Bmatrix}_u$$

$$\mathbf{u}_{1000110} = \begin{Bmatrix} 0.4363 \\ 0.4363 \\ -0.5236 \\ -0.5236 \\ 0.5236 \\ 0.5236 \\ -0.5236 \end{Bmatrix}_u \quad \mathbf{u}_{1100110} = \begin{Bmatrix} 0.4363 \\ -0.9599 \\ -0.5236 \\ -0.5236 \\ 0.5236 \\ 0.5236 \\ -0.5236 \end{Bmatrix}_u$$

(6.163)

The control that generates $\mathbf{u}^*$ is

$$\mathbf{u}^* = \mathbf{u}_1 + b(\mathbf{u}_2 - \mathbf{u}_1) + c(\mathbf{u}_3 - \mathbf{u}_1) = \begin{Bmatrix} -0.2618 \\ -0.2618 \\ -0.5236 \\ -0.5236 \\ 0.5236 \\ 0.5236 \\ -0.5236 \end{Bmatrix}_u$$

(6.164)

One solution to the control allocation problem is $\mathbf{u}^*/a = \mathbf{u}^*/2$

$$\mathbf{u}_{DA} = \mathbf{u}^*/2 = \begin{Bmatrix} -0.1309 \\ -0.1309 \\ -0.2618 \\ -0.2618 \\ 0.2618 \\ 0.2618 \\ -0.2618 \end{Bmatrix}$$

(6.165)

6.11.3   Edge and Facet Searching

Example 6.8

Edge Searching For this example we take the first and second rows of the control effectiveness data from the ADMIRE simulation. The data are not representative of any particular problem. However, the two-dimensional edge search is a component of the three-dimensional facet search, and the removal of the third row of the B matrix is akin to the projection into a plane that occurs in three-dimensional searches. The control effector limits are unchanged.

$$B = \begin{bmatrix} 0.7073 & -0.7073 & -3.4956 & -3.0013 \\ 1.1204 & 1.1204 & -0.7919 & -1.2614 \\ & 3.0013 & 3.4956 & 2.1103 \\ & -1.2614 & -0.7919 & 0.0035 \end{bmatrix} \tag{6.166}$$

We take the $\dot{p}$ and $\dot{q}$ components of Eq. (6.151) in Section 6.11.2 (Example 6.7) for our $\mathbf{m}_{des}$.

$$\mathbf{m}_{des} = \left\{ \begin{array}{c} 2.8493 \\ -0.2942 \end{array} \right\} \tag{6.167}$$

First we set up a coordinate system with x aligned with $\mathbf{m}_{des}$:

$$\theta = \arctan\left(\frac{-0.2942}{2.8493}\right) = -0.1029 \text{ rad} = -5.895° \tag{6.168}$$

$$T = \begin{bmatrix} \cos\theta & \sin\theta \\ -\sin\theta & \cos\theta \end{bmatrix} = \begin{bmatrix} 0.9947 & -0.1027 \\ 0.1027 & 0.9947 \end{bmatrix} \tag{6.169}$$

As a quick check,

$$T\mathbf{m}_{des} = \left\{ \begin{array}{c} 2.8644 \\ 0 \end{array} \right\} \tag{6.170}$$

$$B_0 = TB = \begin{bmatrix} 0.5885 & -0.8186 & -3.3958 \\ 1.1871 & 1.0418 & -1.1467 \\ -2.8559 & 3.1150 & 3.5584 & 2.0988 \\ -1.5630 & -0.9465 & -0.4287 & 0.2202 \end{bmatrix} \tag{6.171}$$

The control effector for the initial vertex is determined from the signs in the first row of B_0,

$$
\mathbf{u}_0 = \mathbf{u}_{1000111} = \begin{Bmatrix} 0.4363 \\ -0.9599 \\ -0.5236 \\ -0.5236 \\ 0.5236 \\ 0.5236 \\ 0.5236 \end{Bmatrix} \tag{6.172}
$$

Next, the y component of $B_0\mathbf{u}_0$ is determined:

$$
B_0\mathbf{u}_0 = \begin{Bmatrix} 8.9091 \\ 0.3320 \end{Bmatrix} \tag{6.173}
$$

We note $y_0 = 0.3320 > 0$ so the first rotation will be clockwise $(-\pi/2)$.

$$
T_0 = \begin{bmatrix} \cos(-\pi/2) & \sin(-\pi/2) \\ -\sin(-\pi/2) & \cos(-\pi/2) \end{bmatrix} = \begin{bmatrix} 0 & -1 \\ 1 & 0 \end{bmatrix} \tag{6.174}
$$

$$
B_1 = T_0 B_0 = \begin{bmatrix} -1.1871 & -1.0418 & 1.1467 \\ 0.5885 & -0.8186 & -3.3958 \end{bmatrix}
$$

$$
\begin{bmatrix} 1.5630 & 0.9465 & 0.4287 & -0.2202 \\ -2.8559 & 3.1150 & 3.5584 & 2.0988 \end{bmatrix} \tag{6.175}
$$

$$
\mathbf{u}_1 = \mathbf{u}_{0011110} = \begin{Bmatrix} -0.9599 \\ -0.9599 \\ 0.5236 \\ 0.5236 \\ 0.5236 \\ 0.5236 \\ -0.5236 \end{Bmatrix} \tag{6.176}
$$

$$
B_0\mathbf{u}_1 = \begin{Bmatrix} -0.6572 \\ -4.3937 \end{Bmatrix} \tag{6.177}
$$

Here $y_1 = -4.3937$ and y_0 and y_1 are of different signs, so we check to see if $\mathbf{u}_0$ and $\mathbf{u}_1$ constitute an edge. We have $\mathbf{u}_0 = \mathbf{u}_{1000111}$ and $\mathbf{u}_1 = \mathbf{u}_{0011110}$. They differ in four positions, and so are not vertices of the same edge.

Since $y_1 < 0$ the second rotation will be through $(+\pi/4)$.

$$
T_1 = \begin{bmatrix} \cos(\pi/4) & \sin(\pi/4) \\ -\sin(\pi/4) & \cos(\pi/4) \end{bmatrix} = \begin{bmatrix} 0.7071 & 0.7071 \\ -0.7071 & 0.7071 \end{bmatrix} \tag{6.178}
$$

$$B_2 = T_1 B_1 = \begin{bmatrix} -0.4233 & -1.3155 & -1.5903 \\ 1.2555 & 0.1578 & -3.2120 \\ -0.9142 & 2.8719 & 2.8193 & 1.3283 \\ -3.1246 & 1.5334 & 2.2131 & 1.6398 \end{bmatrix} \tag{6.179}$$

$$\mathbf{u}_2 = \mathbf{u}_{0000111} = \begin{Bmatrix} -0.9599 \\ -0.9599 \\ -0.5236 \\ -0.5236 \\ 0.5236 \\ 0.5236 \\ 0.5236 \end{Bmatrix} \tag{6.180}$$

$$B_2 \mathbf{u}_2 = \begin{Bmatrix} 8.0874 \\ -1.3255 \end{Bmatrix} \tag{6.181}$$

Continuing, the third rotation is clockwise through $+\pi/8$.

$$T_2 = \begin{bmatrix} \cos(\pi/8) & \sin(\pi/8) \\ -\sin(\pi/8) & \cos(\pi/8) \end{bmatrix} = \begin{bmatrix} 0.9239 & 0.3827 \\ -0.3827 & 0.9239 \end{bmatrix} \tag{6.182}$$

$$B_3 = T_1 B_2 = \begin{bmatrix} 0.0894 & -1.1550 & -2.6985 \\ 1.3220 & 0.6492 & -2.3590 \\ -2.0404 & 3.2401 & 3.4516 & 1.8547 \\ -2.5369 & 0.3176 & 0.9657 & 1.0066 \end{bmatrix} \tag{6.183}$$

$$\mathbf{u}_3 = \mathbf{u}_{1000111} = \begin{Bmatrix} 0.4363 \\ -0.9599 \\ -0.5236 \\ -0.5236 \\ 0.5236 \\ 0.5236 \\ 0.5236 \end{Bmatrix} \tag{6.184}$$

$$B_3 \mathbf{u}_3 = \begin{Bmatrix} 8.9091 \\ 0.3320 \end{Bmatrix} \tag{6.185}$$

We have $y_2 = -1.3255$ and $y_3 = 0.3320$, so the last two vertices are of different sign. Comparing their respective controls, $\mathbf{u}_{0000111}$ and $\mathbf{u}_{1000111}$, they differ only in u_1, so we have identified the correct edge, $\mathbf{o}_{2000111}$.

We can determine $\mathbf{u}^*$ using Eqs. (6.55) and (6.57). The moments we just determined are in the rotated moment space, so we use the rotated desired moment,

$$T\mathbf{m}_{des} = \begin{Bmatrix} 2.8644 \\ 0 \end{Bmatrix} \tag{6.186}$$

$$\begin{Bmatrix} a \\ b \end{Bmatrix} = \begin{bmatrix} \mathbf{m}_{des} & -(\mathbf{m}_2 - \mathbf{m}_1) \end{bmatrix}^{-1} \mathbf{m}_1 \tag{6.187}$$

Applied to this example, we take

$$\mathbf{m}_1 = B_2 \mathbf{u}_2 = \begin{Bmatrix} 8.0874 \\ -1.3255 \end{Bmatrix} \tag{6.188}$$

and

$$\mathbf{m}_2 = B_3 \mathbf{u}_3 = \begin{Bmatrix} 8.9091 \\ 0.3320 \end{Bmatrix} \tag{6.189}$$

$$\begin{Bmatrix} a \\ b \end{Bmatrix} = \begin{bmatrix} 2.8644 & -0.8216 \\ 0 & -1.6575 \end{bmatrix}^{-1} \begin{Bmatrix} 8.0874 \\ -1.3255 \end{Bmatrix} = \begin{Bmatrix} 3.0528 \\ 0.7997 \end{Bmatrix} \tag{6.190}$$

$$\mathbf{u}^* = \mathbf{u}_2 + b(\mathbf{u}_3 - \mathbf{u}_2)$$

$$= \begin{Bmatrix} -0.9599 \\ -0.9599 \\ -0.5236 \\ -0.5236 \\ 0.5236 \\ 0.5236 \\ 0.5236 \end{Bmatrix} - 0.7997 \left(\begin{Bmatrix} 0.4363 \\ -0.9599 \\ -0.5236 \\ -0.5236 \\ 0.5236 \\ 0.5236 \\ 0.5236 \end{Bmatrix} - \begin{Bmatrix} -0.9599 \\ -0.9599 \\ -0.5236 \\ -0.5236 \\ 0.5236 \\ 0.5236 \\ 0.5236 \end{Bmatrix} \right)$$

$$= \begin{Bmatrix} 0.1567 \\ -0.9599 \\ -0.5236 \\ -0.5236 \\ 0.5236 \\ 0.5236 \\ 0.5236 \end{Bmatrix} \tag{6.191}$$

(Since we know that $\mathbf{u}_2$ and $\mathbf{u}_3$ differ in only their first element, Eq. (6.191) could have been reduced to the determination of the first element of $\mathbf{u}^*$, $\mathbf{u}^*(1) = \mathbf{u}_2(1) + b(\mathbf{u}_3(1) - \mathbf{u}_2(1))$. The rest of $\mathbf{u}^*$ is the same as $\mathbf{u}_2$ and $\mathbf{u}_3$.)

As confirmation,

$$B\mathbf{u}^*/a = \begin{Bmatrix} 2.8493 \\ -0.2942 \end{Bmatrix} = \mathbf{m}_{des} \tag{6.192}$$

Example 6.9

Facet Searching We use the desired moment from Example 6.7, Eq. (6.151):

$$\mathbf{m}_{des} = \begin{Bmatrix} 2.8493 \\ -0.2942 \\ 0.5726 \end{Bmatrix} \tag{6.193}$$

The first step is to rotate the problem so that the x-axis is aligned with $\mathbf{m}_{des}$. A straightforward method involves the use of Euler parameters. First we take the cross product of a unit vector in the x-direction with the normalized desired moment. The resultant vector is perpendicular to the plane of the x-axis and $\mathbf{m}_{des}$. The magnitude of the result is the sine of the angle between the two vectors, and when normalized it serves as an eigenaxis.

$$\begin{Bmatrix} 1 \\ 0 \\ 0 \end{Bmatrix} \times \frac{\mathbf{m}_{des}}{|\mathbf{m}_{des}|} = 0.2204 \begin{Bmatrix} 0 \\ -0.8895 \\ -0.4570 \end{Bmatrix} \tag{6.194}$$

Thus the angle between the vectors $\eta = \arcsin(0.2204) = 0.2222$ rad $= 12.73°$. The direction cosines of the eigenaxis are $\xi = 0$, $\zeta = -.8895$, and $\chi = -0.4570$. Then with the customary definitions of the parameters,

$$\begin{aligned} q_0 &\equiv \cos(\eta/2) \\ q_1 &\equiv \xi \sin(\eta/2) \\ q_2 &\equiv \zeta \sin(\eta/2) \\ q_3 &\equiv \chi \sin(\eta/2) \end{aligned} \tag{6.195}$$

the transformation matrix becomes:

$$\begin{aligned} T &= \begin{bmatrix} (q_0^2 + q_1^2 - q_2^2 - q_3^2) & 2(q_1 q_2 + q_0 q_3) & 2(q_1 q_3 - q_0 q_2) \\ 2(q_1 q_2 - q_0 q_3) & (q_0^2 - q_1^2 + q_2^2 - q_3^2) & 2(q_2 q_3 + q_0 q_1) \\ 2(q_1 q_3 + q_0 q_2) & 2(q_2 q_3 - q_0 q_1) & (q_0^2 - q_1^2 - q_2^2 + q_3^2) \end{bmatrix} \\ &= \begin{bmatrix} 0.9754 & -0.1007 & 0.1960 \\ 0.1007 & 0.9949 & 0.0100 \\ -0.1960 & 0.0100 & 0.9805 \end{bmatrix} \end{aligned} \tag{6.196}$$

The starting matrix then is B_0,

$$B_0 = TB = \begin{bmatrix} 0.5122 & -0.7379 & -3.3594 \\ 1.1826 & 1.0467 & -1.1414 \\ -0.4519 & 0.4743 & 0.5295 \end{bmatrix}$$

$$\begin{bmatrix} -2.8610 & 3.1151 & 3.5190 & 1.8095 \\ -1.5603 & -0.9496 & -0.4343 & 0.2033 \\ 0.2729 & -0.2981 & -0.5454 & -1.6570 \end{bmatrix} \tag{6.197}$$

Now, we take the first two rows of B_0 and repeat Example 6.8. As a consequence we find that the x-axis intersects the edge $\mathbf{o}_{2000111}$. Back in three-space,

$$B_0 \mathbf{u}_{0000111} = \begin{Bmatrix} 7.8947 \\ -1.3435 \\ -1.7509 \end{Bmatrix}$$

$$B_0 \mathbf{u}_{1000111} = \begin{Bmatrix} 8.6099 \\ 0.3077 \\ -2.3818 \end{Bmatrix} \tag{6.198}$$

Both vertices of the edge have negative z components, so we need a positive rotation to bring the edge toward the front. A rotation through $\pi/2$ about the x-axis is

$$T_0 = \begin{bmatrix} 1 & 0 & 0 \\ 0 & 0 & 1 \\ 0 & -1 & 0 \end{bmatrix} \tag{6.199}$$

$$B_1 = T_0 B_0 = \begin{bmatrix} 0.5122 & -0.7379 & -3.3594 \\ -0.4519 & 0.4743 & 0.5295 \\ -1.1826 & -1.0467 & 1.1414 \end{bmatrix}$$

$$\begin{bmatrix} -2.8610 & 3.1151 & 3.5190 & 1.8095 \\ 0.2729 & -0.2981 & -0.5454 & -1.6570 \\ 1.5603 & 0.9496 & 0.4343 & -0.2033 \end{bmatrix} \tag{6.200}$$

We take the first two rows of B_1 and repeat Example 6.8. Now we find that the x-axis intersects the edge $\mathbf{o}_{0200110}$. Again in three-space,

$$B_1 \mathbf{u}_{0000110} = \begin{Bmatrix} 4.9695 \\ 0.6465 \\ 0.0949 \end{Bmatrix}$$

$$B_1 \mathbf{u}_{0100110} = \begin{Bmatrix} 5.9998 \\ -0.0157 \\ 1.5563 \end{Bmatrix} \tag{6.201}$$

This edge has positive z components. We also note that the edge found in the initial (B_0) orientation was $\mathbf{o}_{2000111}$, and the edge found after one rotation through $\pi/2$ was $\mathbf{o}_{0200110}$. These two edges differ in the first, second, and seventh position, so their common object is $\mathbf{o}_{2200112}$, a three-dimensional figure and not a facet.

The algorithm proceeds, until after the fifth ($\pi/32$) and sixth ($-\pi/64$) three-dimensional rotations the criteria for a facet are met.

After the fifth rotation the edge found is $\mathbf{o}_{1200110}$, and the defining moments are

$$B_4 \mathbf{u}_{1000110} = \begin{Bmatrix} 6.7149 \\ -0.6343 \\ -0.1577 \end{Bmatrix}$$

$$B_4 \mathbf{u}_{1100110} = \begin{Bmatrix} 5.6847 \\ 0.1680 \\ -1.5472 \end{Bmatrix} \tag{6.202}$$

The sixth rotation yields edge $\mathbf{o}_{0200110}$ and defining moments

$$B_5\mathbf{u}_{0000110} = \begin{Bmatrix} 5.9998 \\ -0.0921 \\ 1.5537 \end{Bmatrix}$$

$$(6.203)$$

$$B_5\mathbf{u}_{0100110} = \begin{Bmatrix} 4.9695 \\ 0.6411 \\ 0.1265 \end{Bmatrix}$$

Thus, the two edges found are $\mathbf{o}_{1200110}$ with a negative z component, and $\mathbf{o}_{0200110}$ with a positive z component. The two edges differ in only the first and second components, so they are both parts of the facet $\mathbf{o}_{2200110}$, which, as we knew all along, is the correct answer.

6.11.4   Banks' Method

For this example we use the same data as in Examples 6.7 and 6.9. To help the reader visualize the process, Figure 6.17 shows each of the vertices found along the way through the algorithm. The notation used to identify each vertex is the first reference, then any renamings, and finally an x if the vertex was not used in the solution. Thus, '$\mathbf{y}_1$, $\mathbf{y}_2$, x' means this vertex was first identified as $\mathbf{y}_1$, later became $\mathbf{y}_2$, and finally was replaced by another vertex (one of the $\mathbf{y}_4$s).

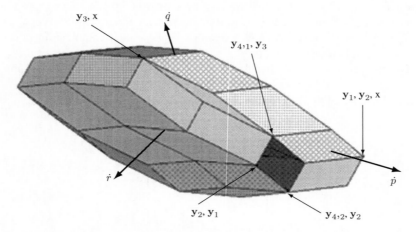

Figure 6.17   The procession of vertices found using Banks' method. The solution facet has been removed. The view is from $(\dot{p}\ \dot{q}\ \dot{r}) = (1\ 1\ 1)$

Example 6.10

Banks' Method for the ADMIRE Simulation We find the first vertex from the signs of $\mathbf{m}_{des}^T B$, which yields the maximum vertex in the direction of $\mathbf{m}_{des}$. To be consistent with the text in Section 6.8 we assign $\mathbf{y} \equiv \mathbf{m}_{des}$.

$$\mathbf{y} = \left\{ \begin{array}{c} 2.8493 \\ -0.2942 \\ 0.5726 \end{array} \right\} \tag{6.204}$$

$$\mathbf{y}^T B = \left[\begin{array}{ccc} 1.4962 & -2.1555 & -9.8133 \\ -8.3573 & 9.0995 & 10.2793 & 5.2858 \end{array} \right] \tag{6.205}$$

The control effectors at the maximum vector are $\mathbf{u}_{1000111}$,

$$\mathbf{u}_{1000111} = \left\{ \begin{array}{c} 0.4363 \\ -0.9599 \\ -0.5236 \\ -0.5236 \\ 0.5236 \\ 0.5236 \\ 0.5236 \end{array} \right\} \tag{6.206}$$

$$\mathbf{y}_1 = B\mathbf{u}_{1000111} = \left\{ \begin{array}{c} 8.8960 \\ -0.5848 \\ -0.6447 \end{array} \right\} \tag{6.207}$$

In Figure 6.17 this vertex is labeled '$\mathbf{y}_1, \mathbf{y}_2, \mathbf{x}$', meaning that it will later be swapped with $\mathbf{y}_2$ (to preserve the handedness of the order of vertices), and finally discarded in favor of another vertex.

The next vertex is found by looking in a direction $\mathbf{d}$ perpendicular to $\mathbf{y}_1$, and in the plane of $\mathbf{y}$ and $\mathbf{y}_1$.

$$\mathbf{d} = \mathbf{y} - \frac{\mathbf{y}^T \mathbf{y}_1}{\mathbf{y}_1^T \mathbf{y}_1} \mathbf{y}_1 = \left\{ \begin{array}{c} 0.0490 \\ -0.1101 \\ 0.7756 \end{array} \right\}_u \tag{6.208}$$

$$\mathbf{d}^T B = \left[\begin{array}{ccc} -0.3454 & 0.0986 & -0.2009 \\ -0.2476 & 0.5254 & 0.3753 & -0.8804 \end{array} \right] \tag{6.209}$$

$$\mathbf{u}_{0100110} = \left\{ \begin{array}{c} -0.9599 \\ 0.4363 \\ -0.5236 \\ -0.5236 \\ 0.5236 \\ 0.5236 \\ -0.5236 \end{array} \right\}_u \tag{6.210}$$

$$\mathbf{y}_2 = B\mathbf{u}_{0100110} = \begin{Bmatrix} 4.7111 \\ -0.5885 \\ 1.6071 \end{Bmatrix} \tag{6.211}$$

In Figure 6.17 this vertex is labeled '$\mathbf{y}_2$, $\mathbf{y}_1$', meaning that it will later be swapped with $\mathbf{y}_1$ (to preserve the handedness of the order of vertices), but not discarded, and thus will be part of the solution.

A third vertex is found in a direction $\mathbf{d}$ perpendicular to the plane of $\mathbf{y}_1$ and $\mathbf{y}_2$, on the same side of the plane $\mathbf{y}$ is on. First the cross product of $\mathbf{y}_1$ and $\mathbf{y}_2$:

$$\mathbf{d} = \mathbf{y}_1 \times \mathbf{y}_2 = \begin{Bmatrix} -1.3193 \\ -17.3344 \\ -2.4800 \end{Bmatrix} \tag{6.212}$$

We calculate $\mathbf{d}^T\mathbf{y} = -0.0793 < 0$, so we swap $\mathbf{y}_1$ and $\mathbf{y}_2$ and calculate $\mathbf{y}_3$ from $-\mathbf{d}$.

$$\mathbf{y}_1 = \begin{Bmatrix} 4.7111 \\ -0.5885 \\ 1.6071 \end{Bmatrix} \tag{6.213}$$

$$\mathbf{y}_2 = \begin{Bmatrix} 8.8960 \\ -0.5848 \\ -0.6447 \end{Bmatrix} \tag{6.214}$$

$$-\mathbf{d}^T B = \{ 19.5339 \quad 19.3090 \quad -18.7125 \\ -26.5909 \quad -17.1402 \quad -8.7417 \quad -0.2999 \} \tag{6.215}$$

$$\mathbf{u}_{1100000} = \begin{Bmatrix} 0.4363 \\ 0.4363 \\ -0.5236 \\ -0.5236 \\ -0.5236 \\ -0.5236 \\ -0.5236 \end{Bmatrix}_u \tag{6.216}$$

$$\mathbf{y}_3 = \begin{Bmatrix} -1.1050 \\ 3.1260 \\ 0.6639 \end{Bmatrix} \tag{6.217}$$

In Figure 6.17 this vertex is labeled '$\mathbf{y}_3$, x', meaning that it will later be discarded in favor of another vertex. It is not hard to see visually why this vertex will not survive, since it is very far from the solution facet.

At this point the three vertices are $\mathbf{m}_{0100110}$, $\mathbf{m}_{1000111}$, and $\mathbf{m}_{1100000}$, which combine to form $\mathbf{o}_{2200222}$, a five-dimensional object and not a facet. The three vertices in Figure 6.17 are '$\mathbf{y}_1$, $\mathbf{y}_2$, x', '$\mathbf{y}_2$, $\mathbf{y}_1$', and '$\mathbf{y}_3$, x'.

We next look along the normal to the plane of $\mathbf{y}_1$, $\mathbf{y}_2$, and $\mathbf{y}_3$.

$$\mathbf{d} = (\mathbf{y}_2 - \mathbf{y}_1) \times (\mathbf{y}_3 - \mathbf{y}_1) = \begin{Bmatrix} 8.3611 \\ 17.0441 \\ 15.5665 \end{Bmatrix} \tag{6.218}$$

The usual application of Eq. (6.72) yields the control vector at this maximum vertex, $\mathbf{u}_{1100110}$ and

$$\mathbf{y}_{4,1} = \begin{Bmatrix} 5.6986 \\ 0.9758 \\ 1.1451 \end{Bmatrix} \tag{6.219}$$

To see which vertex is replaced by $\mathbf{y}_{4,1}$ we perform the determinant tests,

```
if det ([y,y4,y1]) > 0 % Case A, B, or C
  if det ([y,y4,y2]) > 0  % Case A
    y1 = y4;
  else % Case B or C
    y3 = y4;
  end
else & Case D, E, or F
  if det ([y,y4,y3]) > 0 % Case D or E
    y2 = y4;
  else % Case F
    y1 = y4;
  end
end
```

Using the current data,

$$\det \begin{bmatrix} \mathbf{y} & \mathbf{y}_{4,1} & \mathbf{y}_1 \end{bmatrix} = 2.9432 > 0,$$

$$\det \begin{bmatrix} \mathbf{y} & \mathbf{y}_{4,1} & \mathbf{y}_2 \end{bmatrix} = -10.8415 < 0,$$

so $\mathbf{y}_3$ was replaced with $\mathbf{y}_{4,1}$.

In Figure 6.17 this vertex is labeled '$\mathbf{y}_{4,1}$, $\mathbf{y}_3$', meaning that it was created and then replaced the original $\mathbf{y}_3$. The original $\mathbf{y}_3$, as noted above, did not appear in the figure to be much of a candidate for the solution, and so it is not surprising that it is the first to be replaced.

Now the three vertices are $\mathbf{m}_{0100110}$, $\mathbf{m}_{1000111}$, and $\mathbf{m}_{1100110}$, which combine to form $\mathbf{o}_{2200112}$, a three-dimensional object and not a facet.

We repeat the last step with the new set of vertices.

$$\mathbf{d} = (\mathbf{y}_2 - \mathbf{y}_1) \times (\mathbf{y}_3 - \mathbf{y}_1) = \begin{Bmatrix} 3.5209 \\ -0.2903 \\ 6.5429 \end{Bmatrix} \tag{6.220}$$

The control vector at this maximum vertex is $\mathbf{u}_{1000110}$ and

$$\mathbf{y}_{4,2} = \begin{Bmatrix} 6.6861 \\ -0.5885 \\ 0.6831 \end{Bmatrix} \tag{6.221}$$

Using the current data,

$$\det \begin{bmatrix} \mathbf{y} & \mathbf{y}_{4,2} & \mathbf{y}_1 \end{bmatrix} = -3.2894e - 04 < 0,$$

$$\det \begin{bmatrix} \mathbf{y} & \mathbf{y}_{4,2} & \mathbf{y}_3 \end{bmatrix} = 2.9440 > 0,$$

so $\mathbf{y}_2$ was replaced with $\mathbf{y}_{4,2}$.

In Figure 6.17 this vertex is labeled '$\mathbf{y}_{4,2}, \mathbf{y}_2$', meaning that it was created and then replaced the current $\mathbf{y}_2$.

Now the three vertices are $\mathbf{m}_{0100110}$, $\mathbf{m}_{1000111}$, and $\mathbf{m}_{1000110}$, which combine to form $\mathbf{o}_{2200110}$, a two-dimensional object that is the solution facet. It is easy to see in Figure 6.17 that the last three vertices are all parts of the same facet.

6.11.5   Linear Programming

The next two examples continue to use the data for the ADMIRE simulation in Eqs. (6.129) and (6.130), with the desired moment from the direct allocation example, Section 6.11.2.

$$\mathbf{m}_{des} = \begin{Bmatrix} 2.8493 \\ -0.2942 \\ 0.5726 \end{Bmatrix} \tag{6.222}$$

Example 6.11

Direction-preserving Formulation for the ADMIRE Simulation   The problem was solved using the MATLAB® function linprog. The calling arguments were defined using Eqs. (6.87):

```
A=[B  -md];
b=-B*umin;
ct=[0 0 0 0 0 0 0 -1];
h=[umax-umin;1];
```

The function linprog(f,A,b,Aeq,beq,LB,UB) was called, with c as the vector named f in MATLAB®. The next two arguments A,b are for an inequality constraint, and so were set as empty matrices. The arguments Aeq,beq are equality constraints corresponding to our A and b. LB,UB are lower and upper bounds. The MATLAB® function optimoptions was used to specify that linprog use the simplex algorithm.

```
op=optimoptions('linprog','Algorithm','simplex');
[X] = linprog(ct',[],[],A,b,zeros(8,1),h);
u=X(1:7)+umin;
lambda=-X(8);
```

This resulted in a correct solution,

$$\mathbf{u}_{DP} = \begin{Bmatrix} -0.2719 \\ -0.1331 \\ -0.2886 \\ -0.3126 \\ 0.2220 \\ 0.2305 \\ -0.2235 \end{Bmatrix} \tag{6.223}$$

The solution is different from that achieved with direct allocation,

$$\mathbf{u}_{DA} = \begin{Bmatrix} -0.1309 \\ -0.1309 \\ -0.2618 \\ -0.2618 \\ 0.2618 \\ 0.2618 \\ -0.2618 \end{Bmatrix} \tag{6.224}$$

However, as the desired moment nears the boundary of the set of attainable moments the two solutions become more and more similar until, on the boundary, they are the same.

For unattainable moments this algorithm does indeed preserve the direction, returning the solution that is on the boundary.

Example 6.12

Reduced-size Direction-preserving Formulation for the ADMIRE Simulation The problem was also solved using the MATLAB® function linprog, calling arguments from Eqs. (6.107):

```
M=[md(2:3,1)  -md(1)*eye(2)];
A=M*B;
b=-A*umin;
ct=-md'*B;
h=umax-umin;
```

The function `linprog` was called, again specifying the simplex algorithm.

```
op=optimoptions('linprog','Algorithm','simplex');
[X] = linprog(ct',[],[],A,b,zeros(7,1),h);
u=(X+umin);
rho=B(1,:)*u/md(1)
u=u/rho;
```

This resulted in the same solution as obtained using direct allocation,

$$
\mathbf{u}_{RSDP} = \begin{Bmatrix} -0.1309 \\ -0.1309 \\ -0.2618 \\ -0.2618 \\ 0.2618 \\ 0.2618 \\ -0.2618 \end{Bmatrix} \tag{6.225}
$$

6.11.6   Convex-hull Volume Calculations

Example 6.13

Volume Calculation for Daisy-chain Solution   This example will use a daisy-chaining solution with direction preserving in the first step (see step 1 on p. 86). We continue to use Eqs. (6.129) and (6.130). We intend to use the directions of the vertices of Φ as directions, find the maximum moment in each direction that the daisy chain attains, and fit a convex hull to those points.

The daisy chain will consist of the following sub-matrices, with $\mathbf{b}_i$ denoting a column of B:

$$
\begin{aligned}
B_1 &= \begin{bmatrix} \mathbf{b}_4 & \mathbf{b}_5 & \mathbf{b}_7 \end{bmatrix} \\
B_2 &= \begin{bmatrix} \mathbf{b}_1 & \mathbf{b}_2 & \mathbf{b}_3 & \mathbf{b}_6 \end{bmatrix}
\end{aligned} \tag{6.226}
$$

We first continue as in Section 5.4.1 and continue to find the vertices as we did in Eq. (5.49). The complete set of vertices number 44, and are entered into MATLAB® as a matrix `Verts` whose rows are the vertices.

```
Verts=[
    5.6989    -2.1520     1.1451
    4.7113    -0.5883     1.6071
    5.6989     0.9755     1.1451
    6.6864    -0.5883     0.6831
   -5.6989    -2.1483    -1.1451
   -6.6864    -0.5846    -0.6831
   -5.6989     0.9791    -1.1451
```

```
       -4.7113    -0.5846    -1.6071
       -2.0923     1.5615     1.1259
       -5.7533     0.7322     0.9681
       -4.7658     2.2960     0.5061
       -1.1048     3.1253     0.6639
        4.7658    -3.4689    -0.5061
        1.1048    -4.2981    -0.6639
        2.0923    -2.7344    -1.1259
        5.7533    -1.9051    -0.9681
       -8.8960    -0.5883     0.6447
       -7.9085     0.9755     0.1827
        7.9085    -2.1483    -0.1827
        8.8960    -0.5846    -0.6447
       -2.0379    -2.9776    -0.9873
       -1.0503    -1.4139    -1.4493
        1.0503     0.2410     1.4493
        2.0379     1.8048     0.9873
        2.0379    -2.9813     0.9873
        1.0503    -1.4175     1.4493
       -1.0503     0.2447    -1.4493
       -2.0379     1.8084    -0.9873
       -1.1048    -4.3018     0.6639
       -2.0923    -2.7381     1.1259
        2.0923     1.5652    -1.1259
        1.1048     3.1289    -0.6639
        5.7533     0.7359    -0.9681
        7.9085     0.9791    -0.1827
        4.7658     2.2997    -0.5061
       -7.9085    -2.1520     0.1827
       -4.7658    -3.4725     0.5061
       -5.7533    -1.9088     0.9681
        2.5562    -3.4725     0.8217
       -2.5562     2.2997    -0.8217
       -2.6107    -0.5883     1.2915
        2.6107    -0.5846    -1.2915
        2.5562     2.2960     0.8217
       -2.5562    -3.4689    -0.8217];
```

The matrices B_1 and B_2 are defined as b1 and b2. The inverse of B_1, b1i, and the pseudo-inverse of B_2, p2, are calculated.

```
b1=[B(:,4:5) B(:,7)];b1i=inv(b1);
b2=[B(:,1:3) B(:,6)];p2=pinv(b2);
```

An initially empty matrix M will be used to accumulate the results of the daisy-chaining operation. An index vector ix is defined, corresponding to the partitions of **u**, for ease in manipulating the controls without defining a new control vector.

```
M=[];
ix=[4 5 7 1 2 3 6];
```

We then take each vertex m (the transpose of a row of Verts) and try to allocate using b1i. Two scale factors—for the first group, scale457 and the second, scale1236—are initialized to a value of 1. Next we check u_4, u_5, and u_7 against their limits and adjust scale457 as necessary, and repeat for the other controls and scale1236. The two control vectors are u457 and u1236. Since we preserve the direction in each case, these two operations are independent of each other.

```
for iV=1:44 % 44 vertices
 scale457=1;scale1236=1; % Initialize scale factors
 m=Verts(iV,:)'; % Pick a row
 u457=b1i*m;u1236=p2*m; % Sol'ns without constraints
 for iu=1:3
   if(scale457*u457(iu)>umax(ix(iu)))
     scale457=umax(ix(iu))/u457(iu);
   end
   if(scale457*u457(iu)<umin(ix(iu)))
     scale457=umin(ix(iu))/u457(iu);
   end
 end % for iu=1:3
 for iu=4:7
   if(scale1236*u1236(iu-3)>umax(ix(iu)))
     scale1236=umax(ix(iu))/u1236(iu-3);
   end
   if(scale1236*u1236(iu-3)<umin(ix(iu)))
     scale1236=umin(ix(iu))/u1236(iu-3);
 end for iu=4:7
```

We then scale the vectors, assemble the original control vector in U, and add BU as a new column to M. After all 44 vertices are processed, M will contain 44 points on the convex hull of Π for this daisy chain.

```
 u457=scale457*u457;
 u1236=scale1236*u1236;
 U=[u1236(1:3);u457(1:2);u1236(4);u457(3)];
 M=[M B*U];
end % for iV=1:44
```

Finally we pass the matrix of points to convhull, invoking the volume V option,

```
[K,V]=convhull(M');
```

The transpose of M is necessary because of the shape of the matrix. We find the volume of this approximation to the convex hull of Π to be 66.15 compared to 177.10 for Φ. Since this B matrix was obtained by linearizing the ADMIRE simulation using angular acceleration (rad/s^2), the units of Π and Φ are the cube of the angular accelerations, but the actual values are of less importance than the volume relative to the maximum attainable.

6.12 Afterword

In this chapter we have described and in most cases given numerical examples of several methods of control allocation. Despite the apparent complexity of some of the algorithms, they have all been implemented and flown real-time in a manned flight simulator (see Scalera and Durham (1998)). The results of these experiments are described in the MS theses of Virginia Polytechnic Institute & State University students listed in the bibliography, Appendix C.

The simplest and least computationally-intensive method by far is the generalized inverse (Section 6.5). However, the method is often far from optimal in the sense of generating admissible solutions for all attainable moments, or even most of them. The use of generalized inverses in the cascaded generalized inverse (Section 6.5.5.2) appears to overcome that deficiency with only a small price to be paid of complexity. There is no proof or compelling evidence that the cascaded general inverse is thus efficient, or a guarantee that in some applications it will not perform poorly.

The remaining methods are almost equivalent in that they aim to provide admissible solutions for every attainable moment. The fastest is probably Banks' method (Section 6.8) and the slowest the direct method (Section 6.6).

So far as ease of understanding, the direct method is generally just finding intersections of lines with plane surfaces. On the other end of the spectrum, linear programming (Section 6.9 and Appendix A) may be the most challenging. However, linear programming offers several proven pre-packaged solvers that may be relied upon. Certainly if one takes up the control allocation problem with a strong background in linear programming, then it is the easiest to understand.

So far as versatility goes, many formulations of linear programming have a secondary function that permits the optimization of some other criteria than satisfying moment demands. This is not necessarily unique. If the secondary criterion can be formulated by its dependence of the control effectors, then one may move through the null space of the control effectiveness (B) matrix to satisfy it. This will be addressed further in Chapter 7.

References

Acosta, DM, Yildiz, Y, Craun, RW, Beard, SD, Leonard, MW, Hardy, GH, and Weinstein, M 2015 'Piloted evaluation of a control allocation technique to recover from pilot-induced oscillations,' *J. Aircraft*, **52** (1), 130–140.

Beck, RE 2002 *Application of Control Allocation Methods to Linear Systems with Four or More objectives*. PhD Thesis, Virginia Polytechnic Institute & State University.

Bodson, M 2002 'Evaluation of optimization methods for control allocation' *AIAA J. Guidance, Control, and Dynamics*, **25** (4), 703–711.

Bordignon, KA 1996 *Constrained Control Allocation for Systems with Redundant Control Effectors*. PhD Thesis, Virginia Polytechnic Institute & State University.

Bordignon, K and Bessolo, J 2002 'Control allocation for the X-35B,' AIAA-2002-6020 in *Proceedings of the 2002 Biennial International Powered Lift Conference and Exhibit.*

Hausner, MA 1965 *A Vector Space Approach to Geometry.* Prentice-Hall, p.234.

Lallman, FJ 1985 'Preliminary Design Study of a Lateral-Directional Control System Using Thrust Vectoring,' NASA TM 86425.

Mulkens, JM and Ormerod, AO 'Measurements of aerodynamic rotary stability derivatives using a whirling arm facility,' *J. Aircraft*, **30** (2), 178–183.

Sommerville, D 1929 *An Introduction to the Geometry of n Dimensions,* Methuen. Republished by Dover Publishing, New York, 1958: pp.9–10.

Scalera, KR and Durham, W 1998 'Modification of a surplus Navy 2F122A A-6E OFT for flight dynamics research and instruction,' AIAA-98-4180 in *AIAA Modeling And Simulation Technologies Conference And Exhibit.*

Stevens, B and Lewis, F 2003 *Aircraft Control and Simulation,* 2nd edn. John Wiley & Sons, pp. 646–647.

7

Frame-wise Control Allocation

7.1 General

The expression *frame-wise control allocation* arose when considering that modern flight control computers operate by determining the required control effector deflections several times per second, around 100 in current tactical aircraft design. At the beginning of each cycle of operation, or frame, all the inputs to the control law are evaluated, all the control laws are applied, and the required control deflections calculated.

If the flight control computer is treated as a continuous process, then the control allocator calculates control effector positions based on a continuously varying demand for a given moment (or set of objectives) and calculates the global positions of the effectors required to satisfy that demand. It is easy when viewing the problem in this manner to expect that the control effectors will continuously and instantaneously move to these commanded positions.

This view will not hold when the physical limitations of the effectors and their actuators are considered. We have extensively considered the global position limits, for example, an elevator that cannot exceed some minimum or maximum position. But control effectors and their actuation systems also have *rate limits*.

The maximum rate at which a flapping control surface can move is a nonlinear function of its deflection angle and direction of travel, which generate hinge moments transmitted by air loads on the surface. When a surface moves away from its free-float (zero hinge moment) position it opposes the hinge moment, and its maximum rate is usually lower (sometimes much lower) than when moving toward its free-float position.

Figures cited for maximum rate capabilities are usually those that obtain with no air loads on the surface. During low-speed, high angle-of-attack maneuvering the dynamic pressure is relatively low, and the cited no-load maximum rate capabilities are assumed adequate for analysis.

Rate limiting in a control system can have serious, even catastrophic consequences. The effect of rate limiting on a pilot's control of the airplane is to effectively introduce a time delay. In general, the pilot applies an input to the inceptors expecting a certain response from the airplane. When that response is delayed, he applies a greater input. Finally, however, the airplane response catches up to the original desired, and overshoots it. Perceiving this, the pilot

Aircraft Control Allocation, First Edition. Wayne Durham, Kenneth A. Bordignon and Roger Beck.
© 2017 John Wiley & Sons, Ltd. Published 2017 by John Wiley & Sons, Ltd.
Companion website: www.wiley.com/go/durham/aircraft_control_allocation

applies opposite control, with the response again delayed, until finally the pilot's inputs are far out of phase with the response, a condition known as *pilot-induced oscillation.*

We will incorporate these rate limits into the control allocation problem by considering how far the control effectors are able to move within one frame, the time it takes the flight control computer to process all the information and arrive at new commanded positions for the effectors. For example, if a certain effector can move no faster than $100°/s$ and the frame length is 0.01 s (100-Hz computation speed), then the effector cannot move more than $1°$ during the interval. That is, unless the effector comes up against a global physical limit: the ones we have been discussing up until now.

There are other limitations that will affect the distance an effector can travel in a given frame but for now we will assume that each effector can move at its maximum rate in either direction at any time.

We therefore consider the frame-wise problem. Given a control effector u_i with positive and negative rate limits $\dot{u}_{Max}$, $\dot{u}_{Min}$ and a computational frame length Δt, the maximum limits of travel (based on rate alone) of that effector during the frame are $\dot{u}_{Max}\Delta t$, $\dot{u}_{Min}\Delta t$.

It is possible that, during a frame, a positive or negative movement of the effector may encounter a global position limit for that effector, u_{iMax} or u_{iMin}. Thus the actual distance a control effector can travel during Δt is the more restrictive of the rate-limited travel or the global limit:

$$\Delta u_{iMax} = \min(\dot{u}_{Max}\Delta t, u_{iMax})$$
$$\Delta u_{iMin} = \max(\dot{u}_{Min}\Delta t, u_{iMin}) \tag{7.1}$$

Considering all m control effectors, we have the frame-wise set of admissible controls $\Delta\Omega$, completely analogous to the set of globally admissible controls, Ω.

During one frame of computation the moment generated by the controls will change according to the actual control effectiveness, which we model as the control effectiveness matrix, B.

But now we have a choice of selecting B: it could be global, B_{Global}, based on the control effectiveness based on changes in deflection from the origin of control space (the one assumed in Chapter 4 and implied throughout Chapter 5), or we may calculate a local control effectiveness matrix based on the current state (including effector positions) of the airplane, B_{Local}. We will discuss this further in Section 7.3.

We keep track of conditions at the beginning and end of frames with subscripts k at the beginning, $k + 1$ at the end of the first frame, and if necessary by $k + 2$ and so on for subsequent frames with reference to the first.

The current moment (or objective) at the beginning of the frame is $\mathbf{m}_k$. The flight control computer has computed a *global* desired moment, $\mathbf{m}_{d_{k+1}}$ to be obtained by the effectors by the end of the next frame. The change in desired moment from the beginning to the end of the frame is

$$\Delta\mathbf{m}_{des} = \mathbf{m}_{d_{k+1}} - \mathbf{m}_k \tag{7.2}$$

The current effector positions at the beginning of the frame are $\mathbf{u}_k$. The control allocator must calculate a change in control effector positions, $\Delta\mathbf{u}$, that generates the change in moment $\Delta\mathbf{m}_{des}$. We approximate the actual control-generated moment at the beginning of the frame, $\mathbf{m}_k$, as $B_{Global}\mathbf{u}_k$.

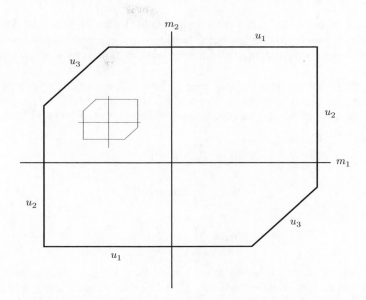

Figure 7.1   Two-dimensional set of attainable moments, Φ

The instantaneous set of changes in moment that can be attained using permissible control deflections is the $\Delta\Phi$ (ΔAMS). The origin of the $\Delta\Phi$ is the current global moment being generated by the control effectors, and its convex hull is generated by the range of motion of the control effectors from their current position. This range of motion will be determined by deflection at the effectors' maximum rate for the duration of the frame, or by global limits, whichever is more restrictive.

Figure 7.1 illustrates the frame-wise problem. It is borrowed from Figure 5.4. The large object is the global Φ, and the controls that define the edges on $\partial(\Phi)$ are shown.

The smaller figure inside Φ is $\Delta\Phi$, the amount the moment can be changed in one frame. Its origin is at the current global moment, $\mathbf{m}_k$, and the controls defining the edges are $\Delta\mathbf{u}_1$, $\Delta\mathbf{u}_2$, and $\Delta\mathbf{u}_3$. As shown, $\Delta\Phi$ does not abut a global limit, so the the minimum and maximum $\Delta\mathbf{u}$ is determined by the rate limits of the effectors. If the next desired moment, $\mathbf{m}_{k+1}$, lies within $\Delta\Phi$, then $\Delta\mathbf{m}$ may be achieved; otherwise control rate-limiting will occur.

7.2   Path Dependency

One quite serious effect of frame-wise control allocation is that after maneuvering in a general manner, the controls that generate the moment that results at the end of the maneuver depend on the sequence of moments generated in between the start and finish. For instance, starting from level flight and maneuvering to achieve the same level flight condition, the control configuration at the end will be different if the path taken was via a loop than if the path taken were a barrel roll.

It is very difficult to generalize this path dependency for any but simple contrived cases. Instead we will present an illustrative example to convince the reader that path dependency is real, then propose a method to eliminate it, whatever the cause.

We consider a sequence of desired moments generated by the flight control computer, each calculated according to the control law being employed at the beginning of a frame of computation. Each change in desired moment from the current $\mathbf{m}$, $\Delta\mathbf{m}$, is to be accomplished during the next frame.

Beginning ($k = 0$) from the origin of both Φ and Ω ($\mathbf{m}_0 = \mathbf{0}$, $\mathbf{u}_0 = \mathbf{0}$), there will be a sequence of $\Delta\mathbf{u}$ calculated according to the control allocation scheme in use. Assuming that the allocation at each frame was successful, then after a period of progress to the nth frame ($k = n$) we should have

$$B\Delta\mathbf{u}_k = \Delta\mathbf{m}_k, k = 1 \cdots n \tag{7.3}$$

$$\mathbf{m} = \sum_{k=0}^{n} \Delta\mathbf{m}_k = \mathbf{m}_{desn} \tag{7.4}$$

$$\mathbf{u}_n = \sum_{k=0}^{n} \Delta\mathbf{u}_k \tag{7.5}$$

Here, $\mathbf{m}_{desn}$ is the desired moment at the end of the nth frame. Equation (7.3) says the allocation was successful at each stage, Eq. (7.4) says the moment being generated after n successful frame-wise allocations is $\mathbf{m}_{desn}$, while Eq. (7.5) merely states the obvious.

The issue with path dependency is, if $\mathbf{m}_{desn} = \mathbf{0}$, does $\mathbf{u}_n$ necessarily $= \mathbf{0}$?

We will assert that a discontinuity such as a change in the control effectiveness matrix during the maneuver will result in a non-zero control vector when returning to the origin of moment space. The problem with nonlinearities is that they are often hard or impossible to model for analysis. Instead we will demonstrate this assertion and, absent someone's counter-demonstration, consider the point made.

7.2.1 Examples of Path Dependency

We will use the simple 3×2 system from Chapter 5, given in Eq. (5.3). The geometry of this problem was described in Figure 7.1.

$$B = \begin{bmatrix} 1 & 0 & -0.5 \\ 0 & 1 & -0.5 \end{bmatrix} \tag{7.6}$$

The global position limits were

$$-0.5 \le u_i \le 0.5, i = 1 \cdots 3 \tag{7.7}$$

No units were specified in the original formulation, so we will take them to be feet-pound for moments and radians for deflection. We will also adopt seconds as our time units and implement our moment-rate allocator in a 10-Hz computer (0.1-s frame length).

We take for the rate limits

$$-2 \le \dot{u}_i \le 2, i = 1 \cdots 3 \tag{7.8}$$

With units of radians per second, each effector can travel a maximum of 0.2 rad in either direction in one frame, or from limit to limit in five frames.

Example 7.1

Path Dependency of Direct Allocation As used here, the term *direct allocation* includes any allocation method that achieves the maximum capabilities and preserves direction. This includes at least the method described as direct allocation (Section 6.6), edge- and facet-searching (6.7), Banks' method (Section 6.8), and the linear programming methods presented in Section 6.9 (but not necessarily those in Appendix A).

For our moment time history we will command the following sequence of moments, each to be achieved in sequential frames of the flight control computer:

$$\mathbf{m}_0 = \mathbf{0}, \mathbf{m}_1 = \begin{Bmatrix} 0.2 \\ 0 \end{Bmatrix}, \mathbf{m}_2 = \begin{Bmatrix} 0.2 \\ 0.2 \end{Bmatrix}, \mathbf{m}_3 = \mathbf{0} \tag{7.9}$$

This sequence is shown in Figure 7.2.

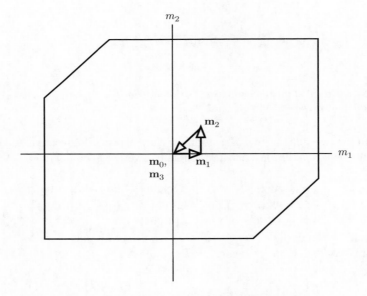

Figure 7.2   Sequence of moments in Φ. Not to scale

At the beginning of each frame we will center the problem at the currently attained moment and consider a $\Delta\Phi$ and the $\Delta\mathbf{m}$ required to advance to the next moment in the sequence. Figure 7.3 illustrates the three copies of $\Delta\Phi$ required. If all is well then the currently attained moment will be the most recent moment in the sequence, and that is assumed in Figure 7.3.

We will use a constant control effectiveness matrix throughout, as this is sufficient to demonstrate the path dependency of direct rate-allocation methods. We will pose three allocation problems with $\Delta\Phi$ centered consecutively at $\mathbf{m}_0$, $\mathbf{m}_1$, and $\mathbf{m}_2$ (Eq. (7.9)), as shown in Figure 7.3.

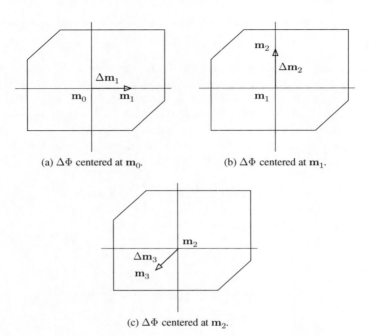

(a) $\Delta\Phi$ centered at $\mathbf{m}_0$.   (b) $\Delta\Phi$ centered at $\mathbf{m}_1$.

(c) $\Delta\Phi$ centered at $\mathbf{m}_2$.

Figure 7.3 $\Delta\Phi$ centered at the three moments, showing $\Delta\mathbf{m}_{k+1} = \mathbf{m}_k - \mathbf{m}_{k-1}$ $k = 1 \cdots 3$. The $\Delta\mathbf{m}$ is the vector with the arrowhead. Not to scale.

The steps in the two-dimensional allocation are easy, especially given that the intersecting edges are easily identified. The last one turns out to be a vertex (seen in Figure 7.3c). The three $\Delta\mathbf{u}$ are

$$\Delta\mathbf{u}_1 = \left\{ \begin{array}{c} 0.1333 \\ -0.0667 \\ -0.1333 \end{array} \right\}, \Delta\mathbf{u}_2 = \left\{ \begin{array}{c} -0.0667 \\ 0.1333 \\ -0.1333 \end{array} \right\}, \Delta\mathbf{u}_3 = \left\{ \begin{array}{c} -0.1333 \\ -0.1333 \\ 0.1333 \end{array} \right\} \tag{7.10}$$

The final control configuration is the sum of these three $\Delta\mathbf{u}$,

$$\mathbf{u}_{DA} = \Delta\mathbf{u}_1 + \Delta\mathbf{u}_2 + \Delta\mathbf{u}_3 = \left\{ \begin{array}{c} -0.0667 \\ -0.0667 \\ -0.1333 \end{array} \right\} \tag{7.11}$$

The solution is not the zero vector, but rather a vector in the null space of B, $B\mathbf{u}_{DA} = \mathbf{0}$.

We have shown only that direct allocation does not necessarily return to the origin of control space at zero moment. To show that it does depend on the path, consider altering our sequence of moment to take a different path,

$$\mathbf{m}_0 = \mathbf{0}, \mathbf{m}_1 = \left\{ \begin{array}{c} 0.2 \\ 0 \end{array} \right\}, \mathbf{m}_2 = \left\{ \begin{array}{c} 0.2 \\ 0.2 \end{array} \right\}, \mathbf{m}_3 = \left\{ \begin{array}{c} 0 \\ 0.2 \end{array} \right\}, \mathbf{m}_4 = \mathbf{0} \tag{7.12}$$

Instead of going to the origin on the third leg, we go straight to the m_2 axis and then down to the origin. The four $\Delta\Phi$ and $\Delta\mathbf{m}$ are shown in Figure 7.4.

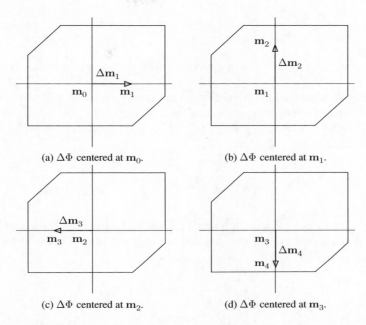

Figure 7.4 $\Delta\Phi$ centered at the four moments, showing $\Delta\mathbf{m}_k = \mathbf{m}_k - \mathbf{m}_{k-1}$ $k = 1 \cdots 4$. The $\Delta\mathbf{m}$ is the vector with the arrowhead. Nothing is to scale

It is obvious that Figures 7.3a and 7.3c are mirrors of each other, both defined by the same edge ($\Delta\mathbf{u}_2$). So are Figures 7.3b and 7.3d, both defined by $\Delta\mathbf{u}_1$. Because of the symmetries built into this problem, the direct-allocation solution of one will be the negative of the solution to its mirror.

$$\Delta\mathbf{u}_1 = \left\{ \begin{array}{c} 0.1333 \\ -0.0667 \\ -0.1333 \end{array} \right\}, \Delta\mathbf{u}_2 = \left\{ \begin{array}{c} -0.0667 \\ 0.1333 \\ -0.1333 \end{array} \right\}, \Delta\mathbf{u}_3 = -\Delta\mathbf{u}_1, \quad \Delta\mathbf{u}_4 = -\Delta\mathbf{u}_2 \qquad (7.13)$$

$$\mathbf{u}_{DA} = \Delta\mathbf{u}_1 + \Delta\mathbf{u}_2 + \Delta\mathbf{u}_3 + \Delta\mathbf{u}_4 = \mathbf{0} \qquad (7.14)$$

This solution is different from that that resulted from the first path, thus demonstrating path dependency. Not too much should be made of the fact that the controls returned to the origin since the problem was very artificial.

Example 7.2

Path Dependency of the Generalized Inverse Generalized inverses do not exhibit path dependency in the strict sense, but something very much like it. First, consider control allocation based on frame-wise use of a *constant* generalized inverse P. Equation (7.5) becomes

$$\mathbf{u}_n = \sum_{k=0}^{n} \Delta\mathbf{u}_k = \sum_{k=0}^{n} P\Delta\mathbf{m}_k = P \sum_{k=0}^{n} \Delta\mathbf{m}_k = P\mathbf{m}_{desn} \qquad (7.15)$$

Clearly, the use of a constant generalized inverse will always return the solution to the origin, $\mathbf{m}_{desn} = \mathbf{0} \Rightarrow \mathbf{u}_n = \mathbf{0}$.

Depending on the changing flight conditions and state of the airplane, control effectiveness may change substantially during a maneuver. If the control effectiveness matrix changes during the evolution then in general $\mathbf{m}_{desn} = \mathbf{0} \not\Rightarrow \mathbf{u}_n = \mathbf{0}$. If we begin with $B_1 = B$ and after q steps change to some different $B = B_2$ then

$$\mathbf{u} = \sum_{k=0}^{n} \Delta\mathbf{u}_k = \sum_{k=0}^{q} P_1\Delta\mathbf{m}_k + \sum_{k=q}^{n} P_2\Delta\mathbf{m}_k = ? \qquad (7.16)$$

Here, P_1 and P_2 are associated with B_1 and B_2, respectively. To demonstrate what happens we return to the three moments of Eq. (7.9), shown in Figures 7.2 and 7.3. We will use Eq. (7.6) for the first two frames,

$$B_1 = \begin{bmatrix} 1 & 0 & -0.5 \\ 0 & 1 & -0.5 \end{bmatrix} \qquad (7.17)$$

and with a change in the (1,3) entry for the last frame:

$$B_2 = \begin{bmatrix} 1 & 0 & -0.25 \\ 0 & 1 & -0.5 \end{bmatrix} \qquad (7.18)$$

For this demonstration we will use the pseudo-inverse as representative of generalized inverses. The two pseudo-inverse matrices are

$$P_1 = \begin{bmatrix} 0.8333 & -0.1667 \\ -0.1667 & 0.8333 \\ -0.3333 & -0.3333 \end{bmatrix} \qquad (7.19)$$

$$P_2 = \begin{bmatrix} 0.9524 & -0.0952 \\ -0.0952 & 0.8095 \\ -0.1905 & -0.3810 \end{bmatrix} \qquad (7.20)$$

From the origin to $\mathbf{m}_1$,

$$\Delta\mathbf{m}_1 = \mathbf{m}_1 - \mathbf{m}_0 = \begin{Bmatrix} 0 \\ 0.2 \end{Bmatrix}$$

$$\Delta\mathbf{u}_1 = P1\Delta\mathbf{m}_1 = \begin{Bmatrix} -0.0333 \\ 0.1667 \\ -0.0667 \end{Bmatrix} \qquad (7.21)$$

We note that none of the effectors has been commanded more than ± 0.2 rad during this 0.1-s frame length, so the rate limits have not been violated.

Continuing,

$$\Delta \mathbf{m}_2 = \mathbf{m}_2 - \mathbf{m}_1 = \begin{Bmatrix} 0.2 \\ 0 \end{Bmatrix}$$

$$\Delta \mathbf{u}_2 = P1\Delta \mathbf{m}_2 = \begin{Bmatrix} 0.1667 \\ -0.0333 \\ -0.0667 \end{Bmatrix} \tag{7.22}$$

We now switch to B_2 and P_2. This means that the current moment is not $\mathbf{m}_2$, since that was arrived at using B_1 and P_1, but rather it is

$$\mathbf{m}_2^* = B_2(\Delta \mathbf{u}_1 + \Delta \mathbf{u}_2) = \begin{Bmatrix} 0.1667 \\ 0.2000 \end{Bmatrix} \tag{7.23}$$

Now Figure 7.3c is not accurate. The origin of the third $\Delta \Phi$ is not at $\mathbf{m}_2$, but rather at $\mathbf{m}_2^*$. This does not slow us down, since the next $\Delta \mathbf{m}$ is easily calculated:

$$\Delta \mathbf{m}_3 = \mathbf{m}_3 - \mathbf{m}_2^* = \begin{Bmatrix} -0.1667 \\ -0.20 \end{Bmatrix}$$

$$\Delta \mathbf{u}_3 = P2\Delta \mathbf{m}_3 = \begin{Bmatrix} -0.1397 \\ -0.1460 \\ 0.1079 \end{Bmatrix} \tag{7.24}$$

The final control configuration, $\mathbf{u}_3$, is

$$\mathbf{u}_3 = \Delta \mathbf{u}_1 + \Delta \mathbf{u}_2 + \Delta \mathbf{u}_3 = \begin{Bmatrix} -0.0063 \\ -0.0127 \\ -0.0254 \end{Bmatrix} \tag{7.25}$$

This is not the zero vector, but the moment (according to B_2) is zero:

$$B_2 \mathbf{u}_3 = \mathbf{0} \tag{7.26}$$

In other words, $\mathbf{u}_3$ is in the null space of B_2, $\mathcal{N}(B_2)$. Winding up in the null space was not a consequence of the path taken, but rather of the changing conditions along the way.

7.3 Global vs. Local Control Effectiveness

Control effectiveness (the B matrix) may be pre-calculated and stored in the flight control computer as functions of the current flight condition. In its most basic form this would result in gain scheduling, the lookup of feedback gains for the control law, which are implicitly calculated using the control effectiveness.

Alternatively, the flight control system may utilize an on-board model (or OBM) consisting of an extensive model of the airplane: at a minimum, aerodynamic forces and moments as

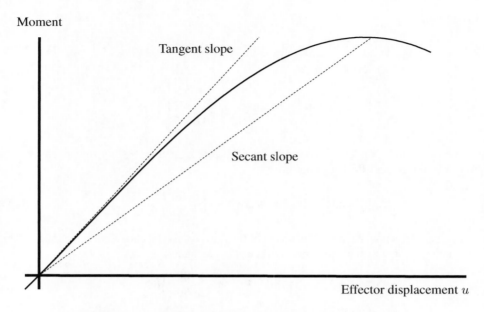

Figure 7.5   Possible interpretations of linear control effectiveness

functions of airplane state, control effector positions, and any other relevant variable. More detailed models may be incorporated, such as the control-actuator dynamics models.

With respect to the global allocation problem, there may be some uncertainty as to how to calculate the global control effectiveness matrix, B_{Global}. If the graph of moment versus deflection is linear for all control deflections, then the answer is easy: the derivative of moment with respect to control deflection is the constant slope of the graph. But it never is.

Consider Figure 7.5. The usual interpretation of control effectiveness determines the tangent slope at the origin. Alternatively, one may use what is sometimes called the secant slope, the slope of a line from the origin to a point on the graph. Shown in the figure is a secant slope to the maximum moment that can be generated by the control effector. Other possibilities are clearly available, but all will introduce errors at some deflection if used as linear approximations for control effectiveness, and the 'best' choice is left to the judgement of the control system designer. A secant slope to the moment at maximum effector deflection is not shown, and would not appear useful in any sense.

In the case of frame-wise allocation we may use the global slope for control effectiveness (B_{Global}), or calculate the slope at the current operating condition, or local slope (B_{Local}). The latter approach will require a detailed OBM, but has the great advantage of generally reducing the error introduced by non-linear control effectiveness. There are, however, pitfalls to be avoided when using local slopes.

Consider Figure 7.6. Three example cases are shown in which the control effector is, at the beginning of the frame, at u^I, u^{II}, and u^{III}. For u^I, if the frame-wise control limit Δu is small then very little error will be introduced using the tangent slope as the effectiveness. u^{II} shows a different problem: the effectiveness of the effector at this condition is zero, and the control allocator will not attempt to employ it. Worse yet is u^{III}: the control allocator will drive it to greater deflection if a decrease in moment is commanded.

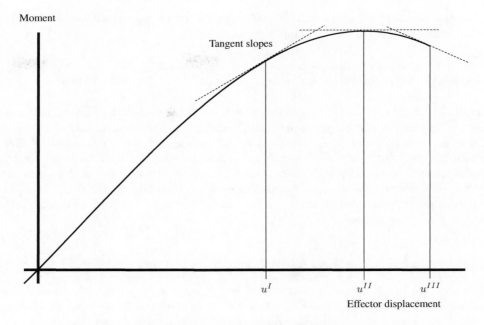

Figure 7.6   Local control effectiveness

It would appear better to establish an arbitrary upper global deflection limit somewhat short of u^{II}, since no good can come from operating at greater deflections. That is, anything that can be accomplished at deflections greater than u^{II} may be accomplished using deflections less that u^{II}, and without special handling.

All this suggests that some special treatment will be required, either in preparing the OBM or in utilizing software that detects when a change in the sign of the slope is about to occur.

In frame-wise control allocation the flight control computer is constantly calculating the change in moment to be used in the allocation in the next frame. The change in moment is that from the current moment being generated to the next moment required to satisfy the dynamics of the control law. This current moment and next moment are obviously with respect to some B_{Global}. The next frame's control allocation problem may be with respect to B_{Global}, but it makes more sense to use B_{Local}, since it will be a more accurate allocation.

All this is will play some part in the discussion of *restoring* (Section 7.4), in which we resolve the problem of the control effectors migrating toward deflections that, in general, will result in non-zero deflections for zero moment demands, as described in Section 7.2 and demonstrated in Section 7.2.1.

7.4   Restoring

Restoring refers to methods to be utilized in response to the path dependency described in Section 7.2. The control effector deflections have migrated to positions that, even though generating the desired moment (through the null space of B), are undesirable for other reasons (see the discussion regarding Figure 5.6).

Since we arrived here through the null space, we will return to desirable deflections through the null space. That is, we seek a vector $\Delta\mathbf{u}^\perp$ such that:

- $\Delta\mathbf{u}^\perp$ is in the null space of B, $\mathcal{N}(B)$
- adding $\Delta\mathbf{u}^\perp$ to the current $\Delta\mathbf{u}$ will move the solution toward a desired configuration.

Originally the idea behind restoring was to drive the effectors toward minimum deflection, zero when possible, while remaining in the null space of B. It was then realized that there may be more desirable goals than just zero deflection. One early success was to restore toward effector configurations that minimize the aerodynamic drag generated by the deflections, which is generally not the same as zero deflection. Other goals were hypothesized, such as minimizing hydraulic power requirements or radar cross-section (the latter is interesting, since the cross-section one wishes to minimize is a function of the threat axis). More generally, we speak of *preferred solutions*.

7.4.1   The Augmented B matrix

Many preferred solutions may be dealt with by adding the scalar function $y(\mathbf{u})$ to the existing moment vector $\mathbf{m}$. This increases the dimension of the problem by one. A two-moment problem becomes a three-objective problem, and a three-moment problem becomes a four-objective problem.

$$\mathbf{m}^* = \left\{ \begin{array}{c} \mathbf{m} \\ y(\mathbf{u}) \end{array} \right\} \tag{7.27}$$

The new function is treated the same as the moments in the original problem. A Taylor series expansion of $y(\mathbf{u})$ is the basis for a new row of B, creating the augmented matrix B^*.

$$B^* = \left[\begin{array}{ccccc} & & B & & \\ \frac{\partial y}{\partial u_1} & \frac{\partial y}{\partial u_2} & \cdots & \frac{\partial y}{\partial u_{n-1}} & \frac{\partial y}{\partial u_n} \end{array} \right] \tag{7.28}$$

For instance, if we are minimizing drag, the augmented B matrix will be

$$B^* = \left[\begin{array}{ccccc} & & B & & \\ \frac{\partial C_D}{\partial u_1} & \frac{\partial C_D}{\partial u_2} & \cdots & \frac{\partial C_D}{\partial u_{n-1}} & \frac{\partial C_D}{\partial u_n} \end{array} \right] \tag{7.29}$$

To minimize the 2-norm of the control vector, we define a related function

$$y = \sum_{i=0}^{m} \frac{1}{2} u_i^2 \tag{7.30}$$

The fraction $1/2$ is in anticipation of a factor 2 that will arise from the derivative of the square. Minimizing y will also minimize the 2-norm, and it is easier to work with. The augmented B matrix is

$$B^* = \left[\begin{array}{ccccc} & & B & & \\ u_1 & u_2 & \cdots & u_{n-1} & u_n \end{array} \right] \tag{7.31}$$

The new row in B^* are the values of the u_i after the step to attain $\Delta\mathbf{m}_{des}$ and before applying some component of $\Delta\mathbf{u}^\perp$ for restoring.

7.4.1.1 The Augmented Set of Attainable Moments

To help visualize the augmented set of attainable moments, let us revisit our 3×2 example using the minimum norm as our goal. Augmented, B^* becomes

$$B^* = \begin{bmatrix} 1 & 0 & -0.5 \\ 0 & 1 & -0.5 \\ u_1 & u_2 & u_3 \end{bmatrix} \tag{7.32}$$

Assume now that we have performed the frame-wise allocation shown in Figure 7.3a using B, but before we leave the frame we wish to minimize y. The augmented B matrix is

$$B^* = \begin{bmatrix} 1 & 0 & -0.5 \\ 0 & 1 & -0.5 \\ 0.1333 & -0.0667 & -0.1333 \end{bmatrix} \tag{7.33}$$

To view the attainable objective set $\Delta\Phi^*$ we rotate the figure $90°$ about Δm_1 so that we are looking at the plane of Δy–Δm_1, with Δy positive upward. Δm_2 will be perpendicular to the plane of Δy–Δm_1. The picture is drawn by removing the second row of B^* (corresponding to Δm_2) resulting in B^{**}. This has the effect of projecting the admissible controls Ω into the plane of Δy–Δm_1, resulting in a side-view of $\Delta\Phi^*$, named $\Delta\Phi^{**}$, shown in Figure 7.7.

$$B^{**} = \begin{bmatrix} 1 & 0 & -0.5 \\ 0.1333 & -0.0667 & -0.1333 \end{bmatrix} \tag{7.34}$$

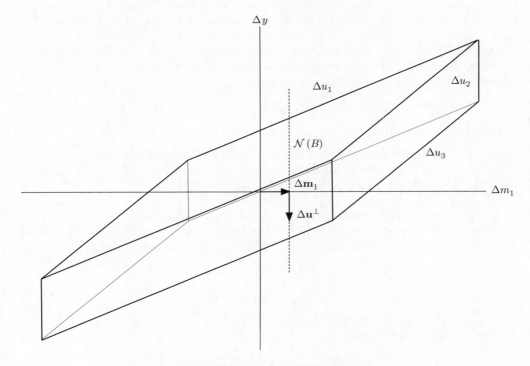

Figure 7.7 $\Delta\Phi^*$ viewed in the plane of Δy–Δm_1

The matrix B^{**} has significance only for constructing Figure 7.7, a way to view $\Delta\Phi^*$.

The line labeled Δm_1 is the edge-on view of Figure 7.3a. The vector with the solid arrowhead is $\Delta\mathbf{m}_1$, the same as in Figure 7.3a. The Δu_i symbols adjacent to three of the edges are the controls that define those edges; all other edges parallel to these three are defined by the same control.

Because we augmented B by adding a row, Figure 7.7 is a three-dimensional figure. All the lines in the figure not vectors or axes are edges, and those hidden from view (into the page, in the direction of positive Δm_2) are drawn as fine dotted lines.

The dashed line through the arrowhead of $\Delta\mathbf{m}_1$ yields admissible control deflections that vary y while still maintaining the $\Delta\mathbf{m}$ (Δm_1, Δm_2) required. It is labeled $\mathcal{N}$ (B), the one-dimensional (in this case) null space of the 2×3 B matrix used, and $\Delta\mathbf{u}^\perp$ must lie along it in the negative Δy direction; that is, down in the figure as shown.

Because this is now a three-dimensional problem, $\Delta\mathbf{u}^\perp_{Max}$ will not intersect an edge, but a two-dimensional facet. We will determine that the facet is defined by $\Delta\mathbf{u}_1$ and $\Delta\mathbf{u}_2$, specifically the facet $\mathbf{o}_{221}$ (recall the discussion on p. 39 for the meaning of $\mathbf{o}_{221}$). Thus we may pick any $\Delta\mathbf{u}^\perp$ and, with suitable scaling, decrease the drag while still solving the original problem.

Some observations on Figure 7.7 follow:

1. The allocation must be done in two steps. There is no *a priori* knowledge of any $\Delta\mathbf{m}^*$ associated with any point along the $\Delta\mathbf{u}^\perp$ shown.
2. We do not necessarily wish to attain minimum Δy in one frame. In the first place the solution may be obsolete in the next frame. Secondly there is no great need to do so, since all we wish is to continually tend toward the minimum 2-norm. Thirdly, taking smaller steps will allow us to use a cheaper allocator for this step, such as a pseudo-inverse solution suitably scaled. And finally, we wish to ameliorate *chattering*, to be discussed in Section 7.4.3.
3. In general if the change in desired moment $\Delta\mathbf{m}_{des}$ ($\Delta\mathbf{m}_1$ in Figure 7.7) is on the boundary $\partial(\Delta\Phi)$ there will be no possibility of restoring because solutions on the boundary are unique (barring degeneracy–see Section 5.2).

7.4.2 Implementation

In order to implement restoring, we first find the effector configuration that satisfies the $\Delta\mathbf{m}_{des}$ allocating as above using the B matrix, then move the origin of the problem to this point in moment space, and then allocate according a desired Δy using B^*.

1. Using B, solve the m-dimensional allocation problem for $\Delta\mathbf{m}_{des}$ to find $\Delta\mathbf{u}$.
2. If the solution is on the boundary of $\Delta\Phi$, stop. No restoring is possible. Otherwise, move the origin to $\Delta\mathbf{m}_{des}$ by changing the upper and lower deflection limits to $\Delta\mathbf{u}'_{Min}$ and $\Delta\mathbf{u}'_{Max}$.

$$\Delta\mathbf{u}'_{Min} = \Delta\mathbf{u}_{Min} - \Delta\mathbf{u}$$

$$\Delta\mathbf{u}'_{Max} = \Delta\mathbf{u}_{Max} - \Delta\mathbf{u} \tag{7.35}$$

3. Augment the B matrix as required (Section 7.4.1).
4. Select an augmented moment vector of the form

$$\Delta\mathbf{m}^* = \begin{Bmatrix} \mathbf{0} \\ a \end{Bmatrix} \ (a < 0) \tag{7.36}$$

The zeros in the first n (2 or 3 for our purposes) components of $\Delta\mathbf{m}^*$ in Eq. (7.36) ensure that $\Delta\mathbf{m}_{des}$ is unchanged, and the negative a indicates that we are minimizing y.

5. Using B^*, solve the $(m + 1)$-dimensional allocation problem for $\Delta\mathbf{m}^*$ to find any $\Delta\mathbf{u}^\perp$.
6. Scale $\Delta\mathbf{u}^\perp$ as necessary.
7. The final control deflection for this frame $\Delta\mathbf{u}'$ is

$$\Delta\mathbf{u}' = \Delta\mathbf{u} + \Delta\mathbf{u}^\perp \tag{7.37}$$

7.4.3 Chattering

Chattering is a phenomenon that can occur with some desired solutions if $\mathbf{u}^\perp$ is not suitably scaled. Consider, for example, the drag coefficient as a function of a moving control surface such as a single horizontal tail surface. There is usually some position at which the control-induced drag is a minimum, generally corresponding to what its free-float position (no hinge moment) would be. The drag then increases with deflections either side of this minimum. The graph of drag vs. deflection is U-shaped, as suggested in Figures 7.8 and 7.9.

A current operating point of u_i is shown in Figure 7.8. The dashed line tangent to the curve is the value of the appropriate entry in B_{Aug}, $\partial D/\partial u_i$. Point (a) represents the deflection following the addition of $\mathbf{u}^\perp$, and point (b) the position on the drag curve following that addition. In this case the drag is reduced substantially, although not by so much as the linear approximation would suggest.

A different operating point of u_i is shown in Figure 7.9. Points (a) and (b) have the same meaning as before. In this case the sign of u_i has changed, and the value of drag has actually increased.

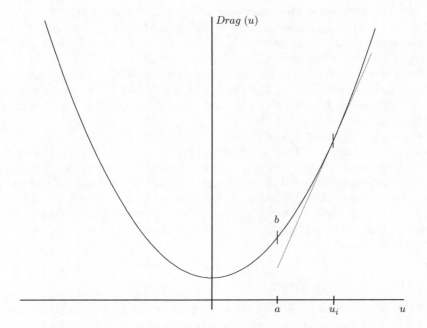

Figure 7.8 Typical drag versus control deflection with restoring

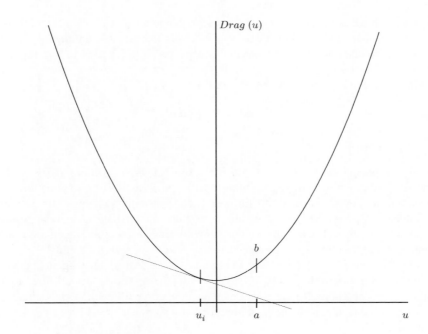

Figure 7.9   Figure 7.8 with different operating point

If we operate in the vicinity of zero deflection for a while, the effector will switch back and forth persistently, a condition called *chattering*. Unless means are taken to prevent it, chattering is an altogether common phenomenon. As we near the origin, we need to scale $\mathbf{u}^{\perp}$. Just how much scaling is required is difficult to determine analytically, and such questions will keep flight control engineers employed for the foreseeable future.

Chattering is an undesirable phenomenon. While it may be that the average effector deflection over time suits the control system's needs, chattering will create wear on the effector actuators, which should be avoided. Aesthetically, it is hard to find appeal in the thought of control effectors buzzing away at 100 Hz.

One way to deal with chattering is to place a low-pass filter in the path to the effectors. In fact, such a filter is normally present naturally, in the form of the dynamic response of the actuators driving the control effectors. Actuators are often modeled as first-order dynamic systems, which, in the their mathematical models, are indistinguishable from electrical low-pass filters. That is to say, while the frame-wise control allocator may command the effector to change direction and assume a new position away from the most recent direction of travel within the next one-hundredth of a second, the actuator–effector system dynamics will not be able to effect the abrupt change required.

7.4.4   Minimum-norm Restoring

Chattering is not a problem when restoring to a minimum-norm solution, as the scaling factor can be determined analytically. Restoring to the minimum-norm solution, or as near to it as possible, may be performed in one step. Any time the pseudo-inverse solution is admissible, minimum-norm restoring will generate that solution. This means, of course, that any time the moment demand is zero, the returned control vector will be zero. Path dependency is removed for any moments that can be achieved by the pseudo-inverse, including the origin.

7.4.4.1 Optimal Scaling for the 2-norm

We wish to determine K that minimizes the norm of $\Delta \mathbf{u}' = \Delta \mathbf{u} + K \Delta \mathbf{u}^\perp$. That is, we wish to minimize

$$J = \frac{1}{2} \Delta \mathbf{u}'^T \Delta \mathbf{u}' \tag{7.38}$$

For the minimum to occur,

$$\frac{\partial J}{\partial K} = \frac{\partial J}{\partial \Delta \mathbf{u}'} \frac{\partial \Delta \mathbf{u}'}{\partial K} = \Delta \mathbf{u}'^T \Delta \mathbf{u}^\perp = 0 \tag{7.39}$$

Substituting $\Delta \mathbf{u}'^T = \Delta \mathbf{u}^T + K \Delta \mathbf{u}^{\perp T}$,

$$\left[\Delta \mathbf{u}^T + K \Delta \mathbf{u}^{\perp T} \right] \Delta \mathbf{u}^\perp = 0 \tag{7.40}$$

The augmented B matrix for minimum-norm restoring was

$$B^* = \begin{bmatrix} B \\ u_1 \ u_2 \ \cdots \ u_{n-1} \ u_n \end{bmatrix} \tag{7.41}$$

This may be written as

$$B^* = \begin{bmatrix} B \\ \Delta \mathbf{u}^T \end{bmatrix} \tag{7.42}$$

We then chose $\Delta \mathbf{u}^\perp$ so that

$$B^* \Delta \mathbf{u}^\perp = \begin{Bmatrix} \mathbf{0} \\ a \end{Bmatrix} \ (a < 0) \tag{7.43}$$

Therefore

$$\Delta \mathbf{u}^T \Delta \mathbf{u}^\perp = a \tag{7.44}$$

The optimal scaling, K_{Opt}, is

$$K_{Opt} = \frac{-a}{\Delta \mathbf{u}^{\perp T} \Delta \mathbf{u}^\perp} \tag{7.45}$$

Example 7.3

Optimal Scaling for Minimum Norm Restoring To illustrate we continue with the simple three-control, two-moment example from Section 7.2.1. We will loosely follow the steps in Section 7.4.2. In the first frame, using direct allocation, we moved from $\mathbf{m}_0$ to $\mathbf{m}_1$.

$$\mathbf{m}_0 = \mathbf{0}, \mathbf{m}_1 = \begin{Bmatrix} 0.2 \\ 0 \end{Bmatrix} \tag{7.46}$$

$$\Delta\mathbf{u}_1 = \begin{Bmatrix} 0.1333 \\ -0.0667 \\ -0.1333 \end{Bmatrix} \tag{7.47}$$

We displace the origin by changing the upper and lower deflection limits using Eq. (7.35):

$$\Delta u_{1_{Min}}' = -0.5 - 0.1333 = -0.6333$$
$$\Delta u_{1_{Max}}' = 0.5 - 0.1333 = -0.3667$$
$$\Delta u_{2_{Min}}' = -0.5 + 0.0667 = -0.4333$$
$$\Delta u_{2_{Max}}' = 0.5 + 0.0667 = -0.5667$$
$$\Delta u_{3_{Min}}' = -0.5 + 0.1333 = -0.3667$$
$$\Delta u_{3_{Max}}' = 0.5 + 0.1333 = -0.6333 \tag{7.48}$$

The augmented B matrix B^* was given in Eq. (7.31) and the augmented moment vector $\mathbf{m}^*$ in Eq. (7.36). We arbitrarily select $a = -1$ for $\mathbf{m}^*$.

The direction of $\mathcal{N}(B)$ is found using any allocation method. All we need is a direction, so the pseudo-inverse may be used. In this case B^* is square, so we simply use its inverse.

$$\mathcal{N}(B) = B^{*-1} \begin{Bmatrix} 0 \\ 0 \\ -1 \end{Bmatrix} = \begin{Bmatrix} 5 \\ 5 \\ 10 \end{Bmatrix} \tag{7.49}$$

The restoring $\Delta\mathbf{u}^\perp$ will be of the form

$$\Delta\mathbf{u}^\perp = K \begin{Bmatrix} 5 \\ 5 \\ 10 \end{Bmatrix} \quad (K > 0) \tag{7.50}$$

The greatest restoring vector, $\Delta\mathbf{u}_{Max}^\perp$ is the greatest admissible vector in the direction of $\mathcal{N}(B)$ from the shifted origin, determined to occur at $K_{Max} = 0.06333$.

$$\Delta\mathbf{u}_{Max}^\perp = \begin{Bmatrix} 0.3166 \\ 0.3166 \\ 0.6333 \end{Bmatrix} \tag{7.51}$$

$$K_{Opt} = \frac{-a}{\Delta\mathbf{u}^{\perp T} \Delta\mathbf{u}^\perp} = \frac{1}{150} = 0.006667 \tag{7.52}$$

If K_{Opt} is greater than K_{Max} previously determined, K_{Max} must be used. In the current problem this is $K_{Opt} = 1/150 = 0.006667 < K_{Max}$, and the $\Delta\mathbf{u}'$ with the smallest norm is:

$$\Delta\mathbf{u}'_{Opt} = \begin{Bmatrix} 0.1666 \\ -0.0334 \\ -0.0666 \end{Bmatrix} \tag{7.53}$$

The norm of this vector is $||\Delta\mathbf{u}'_{Opt}||_2 = 0.1825$, compared to 0.20 with $K = 0$. Figure 7.10 illustrates the process.

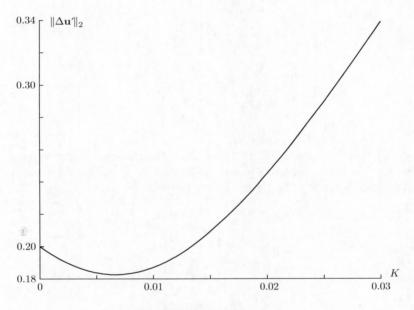

Figure 7.10 $||\Delta\mathbf{u}'||_2$ vs. K

Example 7.4

K_{Opt} versus K_{Max} and the Pseudo-inverse Solution The procedure outlined in the previous section will return exactly the same solution as the pseudo-inverse if that solution is admissible. If the pseudo-inverse solution is inadmissible the pseudo-inverse solution may be uniformly scaled to yield an admissible solution that preserves the direction of $\mathbf{m}_{des}$, but that is different from scaling $\mathbf{u}^\perp$ by the smaller of K_{Opt} and K_{Max}.

To emphasize this result and reinforce the minimum-norm restoring method, we consider a more complicated problem. In Section 5.2 we used a three-moment, seven-control example, reproduced below. We will not bother with the frame-wise problem since the mathematics involved are the same for the global problem.

$$B = \begin{bmatrix} 0.7073 & -0.7073 & -3.4956 & -3.0013 & 3.0013 & 3.4956 & 2.1103 \\ 1.1204 & 1.1204 & -0.7919 & -1.2614 & -1.2614 & -0.7919 & 0.0035 \\ -0.3309 & 0.3309 & -0.1507 & -0.3088 & 0.3088 & 0.1507 & -1.2680 \end{bmatrix} \quad (7.54)$$

$$\mathbf{u}_{Min} = \begin{Bmatrix} -0.9599 \\ -0.9599 \\ -0.5236 \\ -0.5236 \\ -0.5236 \\ -0.5236 \\ -0.5236 \end{Bmatrix}_u \qquad \mathbf{u}_{Max} = \begin{Bmatrix} 0.4363 \\ 0.4363 \\ 0.5236 \\ 0.5236 \\ 0.5236 \\ 0.5236 \\ 0.5236 \end{Bmatrix}_u \qquad (7.55)$$

We take

$$\mathbf{m}_{des} = \begin{Bmatrix} -1.8832 \\ 1.4055 \\ 1.0133 \end{Bmatrix} \qquad (7.56)$$

This vector is 90% of $\mathbf{m}_{0100000}$ in Eq. (5.50), attainable using direct allocation but not by the pseudo-inverse. The direct-allocation solution is 90% of $\mathbf{u}_{0100000}$ in Eq. (5.49) so we can get straight to restoring without getting bogged down in allocation.

$$\mathbf{u}_{DA} = \begin{Bmatrix} -0.8639 \\ 0.3927 \\ -0.4712 \\ -0.4712 \\ -0.4712 \\ -0.4712 \\ -0.4712 \end{Bmatrix} \qquad (7.57)$$

The augmented B matrix is

$$B^* = \begin{bmatrix} 0.7073 & -0.7073 & -3.4956 & -3.0013 & 3.0013 & 3.4956 & 2.1103 \\ 1.1204 & 1.1204 & -0.7919 & -1.2614 & -1.2614 & -0.7919 & 0.0035 \\ -0.3309 & 0.3309 & -0.1507 & -0.3088 & 0.3088 & 0.1507 & -1.2680 \\ -0.8639 & 0.3927 & -0.4712 & -0.4712 & -0.4712 & -0.4712 & -0.4712 \end{bmatrix} \quad (7.58)$$

Arbitrarily selecting $a = -2$ and using P^*, the pseudo-inverse of B^*,

$$\mathbf{u}^\perp = P^* \begin{Bmatrix} \mathbf{0} \\ -2 \end{Bmatrix} = \begin{Bmatrix} 1.5545 \\ 0.0375 \\ 0.6303 \\ 0.3026 \\ 0.4391 \\ 0.4389 \\ -0.3854 \end{Bmatrix} \qquad (7.59)$$

The optimal scaling factor is

$$K_{Opt} = \frac{-a}{\mathbf{u}^{\perp T}\mathbf{u}^{\perp}} = 0.5813 \tag{7.60}$$

The pseudo-inverse solution is $\mathbf{u}_{Pseudo} = \mathbf{u}_{DA} + K_{Opt}\mathbf{u}^{\perp}$

$$\mathbf{u}_{Pseudo} = \begin{Bmatrix} 0.0397 \\ 0.4145 \\ -0.1048 \\ -0.2953 \\ -0.2160 \\ -0.2161 \\ -0.6952 \end{Bmatrix} \tag{7.61}$$

Comparing this with $\mathbf{u}_{Min}$ and $\mathbf{u}_{Max}$ we see that u_7 is commanded past its minimum displacement. Because use of the optimal restoring vector would result in an inadmissible control, we cannot use K_{Opt} but must use K_{Max} instead. The displaced origin is

$$\mathbf{u}'_{Min} = \mathbf{u}_{Min} - \mathbf{u} = \begin{Bmatrix} -0.0960 \\ -1.3526 \\ -0.0524 \\ -0.0524 \\ -0.0524 \\ -0.0524 \\ -0.0524 \end{Bmatrix} \tag{7.62}$$

$$\mathbf{u}'_{Max} = \mathbf{u}_{Max} - \mathbf{u} = \begin{Bmatrix} 1.3002 \\ 0.0436 \\ 0.9948 \\ 0.9948 \\ 0.9948 \\ 0.9948 \\ 0.9948 \end{Bmatrix} \tag{7.63}$$

Comparing $K\mathbf{u}^{\perp}$ to $\mathbf{u}'_{Min}$ and $\mathbf{u}'_{Max}$, we determine that $K_{Max} = 0.1359$ based on u_7.

$$\mathbf{u}_{DA,scaled} = \mathbf{u}_{DA} + K_{Max}\mathbf{u}^{\perp} = \begin{Bmatrix} -0.6527 \\ 0.3978 \\ -0.3856 \\ -0.4301 \\ -0.4116 \\ -0.4116 \\ -0.5236 \end{Bmatrix} \tag{7.64}$$

If we were to use the pseudo-inverse solution we would need to scale the vector so that all elements were admissible.

$$
\mathbf{u}_{Pseudo,scaled} = \left\{ \begin{array}{r} 0.0299 \\ 0.3121 \\ -0.0790 \\ -0.2224 \\ -0.1627 \\ -0.1628 \\ -0.5236 \end{array} \right\} \tag{7.65}
$$

The moment that results from using the scaled pseudo-inverse is, as expected, well short of that desired.

$$
B\mathbf{u}_{Pseudo,scaled} = \left\{ \begin{array}{r} -1.4183 \\ 1.0585 \\ 0.7631 \end{array} \right\} \tag{7.66}
$$

8

Control Allocation and Flight Control System Design

This chapter will illustrate how control allocation fits into a typical flight control system. Dynamic inversion has been chosen as a representative flight control methodology. There are many forms dynamic-inversion control laws may take. One example of a simple dynamic-inversion control law has been created using MATLAB®/Simulink® and is available from the companion website for this book. The files are not required, but are provided should readers wish to further their understanding of this material. The examples in this chapter use a simple linear model for the aircraft.

8.1 Dynamic-inversion Desired Accelerations

Dynamic inversion was introduced in Chapter 3. For the case of a single acceleration controlled by a single control effector we derived Eq. (3.4), repeated here as Eq. (8.1),

$$u^*(t) = \frac{1}{b}(\dot{x}_{des} - f(x(t))) \tag{8.1}$$

There we noted that if we assumed that our approximations and measurements are perfect the result was Eq. (8.2):

$$\dot{x} = \dot{x}_{des} \tag{8.2}$$

8.1.1 The Desired Acceleration: $\dot{x}_{des}$

Figure 8.1 shows the top-level block diagram of the Simulink®file[1] described in Appendix B. Figure 8.1 is a pictorial representation of Eq. (8.1). The process of generating $\dot{x}_{des}$ is broken into two steps. The first step is performed in the 'COMMANDS' block of Figure 8.1. The pilot input, which may be in terms of a sensed force or displacement of the inceptor, is converted to a commanded response, such as roll rate. The second step determines the desired acceleration using the commanded state and sensed values. This is performed in the 'REGULATORS'

---

[1] This particular model is 'Linear_ADMIRE_sim.mdl'.

Aircraft Control Allocation, First Edition. Wayne Durham, Kenneth A. Bordignon and Roger Beck.
© 2017 John Wiley & Sons, Ltd. Published 2017 by John Wiley & Sons, Ltd.
Companion website: www.wiley.com/go/durham/aircraft_control_allocation

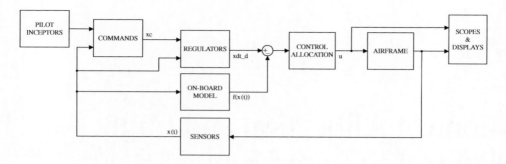

Figure 8.1   Top-level diagram of Eq. (8.1)

block. The 'ON-BOARD MODEL' block provides an estimate of the nominal aircraft accelerations. The nominal accelerations are the part of the total accelerations that are not a function
of the controls to be allocated, the $f(x(t))$ in Eq. (8.1). For the case of a single acceleration with
a single control effector, the 'CONTROL ALLOCATION' block is simply $1/b$.

For the linearized example provided, the airframe dynamics are modeled using the standard
state-space representation:

$$\dot{\mathbf{x}} = A\mathbf{x} + B\mathbf{u} \tag{8.3}$$

These equations are implemented in the 'AIRFRAME' block. This block is expanded and
shown in Figure 8.2.

If the assumptions used to generate Eq. (8.2) are valid – that is, if it is reasonable to take
$\dot{x} = \dot{x}_{des}$ – then Figure 8.2 can be simplified to become Figure 8.3.

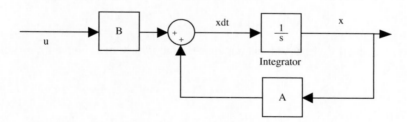

Figure 8.2   State-space representation of the 'AIRFRAME' block in Figure 8.1

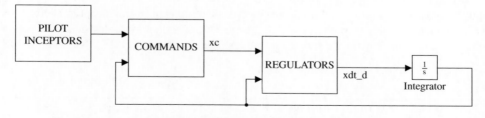

Figure 8.3   Simplified Figure 8.1

With this simplification, the command and regulator parts of the control law can be designed to effect specific transfer functions. That is, by assuming the dynamic inversion and control allocation will function perfectly, the structure of Figure 8.3 may be used for the command and regulator design. Three examples of such designs follow.

8.1.2 Command and Regulator Examples

Example 8.1

Roll-mode Time Constant Usually, the lateral stick is used to command roll rate, p. Pilots expect a fast first-order type of response, and flying qualities documents (for example, Department of Defense (1990)), contain recommended ranges for the roll-mode time constant, τ_R. This can be accomplished using a simple proportional controller, as shown in Figure 8.4. This system can be represented by the transfer function:

$$\frac{p}{p_c} = \frac{K_p}{s + K_p} = \frac{1}{\tau_R s + 1} \tag{8.4}$$

From Eq. (8.4) it can be seen that a desired time constant may be obtained by selecting $K_p = 1/\tau_R$.

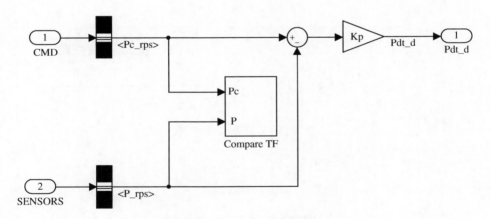

Figure 8.4 Simple proportional controller

A block is included in the simulation that calculates the output of the transfer function of Eq. (8.4) and compares it with the simulation roll rate. See Figure 8.5 for details.

If the assumptions used to simplify the design process are valid, then the simulation roll rate, p_{Sim}, should be identical, or nearly identical, to the output of the transfer function, p_{tf}. Results of running the simulation show the two signals are very close, but are not identical. Figure 8.6 shows the results of the simulation selecting K_p to give a time constant of 0.5 s. The lines for the transfer function response and simulation response are difficult to distinguish at the scale presented. To resolve this distinction, the lower plot shows the difference between the two values.

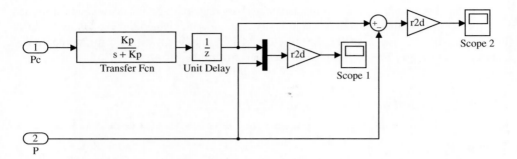

Figure 8.5   Transfer function comparison block

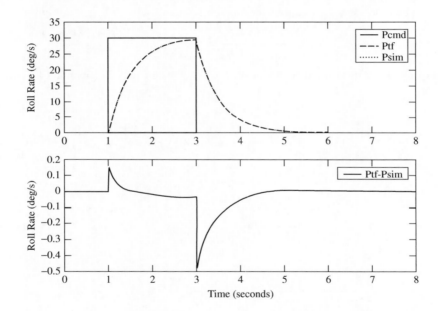

Figure 8.6   Roll rate response

In practice, one will never achieve an identical matching as it is generally impossible for the flight control computer to know the actual aircraft states and nominal accelerations to infinite precision. Moreover, other factors such as actuator and sensor dynamics can impact the response. Even in this simple linear model, the presence of a time delay of one step, 0.01 s, in the 'SENSORS' block of the feedback path is enough to cause minor variations. The gains found from the method described here are a good initial starting point. As a design matures and more information is known about the system, regulator gains are varied to achieve the desired response. This process was documented in a paper describing the dynamic-inversion-based control laws used on the X-35 program:

However, a complication that often occurs in low-order direct synthesis techniques is the impact of higher-order dynamics. To compensate for these effects an offline optimization procedure was wrapped around the dynamic-inversion control law to adjust the regulator gains to achieve the desired flying qualities metrics. (Walker and Allen 2002)

Example 8.2

Pitch-rate Command If a second-order response is desired, a different architecture can be chosen for the regulator. For an aircraft's short-period response, characterized primarily by pitch rate, flying qualities specifications typically recommend a second-order response. Consider a proportional-integral-forward path scheme, shown in Figure 8.7, to control pitch rate, q.

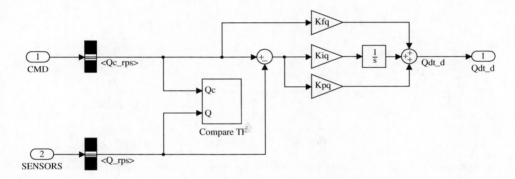

Figure 8.7 Proportional-integral-forward path controller

Performing block diagram algebra, Figure 8.7 reduces to the transfer function:

$$\frac{q}{q_c} = \frac{(K_{fq} + K_{pq})s + K_{iq}}{s^2 + K_{pq}s + K_{iq}} \tag{8.5}$$

A desired response can be expressed in terms of a generic second-order transfer function:

$$G(s) = \frac{\omega_n^2(T_{\theta_2}s + 1)}{s^2 + 2\zeta\omega_n s + \omega_n^2} \tag{8.6}$$

The gains in the regulator can be expressed in terms of the desired response:

$$K_{pq} = 2\zeta\omega_n$$

$$K_{iq} = \omega_n^2$$

$$K_{fq} = T_{\theta_2}\omega_n^2 - 2\zeta\omega_n \tag{8.7}$$

Note that setting the desired T_{θ_2} to zero eliminates the zero in the transfer function. Again, in the simulation, a block is included to compare the results of the desired transfer function with the aircraft response. Figure 8.8 shows the results of the simulation using a desired damping ratio $\zeta = 0.85$, a natural frequency $\omega_n = 2$ rad/s and $T_{\theta_2} = 0$.

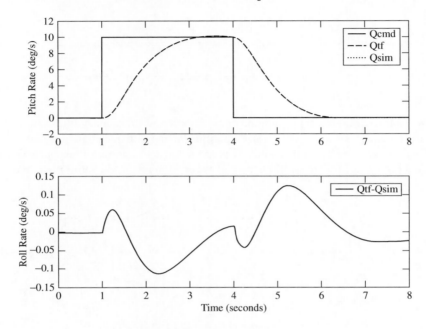

Figure 8.8   Pitch rate response

Example 8.3

Sideslip Controller The method of designing controllers described in this section is not limited to rate command systems. Often, the rudder pedals are used to command sideslip angle. Figure 8.9 shows an architecture to generate desired yaw acceleration, $\dot{r}_{des}$, to control sideslip β. This approach combines proportional control on yaw rate, r, in the regulator block with a proportional-integral-forward path control that generates the yaw-rate command.

Again, performing block diagram algebra, Figure 8.9 reduces to the transfer function:

$$\frac{\beta}{\beta_c} = \frac{K_r(K_{fb} + K_{pb})s + K_r K_{ib}}{s^3 + K_r s^2 + K_r K_{pb} s + K_r K_{ib}}$$

$$= \frac{\omega_n^2 \omega_2 (T_{\theta_2} s + 1)}{(s^2 + 2\zeta \omega_n s + \omega_n^2)(s + \omega_2)} \tag{8.8}$$

If a desired response is known in terms of ζ, ω_n, ω_2, and T_{θ_2}, one can solve for the control law gains that will produce this response. Note this scheme uses the approximation $\dot{\beta} \approx -r$.

This is an oversimplification that results in a noticeable difference between the desired transfer function and the simulation response, as seen in Figure 8.10.

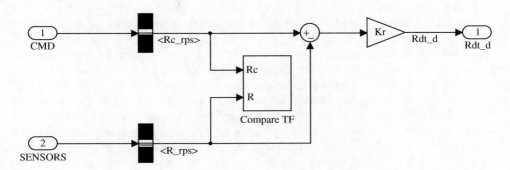

Figure 8.9   Sideslip controller, regulator

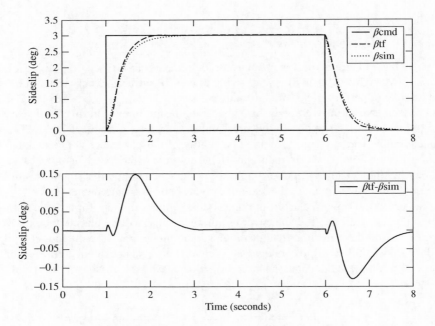

Figure 8.10   Sideslip response

The benefit of using a dynamic-inversion control law is that one can use simple feedback control ideas to determine the command and regulator gains that will produce a desired response. Typically, simplifying assumptions are used early in the design process and then the gains are tuned later on to make the higher-order system behave like a desired lower-order equivalent system.

8.2   The Maximum Set and Control Law Design

In Chapter 5 methods were presented for calculating the AMS, the maximum attainable value in a given direction, and the total volume for the three-dimensional set. This section will detail why this set is important and how it relates to the dynamics and control of the aircraft in general.

The AMS represents the acceleration capability of the aircraft. In Section 8.1 methods were presented for determining the gains of a dynamic-inversion control law that will generate a desired dynamic response. One assumption used in these methods was that the control allocation part of the control law provided the necessary accelerations. This method of calculating gains can be used to design a control system with any dynamic response the designer desires. However, in practice, the range of dynamic responses is limited to those whose required accelerations lie within the AMS. More specifically, the desired accelerations must lie within the subset of the AMS for which a specific method of control allocation will find admissible controls. For this reason, the authors believe the ability of a control allocation method to find admissible controls for every attainable moment is the standard to be used for comparing different methods of control allocation.

8.2.1   In the Design Process

In the early stages of aircraft design, maneuver requirements can be used to specify a minimum size for an AMS. This helps in the initial sizing of control surfaces to ensure that the airplane will be capable of performing desired maneuvers.

Typically, control surfaces are sized by considering the minimum control effectiveness flight regimes. Most aircraft rely upon aerodynamic control surfaces to generate the desired moments for maneuvering. The moment-generating capability of these surfaces is proportional to the dynamic pressure, $\overline{q} = \rho V^2/2$. To reduce available control power, low dynamic pressure flight conditions are often examined. For example, a low-speed condition that might be similar to a landing configuration is often used to examine control power requirements.

Before deciding if a particular set of control effectors is sufficient for a particular design, it is important to determine the maneuver requirements that will be placed upon the vehicle. There are a multitude of methods and ideologies for coming up with such requirements. As it is not the purpose of this text to catalogue all such methods and debate their efficacy, only one method will be discussed in detail. More information is available in Shaw (1985) and Wilson *et al.* (1993).

Example 8.4

Roll Requirements for a Military Aircraft The simulation used as examples for this text is the ADMIRE vehicle, which is a single-engine canard/delta-wing fighter (ADMIRE 2003). If such a vehicle were being designed in response to a proposal by the US government, the designers would be guided by a detailed document that specifies certain performance standards the aircraft is required to meet. This detailed specification would be based upon the guidance of some standard, such as MIL-STD-1797A (Department of Defense 1990). The guidance of 1797A was based on earlier flying qualities requirements such as MIL-F-8785C (Department of Defense 1980). Lacking a detailed specification for the airplane represented by the ADMIRE simulation, we will assume for purposes of discussion that the requirements of 8785C apply to it.

The requirements of 8785C divide airplanes into classes, based on size and maneuverability. The ADMIRE vehicle would be classified as a Class IV airplane, a highly maneuverable fighter. The maximum desired roll-mode time constant, τ_R, for a Class IV airplane is one second:

$$\tau_R \leq 1 \text{ s} \tag{8.9}$$

Recall Eq. (8.4), which related the roll-mode time constant to the desired roll-rate transfer function:

$$\frac{p}{p_c} = \frac{K_p}{s + K_p} = \frac{1}{\tau_R s + 1} \tag{8.10}$$

However, this information alone is insufficient to determine a desired acceleration capability. For a first-order step response, the maximum acceleration will occur at the time of the step input and will be equal to the magnitude of the step divided by the time constant.

$$\dot{p}_{Max} = \frac{p_c}{\tau_R} \tag{8.11}$$

Figure 8.11 shows an example response for the above transfer function for a unit step input and a time constant $\tau_R = 0.5$s.

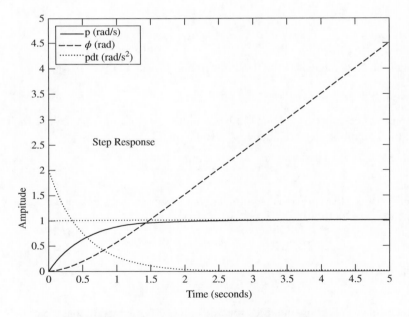

Figure 8.11   Example of a first order roll rate step response

To determine the greatest lateral acceleration required, it is necessary to choose the maximum size of step input for which the system will produce the desired dynamic response. 8785C specifies the maximum allowable time to achieve a change in bank angle in response to a roll input for three different categories of maneuvers, in four different speed ranges from very low (VL), to low (L), to medium (M), and to high (H). The definitions of these speed ranges are specific to each airplane. The data for the ADMIRE simulation used here (Mach 0.22, near sea level) will be assumed to be very low (VL), and the following analyses are based on that assumption.

For initial sizing purposes, the more restrictive requirement is typically selected so that the requirements will be satisfied for all categories. The same analysis would be completed for all the other speed ranges. The more restrictive requirement is identified as that which requires the greatest initial roll acceleration $\dot{p}_{Max} = p_c/\tau_R$.

We choose the input to be a step, $1/s$, of magnitude $p_c(s) = p_c/s$. The transfer function is Eq. (8.10), so

$$p(s) = \left(\frac{p_c}{s}\right)\left(\frac{1}{\tau_R s + 1}\right) \tag{8.12}$$

The inverse LaPlace transform of Eq. (8.12) yields

$$p(t) = p_c(1 - e^{-t/\tau_R}) \tag{8.13}$$

We proceed by assuming the bank angle ϕ is the integral of the roll rate p, and that the roll is initiated from $\phi = 0$. The specification gives values for $\Delta\phi$ and the greatest time in which the maneuver can be performed, T_1. Therefore,

$$\Delta\phi = \phi(T_1) - \phi(0)$$
$$= \int_0^{T_1} p\,dt = p_c T_1 + p_c \tau_R e^{(-T_1/\tau_R)} + C_1 \tag{8.14}$$

Since the initial bank angle, $\phi(0) = 0$ the integration constant $C_1 = -p_c\tau_R$.

$$\Delta\phi = p_c T_1 - p_c \tau_R(1 - e^{-T_1/\tau_R}) \tag{8.15}$$

For a given time constant, τ_R, change in bank angle, $\Delta\phi$, and time to achieve the change in bank, T_1, the size of the step command, p_c, can be found:

$$p_c = \frac{\Delta\phi}{T_1 - \tau_R(1 - e^{-T_1/\tau_R})} \tag{8.16}$$

Applying Eq. (8.16) to the various requirements of 8785C shows the most demanding requirement to be that for $\Delta\phi = 90°$ and $T_1 = 2.0s$. This yields

$$p_c = 79.3 \text{ deg/s} = 1.38 \text{ rad/s} \tag{8.17}$$

Using Eq. (8.11),

$$\dot{p}_{Max} = 79.3 \text{ deg/s}^2 = 1.38 \text{ rad/s}^2 \tag{8.18}$$

For those familiar with flying qualities specifications, it is interesting that the critical requirement was found to be for Category B flight phases. Category B is in general the least demanding, consisting of climb, cruise, and so on. The other two flight phases are A, which includes air combat maneuvering, and C, involving take-off and landings, including shipboard operations. This curious result is probably due to the VL (very low) speed assumed for the flight condition.

Figure 8.12 shows the response that satisfies the MIL-F-8785C time to bank requirements with $\tau_R = 1.0$s.

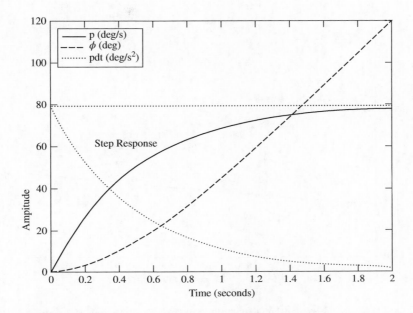

Figure 8.12   Step response meeting time to achieve bank-angle requirements

In practice, the acceleration found in this manner would only be a necessary but not sufficient condition. This simple analysis has neglected several factors, such as actuator dynamics and modeling errors, which could negatively impact the airplane's ability to achieve the desired performance. Having less than the acceleration capability found in this manner would indicate a problem with the design, but having more than this amount does not guarantee a successful design.

One way to increase this initial estimate to account for uncertainties is to change the time constant. Note that the desired time constant was an inequality, $\tau_R \leq 1$s. If the specified time constant is reduced, the acceleration requirement is increased. This will be illustrated by reworking the previous example using a desired time constant of 0.2 s.

Again using Eq. (8.16), now with $\tau_R = 0.2$s, the same flight category (B) is found to be the most demanding. With $\Delta\phi = 90°$ and $T_1 = 2.0$s

$$p_c = 50.0 \text{ deg/s} = 0.87 \text{ rad/s} \tag{8.19}$$

Again using Eq. (8.11),

$$\dot{p}_{Max} = 250 \text{ deg/s}^2 = 4.36 \text{ rad/s}^2 \tag{8.20}$$

Although the step size, p_c, is reduced, the desired acceleration has significantly increased. Figure 8.13 shows the relationship between the required step size and required acceleration, assuming the desired change in bank angle and the time to change remain constant. For this case, as the time constant goes to zero, the step size approaches 45 deg/s and the acceleration goes to infinity.

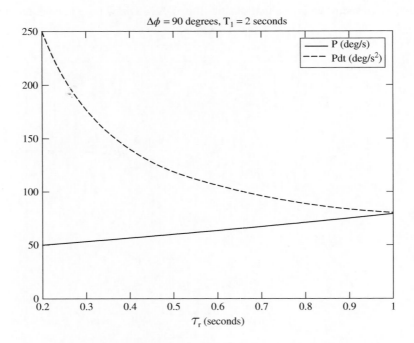

Figure 8.13   Variation in required acceleration

When doing initial sizing estimates, reducing the desired time constant can be used to insure that the vehicle will be able to perform required maneuvers. In the early stages of the design process it is important to make sure there is more than enough acceleration capability so that one can account for uncertainties in the initial estimates and possible failure scenarios.

This example showed only one axis for a single-maneuver requirement. In practice, one would want to generate acceleration requirements for multiple maneuvers, exciting all three axes. The AMS would be calculated for a proposed control effector suite and then checked to insure that all the desired accelerations are contained within the AMS.

8.2.2   In a Mature Design

Once a design is mature, with an established control effector suite and relatively well known control effectiveness data, the control laws can be customized to make use of the available acceleration capabilities. Consider the AMS used in the earlier example shown in Figure 5.11. The maximum roll acceleration from Table 5.5 is ± 7.74 rad/s^2. This is well in excess of the minimum requirement of 1.38 rad/s^2 calculated for a bank angle change of 90 deg in 2s. Recall that this requirement was calculated using a time constant $\tau_R = 1$s and a step size of $p_c = 79.3$ deg/s. With more roll acceleration capability, the performance of the aircraft can be improved by increasing the allowable step size, decreasing the time constant, or some combination of both. The ADMIRE simulation uses a simple mapping of lateral stick force to roll rate command. At the limits, 80 N of force maps to a command of $p_c = 180$ deg/s $= \pi$rad/s. To size the time constant such that the largest possible step input still results in the desired dynamics:

$$\dot{p}_{Max} = \frac{p_c}{\tau_R} \Rightarrow \tau_R = \frac{p_c}{\dot{p}_{Max}} = \frac{\pi \text{ rad/s}}{7.74 \text{ rad/s}^2} = 0.406 \text{ s} \tag{8.21}$$

If we run the linear simulation performing this maneuver, we see that the commanded accelerations lie outside the AMS. This is due to the fact that the aircraft is also trying to control sideslip, and the required yaw accelerations in combination with the roll accelerations are outside of the AMS, as shown in Figures 8.14 and 8.15. The desired accelerations are shown as densely packed vectors from the origin.

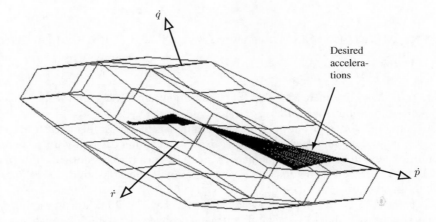

Figure 8.14 AMS and desired accelerations for $p_c = \pi$ rad/s and $\tau_R = 0.406$s. Viewed from $(\dot{p}\ \dot{q}\ \dot{r}) = (1\ 1\ 1)$

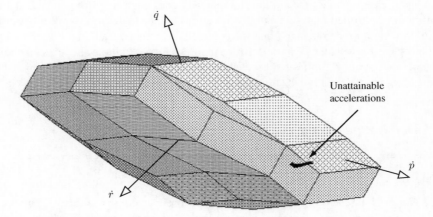

Figure 8.15 AMS and desired accelerations for $p_c = \pi$ rad/s and $\tau_R = 0.406$s. The desired accelerations are attainable except as shown. Viewed from $(\dot{p}\ \dot{q}\ \dot{r}) = (1\ 1\ 1)$

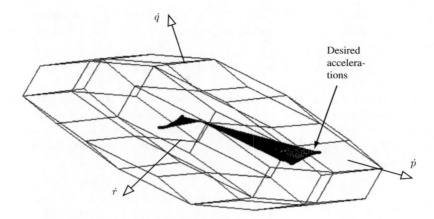

Figure 8.16 AMS and desired accelerations for $p_c = \pi$ rad/s and $\tau_R = 0.5$s. Viewed from $(\dot{p}\ \dot{q}\ \dot{r}) = (1\ 1\ 1)$

Figure 8.16 shows that increasing the desired time constant to 0.5 s allows all the required accelerations to be within the AMS.

The important lesson to take away from this example is that most real aircraft maneuvers will involve multiple axes. Checking the maximum acceleration capability for a single axis is often insufficient to guarantee that a maneuver can be performed as desired. If the desired dynamics were altered for either the roll or yaw axis, it will affect the ability of the aircraft to perform this maneuver as specified.

At this point, it is reasonable to ask what happens if the control law asks for accelerations that are beyond the capability of the airplane (that is, outside the AMS). The response varies depending upon several factors including: how the control allocation algorithm deals with excessive acceleration commands, the flight regime, and the behavior of the airplane. In a best-case scenario, the airplane only flies slightly differently than the specified desired dynamics. In a worst-case scenario, the airplane temporarily goes open-loop when it is unstable and it departs from controlled flight, potentially disastrously. Due to the severity of the worst-case scenario, it is highly recommended that the control law be tuned such that it will not ask the control inceptors for what they cannot deliver. As shown in the example above, good choices in the control law can allow for desired accelerations that use the full capabilities of the airplane while avoiding the potentially hazardous excessive command condition.

8.2.3 Non-optimal Example

Example 8.5

Ganged Controls In Section 6.4 we saw one approach to control effector ganging. That example is continued here, and the effect on the maneuver capabilities of the airplane is shown. To do this we will need the generalized inverse matrix (P) that corresponds to the ganging.

Recall that the control effectors for the ADMIRE vehicle were combined in a conventional pre-ganged arrangement with the following three pseudo-controls:

- Roll effector: outboard elevons only commanded asymmetrically.
- Pitch effector: canards only commanded symmetrically.
- Yaw effector: rudder.

In this scenario, the inboard elevons are deployed symmetrically as a trailing-edge flap device used to improve lift and are scheduled as a function of Mach and angle of attack. As a result, they are not actively used to generate moments.

The effect of ganging can be represented as an invertible 3×3 matrix, as seen in Eq. (6.1). Alternatively we may use a generalized-inverse formulation. Recall Eq. (6.6) for calculating a generalized inverse, $P = N[BN]^{-1}$.

To reflect the ganging, the matrix N becomes

$$N = \begin{bmatrix} 0 & 1 & 0 \\ 0 & 1 & 0 \\ 1 & 0 & 0 \\ 0 & 0 & 0 \\ 0 & 0 & 0 \\ -1 & 0 & 0 \\ 0 & 0 & 1 \end{bmatrix} \tag{8.22}$$

The rows of N represent the seven actual control effectors. The columns represent the three pseudo-controls. The third and sixth rows represent the outboard elevons, which are used asymmetrically to define the first pseudo-control, the roll effector. The first and second rows correspond to the canards, which are summed to become the second pseudo-control, the pitch effector. The seventh row is the rudder, which is also the third pseudo-control, the yaw effector. The fourth and fifth rows are the inboard elevons, which will not be commanded by the control allocation routine and are thus zeroes. The control effectiveness matrix for this example is the same as Eq. (5.43), repeated here for convenience.

$$B = \begin{bmatrix} 0.7073 & -0.7073 & -3.4956 & -3.0013 & 3.0013 & 3.4956 & 2.1103 \\ 1.1204 & 1.1204 & -0.7919 & -1.2614 & -1.2614 & -0.7919 & 0.0035 \\ -0.3309 & 0.3309 & -0.1507 & -0.3088 & 0.3088 & 0.1507 & -1.2680 \end{bmatrix} \tag{8.23}$$

Applying $P = N[BN]^{-1}$,

$$P = \begin{bmatrix} -0.0 & 0.4463 & 0.0011 \\ -0.0 & 0.4463 & 0.0011 \\ -0.1335 & 0 & -0.2221 \\ 0 & 0 & 0 \\ 0 & 0 & 0 \\ 0.1335 & 0 & 0.2221 \\ 0.0317 & 0 & -0.7358 \end{bmatrix} \tag{8.24}$$

However, the cost of this simplification is a loss of acceleration capability. Figure 8.17 shows the AMS for the pre-ganged set of controls inside the original AMS.

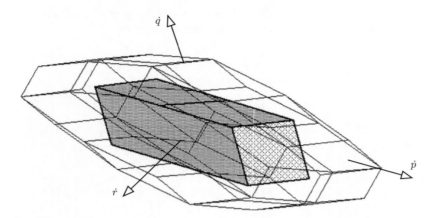

Figure 8.17 AMS for three ganged pseudo-controls inside the original AMS. Viewed from $(\dot{p}\ \dot{q}\ \dot{r}) = (1\ 1\ 1)$

The AMS for the pre-ganged set of controls is significantly smaller than Figure 5.11, 32.6 $(rad/s^2)^3$ compared to 177.1 $(rad/s^2)^3$, less than 19% of the original volume. Figure 8.18 shows the acceleration requirements for a 180 deg/s roll input with a time constant of 0.5s superimposed on the new AMS.

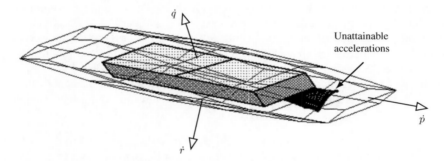

Figure 8.18 AMS for three ganged pseudo-controls with accelerations for $p_c = 180$ deg/s and $\tau_R = 0.5$s. View from $(\dot{p}\ \dot{q}\ \dot{r}) = (1\ 5\ 1)$

Clearly both the roll and yaw accelerations exceed the capabilities of the ganged AMS. To reduce the required accelerations, the maximum step command could be decreased, the time constant could be increased, or some combination of both of these actions could be taken. It was found that the desired accelerations fit to the new AMS only by reducing the magnitude of the roll-rate command to 94.4 deg/s. This represents a loss of over 47% of the original rate command. Alternatively, the desired accelerations fit to the new AMS by increasing the time constant to its maximum allowable value of 1 s and reducing the magnitude of the roll rate command to 110.5 deg/s. This represents a loss of over 38% of the original rate command, with the additional penalty of a response with a time constant that is doubled.

This example illustrates some of the motivation for researching the control allocation problem. As can be seen, one of the traditional methods of solving the control allocation problem, the pre-ganged solution, sacrifices a significant amount of performance.

It should be noted that even the smaller AMS of the ganged effector suite is capable of satisfying the time to bank requirements used in Example 9.4. This brings up another method of doing the initial sizing of the control effector suite. If the ganged effector suite is capable of generating the required accelerations of early design iterations, it is likely that the increase in the AMS due to using the individual effectors will offset any acceleration losses due to unknown or not-modeled effects. When the design matures, an optimal control allocation method can be implemented and the control laws can be changed to take advantage of the increase in available acceleration.

References

Aerodata Model in Research Environment (ADMIRE), Ver. 3.4h, Swedish Defence Research Agency (FOI), Stockholm, Sweden, 2003.

Department of Defense 1980 MIL-F-8785C 'Military Specification, Flying Qualities of Piloted Airplanes'.

Department of Defense 1990 MIL-STD-1797A 'Flying Qualities of Piloted Aircraft'.

Shaw, RL 1985 *Fighter Combat: Tactics and Maneuvering*. Naval Institute Press.

Walker, GP and Allen, DA 2002 'X-35B STOVL flight control law design and flying qualities,' in *AIAA Biennial International Powered Lift Conference*.

Wilson, DJ, Riley, DR, and Citurs, KD 1993 'Aircraft Maneuvers for the Evaluation of Flying Qualities and Agility, Vol. 2: Maneuver Descriptions And Selection Guide,' Wright Laboratory Technical Report WL-TR-93-3082.

9

Applications

9.1 Lessons Learned from the Design of the X-35 Flight Control System

This section will address some of the issues integrating control allocation algorithms as a part of the flight control system. The material presented in this chapter comes from lessons learned while designing the flight control system for the X-35. Not all of the material presented here has its origins in the X-35 program, but all of these issues were discussed at some point in the program and the merits of various options debated. The issues are presented here to inform the reader of difficulties that may arise when attempting to implement various control allocation methodologies. An important point to keep in mind is that the design process is open-ended and does not result in a unique solution. Typically, every design choice has some benefits and some penalties associated with it and trade-offs are made in an attempt to get the best possible final product. It is not the authors' intent to say 'this is the way to fix a problem'; rather we intend to present some of the benefits and penalties associated with various methods so that the reader is more informed of the options available.

9.1.1 Theory vs. Practice

> 'In theory, there is no difference between theory and practice. In practice, this is not the case.'
>
> -Anonymous

Aircraft have issues that are often ignored when developing theoretical solutions using simulation studies. The effects of hardware failures are often initially ignored because if a specific piece of hardware is not to hand, then it cannot be known precisely how that hardware will fail. Creating models to represent multiple possible hardware failures is often not worth the effort.

Prior to flight clearance, rigorous testing must be passed. This testing includes verifying how the system responds to various hardware failure scenarios and checking for bugs in the software to provide assurances that the software will not fail.

Aircraft Control Allocation, First Edition. Wayne Durham, Kenneth A. Bordignon and Roger Beck.
© 2017 John Wiley & Sons, Ltd. Published 2017 by John Wiley & Sons, Ltd.
Companion website: www.wiley.com/go/durham/aircraft_control_allocation

This chapter will address several issues related to hardware and software implementation. These issues are not typically primary considerations when developing control allocation algorithms. They are as follows:

1. **Deterministic:** To pass software testing the flight control system is required to have a unique solution of control effectors for a given flight condition. When flight testing the aircraft, it is important to have expected values for the control effectors. That way, if the effectors are not in the expected positions, something is not quite right (either the model used in the prediction is incorrect, or something unforeseen has occurred).
2. **Known gains through the control law for structural coupling issues:** The forces from the effectors cause vibrations in the airframe. These vibrations get back in to the sensor data, which can change the commands to the effectors. If the gain around this loop is greater than one, then the vibrations will increase. This can lead to catastrophic failure of the structure.
3. **Structural load constraints:** The structure of the aircraft is designed to support a limited amount of force. If the loads applied by the effectors exceed these limits, the airframe may break.
4. **Failure accommodation:** The airplane must be able to land safely in the event of probable control effector failures.
5. **Computation in real time:** Note that hardware approved for military and commercial applications is often significantly behind the state of the art. This is due to the time it takes for hardware to go through the rigorous testing process to get approved for high-risk applications.

9.2   Uses of Redundancy

When aircraft are designed with excess effectors, the typical under-determined problem, the extra capability is there to provide for sufficient control power in the event of a failure scenario. If there are no failures present, there are many ways to make use of the extra degrees of freedom. Some methods of control allocation attempt to optimize some cost function in real time in order to make use of these extra degrees of freedom. This uses computational time to find the optimal solutions, and this requires computer processing power, which is not always available. An alternative approach is to pre-compute effector solutions that are optimal for specific conditions and use those as a preferred solution.

9.2.1   Preferred Solutions

Instead of relying on real-time optimization, optimal configurations could be pre-computed and stored. Often the aerodynamics group will provide a flap schedule for leading- and trailing-edge flaps that is optimal. These schedules are a typically a function of Mach number and angle of attack. They could also include different schedules for different modes, such as a landing configuration or aerial refueling. In this method of implementation the controls are restored through the null space towards the preferred solution in an attempt to minimize the distance between the required controls and the preferred controls. An advantage of this method is the reduction in computational requirements, as the optimal solution does not need to be recomputed each frame. During failure scenarios, the optimal solution would likely change. This method would not find the new optimal solution. It is possible to pre-compute optimal solutions for failure scenarios. However, this may require a very large number of such scenarios to be computed and stored, costing valuable memory space in the on-board computer.

On the X-35 program, pre-computed preferred solutions were used. The preferred values for the aerodynamic effectors were based on aerodynamic efficiency and were provided by the aerodynamics group. The preferred values of the propulsion effectors were based on engine efficiencies and were provided by the propulsion group.

9.2.2   Resolving Path-dependency Issues

One method of dealing with nonlinear functions is to reduce the problem to a locally linear problem and command a change in control effectors for a desired change in acceleration. This method introduces discrete integration into the control allocation problem. Unfortunately, this integral is path dependent.

This path dependency is highly undesirable for flight vehicles. The ability to predict what the control deflections should be at a particular flight condition is extremely important during the test and validation of the flight control software. One method of reducing the path dependency is to use the redundancy to optimize a function, so that the control effectors will be driven towards an optimal solution in the long term. Note that this may not guarantee the desired results of a single solution for a given flight condition. First, the function to be optimized should be chosen such that is has a unique global solution. If not, the optimization may result in finding different local minimums based on the path taken to arrive at a specified condition. Also, being driven towards a unique global minimum still allows for temporary variations, which become small in the long term.

On the X-35 program, discrete-time, or frame-wise control allocation (Chapter 7) was used to command a change in the control effectors. The method of using preferred values to drive the solution to a repeatable configuration worked well, and no path-dependency issues arose during the testing of the vehicle. For a given flight condition the effectors were in the predicted configurations within acceptable tolerances.

9.3   Design Constraints

In addition to satisfying the moment generation required, there may be other constraints imposed on the allocation of the control effectors. Several common ones are detailed below.

9.3.1   Axis Prioritization

In the event that the moment demands exceed the capabilities of the control effector suite (that is, the moment demand is outside the AMS) it may be desirable to prioritize one axis over the others. Often the longitudinal axis (pitch) is prioritized over the lateral-directional axes. There are several reasons for doing this. One is that control of the angle of attack is vitally important to the performance of the airplane and this is accomplished through control of the longitudinal axis. Due to the nonlinear nature of the lift curve slope, there tends to be a maximum angle of attack that should not be exceeded. Exceeding this limit can result in undesirable effects such as a significant loss of altitude or a departure from controlled flight as the aerodynamic forces and moments become more chaotic in nature. Another reason to prioritize control of the longitudinal axis is that this is the axis most likely to be unstable or have reduced stability margins. For a supersonic aircraft, such as the X-35 or the ADMIRE simulation (see Appendix B), the aircraft center of pressure will change as the aircraft transitions from subsonic to supersonic flight. At supersonic conditions the center of pressure is in a more stable position. If the aircraft is already stable at the subsonic condition, the aircraft may be too stable supersonically, resulting in large control effector deflections required to trim and maneuver the aircraft. To avoid this issue, airplane designers may choose to have the aircraft be closer to neutrally stable, or possibly even unstable, for a portion of the subsonic flight envelope. For example, the ADMIRE simulation is unstable longitudinally at subsonic conditions.

One method of enforcing this axis prioritization is to redefine the control effectors into longitudinal and lateral-directional pseudo-effectors. Taking the seven aerodynamic surfaces of the ADMIRE vehicle, they may be grouped into the pseudo-effectors shown in Table 9.1.

The pseudo-effector commands are computed using the control allocation routine and then mapped back to the actual effector commands as shown in Table 9.2.

For this example, to prioritize the longitudinal axis over the lateral-directional axes one would solve for the three longitudinal effectors (CS, IS, OS) given a commanded pitch moment first. Once the symmetric values are known, they are used to recompute the control effector limits of the lateral-directional effectors. Then the four lateral-directional effectors are solved to satisfy the roll and yaw moment demands.

Using pseudo-controls as described imposes constraints on the moment-generating ability of the control effector suite. This method does not permit allocation of admissible controls for every moment in the AMS.

Aircraft design, like all design, is about the art of compromise. Using this method gains the desired axis prioritization. The cost for this gain is the loss of performance. Whether or not this trade-off is acceptable will depend upon the specific aircraft and its requirements.

Table 9.1   Pseudo-effectors

Longitudinal	
Symmetric canard	CS = 1/2 (rc + lc)
Symmetric inboard elevons	IS = 1/2 (rie + lie)
Symmetric outboard elevons	OS = 1/2 (roe + loe)
Lateral-directional	
Asymmetric canard	CA = 1/2 (rc − lc)
Asymmetric inboard elevons	IA = 1/2 (rie − lie)
Asymmetric outboard elevons	OA = 1/2 (roe − loe)

Key: CS, symmetric canard; IS, symmetric inboard elevons; OS, symmetric outboard elevons; CA, asymmetric canard; IA asymmetric inboard elevons; OA, asymmetric outboard elevons; rc, right canard; lc, left canard; rie, right inboard elevon; lie, left inboard elevon; roe, right outboard elevon; loe, left outboard elevon.

Table 9.2   Reverse mapping,
pseudo-effectors

rc	=	CS + CA
lc	=	CS − CA
rie	=	IS + IA
lie	=	IS − IA
roe	=	OS + OA
loe	=	OA − OA
r	=	r

Abbreviations as per Table 9.1
and r, rudder.

On the X-35 program, the effectors were divided into a longitudinal suite and a lateral-directional suite. The longitudinal suite satisfied the pitch moment demands, and vertical and forward acceleration commands when the propulsion effectors were active on the STOVL variant. The lateral-directional suite satisfied the roll and yaw moment demands. The longitudinal suite was prioritized over the lateral-direction suite. During the flight testing of the X-35 this method effectively prioritized the longitudinal effectors and the vehicle operated within its specified angle-of-attack ranges. Also, the vehicle satisfied all of the maneuverability requirements set forth by the program office.

9.3.2   Structural Loads

The control effectors generate moments that accelerate the aircraft. Typically when doing aircraft dynamics we treat the aircraft as a rigid body. While this is in general a valid assumption for dynamic analysis, the good people who design the structure care about the loads that the airframe needs to accommodate. When an effector applies a force or moment, the structure of the airframe must be capable of handling the loads or the airframe will rip apart. This is both a problem for maximum effector-induced loads, as well as concerns about repeated load applications causing fatigue failures.

Certain effector combinations tend to be detrimental to the long-term life of the airframe. An example of this is seen in the fuselage loads for the ADMIRE vehicle. If the canards are generating a positive roll moment, and the elevons are generating a negative roll moment, the fuselage is subjected to torque about the x-axis. Due to the fact that weight is a critical element in aircraft design, the structure of the vehicle is made as light (resulting in it generally being weaker) as it can be while still being capable of carrying the necessary loads. The controls described may generate a net rolling moment, but they will cause more internal load on the fuselage structure than another control solution that results in the same net moments.

One method of dealing with structural loads is to use preferred values to drive the controls towards structurally sound conditions and avoid the excessive load conditions. However, often designers will say that 'I prefer that the airplane not rip apart while I'm flying' is not a strong enough statement and they desire some sort of guarantee that the undesirable control solutions will not be used.

One method of preventing the scenario described above is to pre-gang the control effectors to eliminate the possibility of commanding controls that will cause excessive loads on the airframe. For the ADMIRE example, the longitudinal controls could be combined into a single symmetric pitch effector:

```
(pitch)=-CS+IS+OS
```

The sign convention was selected such that positive (pitch) produces negative (nose-down) pitching moment, similar to an elevator.

The lateral-directional asymmetric controls can be combined into a single roll effector:

```
roll=-CA+IA+OA
```

The sign convention was selected such that positive roll produces negative (left-wing-down) rolling moment.

For a yaw effector, the rudder would be used:

```
yaw=rud
```

There are benefits and penalties for every design choice. For this particular solution, one is penalized with a loss of control authority by imposing additional constraints on the system. This is above and beyond the authority lost by separating the controls into longitudinal and lateral-directional effectors. The benefit is a guarantee that certain structurally adverse control solutions will never be commanded. A compromise that may be struck would be to enforce the preceding ganging only at high dynamic pressure flight conditions and allowing the controls to be commanded individually at lower dynamic pressures. For the high dynamic pressure conditions, it is more important to restrict the controls due to the high loads associated with the high dynamic pressures. Also at the high dynamic pressures, the aircraft will typically have more than the required amount of control authority, so some may be sacrificed to reduce the structural loads. Conversely, at the lower dynamic pressure conditions the aircraft may suffer from a lack of control authority and it may be necessary to use the full capabilities of the effectors. However, at the lower dynamic pressures the structural loads due to the effector deflections will be significantly less.

On the X-35 program, the asymmetric horizontal tail command and the asymmetric flap command were ganged together for high speed conditions to eliminate the possibility that they would be commanded in a configuration that would cause excessive torque on the fuselage. At low speed conditions they were commanded independently to provide greater moment-generation capability.

9.3.3   Effector Bandwidth

If some effectors are slower than others or for some other reason it is desirable to use them less often, this can be accomplished by biasing the weighting matrix used in the cost function.

On the X-35 program, weightings were used in blending the use of aerodynamic and propulsive effectors when transitioning to and from hover. In hover, the aerodynamic effectors had limited effect on the moments generated and instead were fixed in the LIDs configuration (Lift Improvement Devices) to better control the airflow around the vehicle and reduce engine suck-down effects. When the vehicle had more airspeed and the aerodynamic controls became more effective, they became part of the active set. When the vehicle had sufficient airspeed to maneuver using only aerodynamic controls, the propulsion effectors were heavily weighted to de-emphasize their use. This saved wear on the engine components. Also, the aerodynamic effectors were better behaved (more linear in their effects) than the propulsive effectors. At higher airspeeds, the propulsive lift system was disabled and the propulsive effectors were inactive.

Some control system architects have also used high-pass and low-pass filters to carve up the moment demands, sending the low-pass values to slowed effectors and the high-pass to the faster effectors.

9.3.4   Gain Limiting and Stability Margins

When clearing an aircraft for flight testing, one of the important pieces of information to know is the total gain through the control law, the sensor-to-effector gain. This is necessary because of structural coupling issues.

Knowing the gains through the control law can be complicated using this type of control methodology. However it is important to know them to avoid potentially catastrophic consequences. If a simple generalized inverse solution is used, then the control allocation gains are known at any particular flight condition. However, for many of the optimal methods discussed, it is difficult to determine the gain through the control allocation portion of the control law.

On the X-35 program, there were a number of structural coupling challenges that needed to be overcome. A significant amount of effort was required to ensure the vehicle was certified flightworthy. See Tauke and Bordignon (2002) for more details.

9.4   Failure Accommodation

One of the main uses of redundant effectors is the ability to overcome the loss of control effectiveness associated with an actuator failure. An advantage of a control-allocation-based control method is that it is relatively easy to accommodate effector failures. A control allocation algorithm will simply use whatever controls are functional to attempt to satisfy the moment demands.

There are several design features that can help improve failure accommodation. The use of hardware which can detect failed effectors can simplify the process of determining which effectors are available to the control allocation algorithm. Also, when an effector fails it is useful to know the failed position. Some systems are designed to fail to a specific location. Other times an effector may become stuck at some random position. If the control law is aware of the position of the failed effector it can better account for the effects of the failure when computing the required moment.

On the X-35 program, the hardware moving the aerodynamic effectors sent signals to the control law, reporting failures and last known position. The software which controlled the propulsion system could detect and analyze the condition of the propulsion effectors. See Wurth et al. (2002) for a more detailed description of how this was done. From a control allocation standpoint, the propulsion control software communicated logical values for each propulsion effector, set to true if a failure was detected. For some of the failure conditions, a failure meant that the effector could still be commanded, but the range of permissible values was changed by the failure. For these cases, the propulsion software communicated new effector limits for the failed effector.

If an effector was failed in a fixed position, the fixed value was used in the OBM to compute the nominal accelerations and that effector was removed from the active effector set. If a failed effector had new limits, the effector was still part of the active effector set, but the new limits were used.

Fortunately, no effector failures occurred during the X-35 flight tests. However, numerous failure scenario simulations were conducted both off-line and in piloted simulations. The failure accommodation use proved to be very effective during these simulated flights.

References

Tauke, G and Bordignon, K 2002 'Structural coupling challenges for the X-35B,' AIAA 2002-6004 in *AIAA 2002 Biennial International Powered Lift Conference and Exhibit*.

Wurth, S, Hart, J, and Baxter, J 2002 'X-35B integrated flight propulsion control fault tolerance development,' AIAA 2002-6019 in *AIAA 2002 Biennial International Powered Lift Conference and Exhibit*.

A

Linear Programming

Linear programming is an extensively studied approach to numerical optimization. Its popularity may be explained because many problems of practical interest have constraints and objectives that are either inherently linear functions of the unknowns or where a linear model is a convenient approximation (Luenberger and Ye 2008). The constraints for the control allocation problem as posed previously (Chapter 4) fall into the latter category if we restrict ourselves to systems that can be approximated with linear control effectiveness and simple bounds on the controls.

Several researchers have considered linear programming as an approach to the control allocation problem. In one of the earlier treatments, Buffington (1999) illustrates the use of linear programming to change the behavior of how the allocator deals with redundant controls, achieving the desired command with control solutions chosen to meet different requirements depending on the flight-phase. This theme has been expanded by other researchers and linear programming allocators have been designed to allocate controls while optimizing a variety of secondary objectives.

Given the straightforward transformation of $\mathbf{Bu} = \mathbf{m}_{des}$ and $\mathbf{u}_l \leq \mathbf{u} \leq \mathbf{u}_u$ into a set of linear constraints, it remains to be seen if we can find suitable linear cost functions that will achieve the objectives of the control allocator. Additional constraints may also need to be formulated in linear form to meet these objectives. A variety of formulations are possible, we will restrict ourselves to those that meet our previous definitions of optimality (with respect to allocating for maximum capabilities) while also gracefully handling unfeasible commands. Rather than prescribe a single formulation, examples are provided that typify different optimal behavior for the allocator.

The linear programming formulations presented not only allocate for maximum capabilities, but also allow for the optimization of secondary objectives when there is redundant control authority. The concept of using a preferred control solution to resolve control redundancy was briefly introduced in Section 9.2.1; the use of excess control authority to deal with secondary objectives was expanded on in Chapter 7. The methods of Chapter 7 can be applied to a linear programming allocator in the same manner as any of the other allocators introduced earlier. However, we will also see that a subset of the previous methods can be incorporated directly into the linear programming formulation, albeit in slightly different forms.

Aircraft Control Allocation, First Edition. Wayne Durham, Kenneth A. Bordignon and Roger Beck.
© 2017 John Wiley & Sons, Ltd. Published 2017 by John Wiley & Sons, Ltd.
Companion website: www.wiley.com/go/durham/aircraft_control_allocation

Approaches for these two goals, dealing with unfeasible commands while maintaining maximum capabilities and optimizing secondary objectives in the presence of control redundancy, can be developed independently. This allows different approaches for each to be compared; the appropriate approaches for a given problem can then be combined.

This appendix starts with an overview of how the control allocation problem relates to linear programming. Methods for dealing with unattainable commands are presented as part of that overview – if one does not care about secondary objectives, or the existence of non-unique solutions for feasible commands, these methods will function as an optimal allocator achieving moments throughout Φ.

After the overview, some discussion of linear programs is presented, illustrating how to go about formulating a problem in a useful standardized form. Some important properties are discussed, including how those properties can be used to make inferences about the expected optimal solution. The discussion herein is by no means as complete as a dedicated text on linear programming, but gives some of the elements that help understand and diagnose the behavior of an allocator built using these techniques.

Options for optimizing feasible commands are presented before providing methods for combining both the infeasible and feasible approaches into a single method suitable for use as a control allocator. The major design choices are all involved in setting up this allocator. The actual numerical solution, while important from a computation standpoint, is a well-studied field in and of itself. Any of the programs developed in this text can be solved using one of the many available solvers.

In the final section, we will discuss the numerical solution of linear programs. A general overview is presented, and then one of the classic algorithms is described in depth, including examples of how it generates solutions to the control allocation problem. This description, along with the sample algorithm accompanying the numerical simulation described in Appendix B, will allow the reader to develop code that solves the example problems. Understanding the operation of a simple solver will also help in optimizing the performance of more complex tools.

A.1   Control Allocation as a Linear Program

Before we examine the solution to linear programs in detail, we first consider how the control allocation problem might be cast as a linear optimization problem. Ultimately, we seek formulations that meet the definitions of optimality discussed in Section 6.2. Two examples of how the problem could be set up to achieve maximum capabilities according those definitions were introduced in Section 6.9.

Both of the previous examples of linear programs treated infeasible commands in a manner equivalent to the other optimal methods discussed, preserving the direction and returning a solution on the boundary of $\partial(\Phi)$. One of the methods, presented in Eq. (6.107), also scaled the solution for feasible commands from that on $\partial(\Phi)$. In the other, developed in Eqs. (6.86)–(6.87), the solution for feasible commands was under-determined, becoming a function of the specific numerical approach taken to solving the program. Linear programming offers a more general approach than either of these specific methods, allowing us to develop allocators with different goals.

In Eqs. (4.5) and (4.6), the control allocation problem was introduced as solving the constrained linear system,

$$\mathbf{Bu} = \mathbf{m}_{des} \tag{A.1}$$

$$\mathbf{u}_l \leq \mathbf{u} \leq \mathbf{u}_u \tag{A.2}$$

to find $\mathbf{u}$ where $\mathbf{u}_l$, $\mathbf{u}_u$, $\mathbf{B}$, and $\mathbf{m}_{des}$ are given. In Chapter 5, we have noted that, for the case where the number of effectors in $\mathbf{u}$ is greater than the number of elements in the commanded objective, $\mathbf{m}_{des}$, the system in Eq. (A.1) is under-determined and there are potentially an infinite number of solutions for $\mathbf{u}$. Enforcing the bounds on our effectors, Eq. (A.2), limits the set of attainable objectives, Φ, to a set bounded by a convex polyhedron.[1]

A.1.1   Optimality for Attainable Commands

For the moment, we will ignore the possibility that the controller could ask for $\mathbf{m}_{des}$ outside of Φ and only consider feasible commands. Unless such a command is on the boundary of the feasible set, the corresponding effector positions are not unique. An optimality statement for this condition, nearest to a preferred solution, was also introduced in Section 6.2.3. The difference between our two previous examples is that, while both allocated for maximum capabilities, one did not include any such preferred solution. The other implicitly included a preference for the solution scaled from the boundary.

One approach to satisfying this optimality criteria is to search for the solution that yields the best value of some metric, $J(\mathbf{u})$, written as a function of the effector positions. The choice of this metric is analogous to choosing secondary objectives to be minimized after the primary objectives are met. Researchers have considered examples chosen to satisfy any of a variety of goals including: minimizing control deflections or the usage of certain effectors; optimizing the control contribution to quantities such as drag, structural loading, or radar cross section; or even tracking a preferred control solution chosen to simplify analysis or to facilitate system identification (see Buffington (1999), Buffington, Chandler, and Pachter (1998), Buffington (1999), and Frost et al. (2010)).

The result is a constrained optimization problem, Eq. (A.3).

$$\begin{aligned} \min_{\mathbf{u}} \quad & J(\mathbf{u}) \\ & \mathbf{Bu} = \mathbf{m}_{des} \\ & \mathbf{u}_l \leq \mathbf{u} \leq \mathbf{u}_u \end{aligned} \tag{A.3}$$

If $J(\mathbf{u})$ is linear, then the problem stated in (A.3), is classified as a linear program.

A.1.2   Optimality for Unattainable Commands

The program in Eq. (A.3) satisfies our optimality criteria from Section 6.2.4 for maximizing capabilities and yielding solutions close to preferred solutions over the entirety of Φ. However, the equality constraint $\mathbf{Bu} = \mathbf{m}_{des}$ cannot not hold for $\mathbf{m}_{des} \notin \Phi$. This program has no solution

---

[1] Typically for aircraft problems it is convenient to consider two- and three-dimensional problems where the attainable objectives are constrained to a polygon or polyhedron. However, the methods presented can also be extended to systems with more objectives, yielding a higher-dimensional polyhedron (Beck 2002). Consider, for example, the problem of reaction control system (RCS) jet selection where a vehicle uses multiple fixed thrusters to control both angular and linear accelerations

in the face of an unattainable command and thus is not suitable as an allocator unless the control system can guarantee that $\mathbf{m}_{des} \in \Phi$.

We will assume that, for the unattainable command, the desired control solution will be on the boundary of Φ. We can then define a second constrained optimization problem for the unattainable case.

$$\min_{\mathbf{u}} \quad J_{infeas}(\mathbf{u})$$

$$\mathbf{Bu} \in \partial(\Phi)$$

$$\mathbf{u}_l \le \mathbf{u} \le \mathbf{u}_u \tag{A.4}$$

Here, the equality constraint on the moment has been relaxed. Instead, we just require that the solution be on the boundary, $\partial(\Phi)$. In the subsequent sections, we will see that, rather than stating this constraint explicitly, the constraints and objective function can be chosen such that the final control solution satisfies this condition.

We are left with three conditions, as shown in Eqs. (A.5)–(A.7), to consider when formulating our allocator as a linear programming problem. The first and second ensure we meet the maximum capability and preferred control conditions. The third allows us to specify an optimal solution and fail gracefully when presented with an unattainable command.

$$\mathbf{u} = \min_{\mathbf{u}} J(\mathbf{u}) \quad \textit{if } \mathbf{m}_{des} \in \Phi \tag{A.5}$$

$$\mathbf{Bu} = \mathbf{m}_{des} \quad \textit{if } \mathbf{m}_{des} \in \Phi \tag{A.6}$$

$$\mathbf{Bu} \in \partial(\Phi) \quad \textit{if } \mathbf{m}_{des} \notin \Phi \tag{A.7}$$

Ultimately, we will describe methods that satisfy all three conditions. Before considering how to handle some secondary objective, Condition (A.5), we first define linear programs satisfying the second two conditions. The preferences used in meeting Condition (A.7) are similar to those discussed in Section 6.2.4, only we will relax the concept of axis prioritization somewhat and instead consider the minimization of the moment error, as shown in Figure A.1.

Direction Preserving

Consider the constraint equations in Eq. (A.3), $\mathbf{Bu} = \mathbf{m}_{des}$ and $\mathbf{u}_l \le \mathbf{u} \le \mathbf{u}_u$. The equality constraint ensures that we achieve the desired solution, but that it cannot be met if $\mathbf{m}_{des}$ is outside of Φ. Recall that previously we developed control allocation solutions that scaled the desired moment so that it lay on $\partial(\Phi)$ but maintained the same direction, (see Figure A.1). A simple modification to the constraints in our problem yields,

$$\mathbf{Bu} = \lambda \mathbf{m}_{des} \mid \{ \ 1 \ge \lambda \ge 0, \mathbf{u}_l \le \mathbf{u} \le \mathbf{u}_u \} \tag{A.8}$$

In Eq. (A.8), the scaling factor, λ, is always positive, ensuring that the resulting moment $\lambda \mathbf{m}_{des}$ is in the same direction as the desired moment, $\mathbf{m}_{des}$. For commands in Φ, we need to ensure $\lambda = 1$ so that the desired moment is achieved. Outside of Φ, λ is allowed to be less than one, scaling the desired moment.

If our linear program maximizes λ, then the upper bound on λ guarantees ensures that we satisfy Condition (A.6). Similarly, the maximum value of λ occurs when $\mathbf{Bu}$ falls on $\partial(\Phi)$ for unattainable $\mathbf{m}_{des}$ (Φ is convex) meeting Condition (A.7).

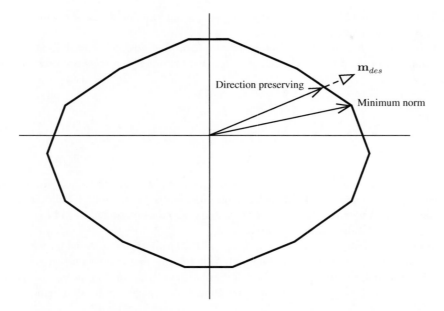

Figure A.1 Φ for a 2×7 B matrix showing two possible linear programming solutions for an unattainable moment command, $\mathbf{m}_{des}$: Preserving the desired moment direction, and minimizing the sum of the absolute error in each axis

By convention, we will be setting up our program to minimize the objective function, thus our cost function is simply, $\min -\lambda = \max \lambda$. Augmenting the unknown control vector with λ and subtracting $\lambda \mathbf{m}_{des}$ from both sides of the constraint in Eq. (A.8), we are left with a problem with $m + 1$ unknowns.

$$\min_{\mathbf{u}, \lambda} \quad -\lambda$$

$$[\mathbf{B} - \mathbf{m}_{des}] \begin{pmatrix} \mathbf{u} \\ \lambda \end{pmatrix} = \mathbf{0}$$

$$0 \leq \lambda \leq 1$$

$$\mathbf{u}_l \leq \mathbf{u} \leq \mathbf{u}_u \tag{A.9}$$

Error Minimization
An alternative approach to satisfying the Conditions (A.6) and (A.7) is to omit the equality constraint entirely from the original linear program in Eq. (A.3). Instead, we will define a cost function that minimizes a linear combination of the absolute value of the error between the commanded moment $\mathbf{m}_{des_i}$, and that achieved $y_i = \mathbf{B}_{r_i} \mathbf{u}$, in each axis.

$$\min_{\mathbf{u}} (w_1 |\mathbf{B}_{r_1} \mathbf{u} - \mathbf{m}_{des_1}| + w_2 |\mathbf{B}_{r_2} \mathbf{u} - \mathbf{m}_{des_2}| + w_3 |\mathbf{B}_{r_3} \mathbf{u} - \mathbf{m}_{des_3}|) \tag{A.10}$$

It is readily apparent that the above function has a minimum of zero when $\mathbf{B}\mathbf{u} = \mathbf{m}_{des}$, ensuring that Condition (A.6) is met when Eq. (A.10) is optimized for $\mathbf{m}_{des}$ inside Φ.

In the case where the command cannot be satisfied, the solution returned will map to one on the boundary. The selection of the weighting terms, w_i, governs the priority of how the individual moment errors are driven to zero. By choosing the magnitude of w_i such that one term always dominates Eq. (A.10), a particular axis of control can be prioritized.

More generally, we can seek to minimize some weighted norm of the errors. The weights on the individual axes have been combined into a single matrix $\mathbf{W_d}$; typically $\mathbf{W_d}$ is chosen to be diagonal.

$$\min_{\mathbf{u}} \quad \| \mathbf{W_d}(\mathbf{Bu} - \mathbf{m}_{des}) \|$$

$$\mathbf{u}_l \leq \mathbf{u} \leq \mathbf{u}_u \tag{A.11}$$

The choice of norm in Eq. (A.11) determines the type of optimization algorithm needed to solve the problem. Typically when presented with minimum-norm-type solutions to engineering problems, we envision a sum-of-squares, Euclidean, type of approach. Use of the ℓ_2-norm presents us with a constrained quadratic programming problem. An allocator using such a program is possible using, for example, quadratic programming, but is not amenable to linear programming techniques.

Fortunately, there are norms that allow us to express the objective function from Eq. (A.11) in a linear form. The absolute value solution of Eq. (A.10) is one such option, corresponding to use of the ℓ_1-norm in Eq. (A.11). Another option is the use of the ℓ_∞-norm of the moment error, minimizing the absolute value of the maximum component.

Because the ℓ_∞ option yields a larger, more complicated, program, the ℓ_1 option is typically utilized for moment error. Later we will consider the use of ℓ_∞-norms of control errors where it has been shown that the added complexity may be justified by the properties of the solution.

Example A.1

Solving a Control Allocation Problem using Linear Programming Recall the example from the ADMIRE simulation above, in Section 6.11.2, where we seek a direction-preserving control solution to the desired moment,

$$\mathbf{m}_{des} = \begin{pmatrix} 2.8493 \\ -0.2942 \\ 0.5726 \end{pmatrix} \tag{A.12}$$

using the control effectiveness and control limits in Eqs. (A.13) and (A.14).

$$B = \begin{bmatrix} 0.7073 & -0.7073 & -3.4956 & -3.0013 & 3.0013 & 3.4956 & 2.1103 \\ 1.1204 & 1.1204 & -0.7919 & -1.2614 & -1.2614 & -0.7919 & 0.0035 \\ -0.3309 & 0.3309 & -0.1507 & -0.3088 & 0.3088 & 0.1507 & -1.2680 \end{bmatrix} \tag{A.13}$$

$$\mathbf{u}_l = \begin{pmatrix} -0.9599 \\ -0.9599 \\ -0.5236 \\ -0.5236 \\ -0.5236 \\ -0.5236 \\ -0.5236 \end{pmatrix} \quad \mathbf{u}_u = \begin{pmatrix} 0.4363 \\ 0.4363 \\ 0.5236 \\ 0.5236 \\ 0.5236 \\ 0.5236 \\ 0.5236 \end{pmatrix} \tag{A.14}$$

Using the data in Eqs. (A.12) – (A.14), we can obtain a solution to the linear program in Eq. (A.9) using any convenient linear programming solver. As an example we will use MATLAB®, invoking x = linprog (f, A, b, Aeq, beq, xMin, xMax), which looks for the solution,

$$\min_{\mathbf{x}} \mathbf{f}^T \mathbf{x}$$

$$\mathbf{A}_{eq}\mathbf{x} = \mathbf{b}_{eq}$$

$$\mathbf{A}\mathbf{x} \leq \mathbf{b}$$

$$\mathbf{x}_{Min} \leq \mathbf{x} \leq \mathbf{x}_{Max}$$

Using MATLAB® we can format our example as required for linprog,

```
f = [zeros(7,1); -1];
A = []; b = [];
Aeq = [B -md];
beq = [zeros(3,1)];
xMin = [ul; 0];
xMax = [uu; 1];
opt=optimoptions('linprog','algorithm','simplex');
x = linprog(f,A,b,Aeq,beq, xMin,xMax,zeros(8,1),opt);
ul = x(1:7)
lambda = x(8)
```

The resulting solution satisfies $\mathbf{Bu} = \mathbf{m}_{des}$, but is different than any of those encountered in previous examples.

$$\lambda = 1, \mathbf{u} = \begin{pmatrix} -0.9599 \\ 0.4363 \\ -0.5236 \\ -0.2913 \\ 0.5236 \\ -0.2150 \\ 0.1479 \end{pmatrix} \tag{A.15}$$

In particular, the magnitude of the control usage appears higher than in the generalized inverse (Eq. (6.3)) or direct allocation (Eq. (6.7)) solutions. An astute reader will note that all of the controls, save $\mathbf{u}_5$ and $\mathbf{u}_6$, are saturated at one of their limits. The cost function given in Eq. (A.9) does not distinguish between equivalent solutions that achieve the same moment inside of Φ, thus the resulting control vector is a function of the choice of solver.[2]

In Chapter 5, a solution on the boundary of Φ was found to be unique and defined by two controls, assuming every 3×3 sub-matrix of **B** is full rank. Thus, our linear program should return the same answer as the direct allocation solution for unattainable commands if it is truly

---

[2] For consistency with the discussion in subsequent sections, linprog was forced to use a variant of Dantzig's (Dantzig, 2002) Simplex algorithm. The reader is invited to examine the output with the 'algorithm' parameter set to 'interior-point', 'dual-simplex', or 'active-set' and compare the magnitude of the resulting controls

returning the maximum moment in the desired direction. We can modify our solution to seek an unattainable command by scaling $\mathbf{m}'_{des} = 4\mathbf{m}_{des}$,

```
md = 4*md; %Note scaling on desired moment
            % to create unattainable command
f = [zeros(7,1); -1];
A = []; b = [];
Aeq = [B -md];
beq = [zeros(3,1)];
xMin = [ul; 0];
xMax = [uu; 1];
opt=optimoptions('linprog','algorithm','simplex');
x = linprog(f,A,b,Aeq,beq, xMin,xMax,zeros(8,1),opt);
ul = x(1:7)
lambda = x(8)
```

The new, larger result below in Eq. (A.16) is identical to that returned by the direct allocation approach (or any of the methods that preserve direction when the command is saturated).

$$\mathbf{u} = \begin{pmatrix} -0.2619 \\ -0.2617 \\ -0.5236 \\ -0.5236 \\ 0.5236 \\ 0.5236 \\ -0.5236 \end{pmatrix} \tag{A.16}$$

$$\lambda = 0.5 \tag{A.17}$$

The solution is found on the facet defined by $\mathbf{u}_2$ and $\mathbf{u}_3$, The multiplier, λ, indicates that the solution is only returning 50% of the desired moment—prior to scaling by four, the original moment demand was exactly halfway to the boundary of the AMS.

A.2 Standard Forms for Linear Programming Problems

The early conception of linear programming was developed as way to address the problem of determining a 'program' for allocating scarce resources (say, personnel or equipment) amongst various tasks by military planners (Dantzig 2002). The introduction of a linear objective function turned this problem into one of optimization, with a set of unknown variables whose constraints could be approximated by linear relationships.

In studying their solutions, it is convenient to restate linear programs in a common, standardized format. Any problem seeking to optimize a linear function of a set of unknown variables that are subject to constraints expressed as linear equalities or inequalities can be rewritten in what is known as the *standard form*.

$$\min_{\mathbf{x}} \mathbf{c}^T \mathbf{x} \mid \mathbf{Ax} = \mathbf{b}, \mathbf{x} \geq \mathbf{0} \tag{A.18}$$

All unknown variables in Eq. (A.18) are assumed to be non-negative; this is reminiscent of the physical interpretation of the unknowns as resources to be allocated in the original linear programs. The remaining constraints have all been converted into a set of linear equalities.[3] This standardized form is frequently adopted in the linear programming literature; it is of interest because it provides us a framework in which to discuss the properties of linear programs and to begin to develop methods for solving them.

Recall the simplistic direction-preserving linear program we previously solved using MATLAB®'s *linprog* function. The program in Eq. (A.9) is not written in standard form, but the process of transforming it into standard form is similar to that required for more complex control allocators. In the subsequent sections, we will transform it into the standard form. Many of the same approaches will be useful later as we examine more complex problems.

A.2.1 Dealing with Negative Unknowns

As a first step in converting a program designed to solve the control allocation problem into the standard form in Eq. (A.18), consider the non-negative constraint on the unknown variables.

$$\mathbf{x} \geq 0 \tag{A.19}$$

Typically, the unknowns in the control allocation problem are bounded, but may have either sign.

$$\mathbf{u}_l \leq \mathbf{u} \leq \mathbf{u}_u \tag{A.20}$$

The simplest transformation we can apply to our effector positions to make them non-negative is to redefine the zero position so that we are measuring each effector deflection with respect to its lower bound.

$$\mathbf{x} = \mathbf{u} - \mathbf{u}_l \tag{A.21}$$

For the direction-preserving control allocation problem in Eq. (A.9), the cost function is a function only of the scaling parameter, $\min - \lambda$, and remains unchanged when we apply this substitution. Using Eq. (A.21) to substitute for $\mathbf{u}$, the constraints are transformed.

$$\mathbf{Bu} = \mathbf{B}\ (\mathbf{x} + \mathbf{u}_l) = \lambda \mathbf{m}_{des}$$

$$\mathbf{Bx} - \lambda \mathbf{m}_{des} = -\mathbf{Bu}_l \tag{A.22}$$

$$[\mathbf{B}\ -\mathbf{m}_{des}] \begin{pmatrix} \mathbf{x} \\ \lambda \end{pmatrix} = -\mathbf{Bu}_l \tag{A.23}$$

$$0 \leq \mathbf{x} \leq \mathbf{u}_u - \mathbf{u}_l$$

$$0 \leq \lambda \leq 1 \tag{A.24}$$

Instead of measuring the effector positions with respect to their lower bound, we can also define them with respect to an arbitrary reference value, $\mathbf{u}_0$. When the reference value falls within the range of the effectors, the difference from the reference, $\mathbf{u} - \mathbf{u}_0$, is still a quantity that

---

[3] The researcher interested in perusing the linear programming literature will quickly learn that there are at least two differing conventions for the *standard* form. In the second, the constraints are all stated as inequalities, $\min_{\mathbf{x}} \mathbf{c}^T \mathbf{x}$ | $\mathbf{Ax} \leq \mathbf{b}, \mathbf{x} \geq 0$

may have either sign. This unknown non-negative quantity can be rewritten as the combination of two non-negative quantities,

$$\mathbf{u} - \mathbf{u}_0 = \mathbf{u}^+ - \mathbf{u}^-$$

$$\mathbf{u} = \mathbf{u}^+ - \mathbf{u}^- + \mathbf{u}_0$$

$$\mathbf{u}^+ \geq \mathbf{0}$$

$$\mathbf{u}^- \geq \mathbf{0} \tag{A.25}$$

It is tempting to think of the new unknown variables, $\mathbf{u}^+$ and $\mathbf{u}^-$ as if they were made up of the magnitude of the positive and negative elements of $\mathbf{u}$,

$$\mathbf{u}_i = \left\{ \begin{array}{l} \mathbf{u}_i^+ + \mathbf{u}_0 \; if \; \mathbf{u}_i - \mathbf{u}_{0i} \geq 0 \\ \mathbf{u}_i^- + \mathbf{u}_0 \; if \; \mathbf{u}_i - \mathbf{u}_{0i} < 0 \end{array} \right\} \; for \; i = 1...m \tag{A.26}$$

$$\tag{A.27}$$

However, unless the constraint in Eq. (A.26) is explicitly included, a constant offset can be added to both $\mathbf{u}^+$ and $\mathbf{u}^-$ without changing the value of $\mathbf{u}$.

Equation (A.25) is applied to the direction-preserving problem, Eq. (A.9), to give a new set of constraints with non-negative variables.

$$\mathbf{B}\mathbf{u} = \mathbf{B}(\mathbf{u}^+ - \mathbf{u}^- + \mathbf{u}_0) = \lambda \mathbf{m}_{des} \tag{A.28}$$

$$[\mathbf{B} \;\; -\mathbf{B}] \begin{pmatrix} \mathbf{u}^+ \\ \mathbf{u}^- \end{pmatrix} = \lambda \mathbf{m}_{des} - \mathbf{B}\mathbf{u}_0$$

$$[\mathbf{B} \;\; -\mathbf{B} \;\; -\mathbf{m}_{des}] \begin{pmatrix} \mathbf{u}^+ \\ \mathbf{u}^- \\ \lambda \end{pmatrix} = -\mathbf{B}\mathbf{u}_0 \tag{A.29}$$

$$0 \leq \lambda \leq 1$$

$$0 \leq \mathbf{u}^+ \leq \mathbf{u}_u - \mathbf{u}_0$$

$$0 \leq \mathbf{u}^- \leq \mathbf{u}_0 - \mathbf{u}_l \tag{A.30}$$

As before, when we wrote $\mathbf{u}$ in terms of $\mathbf{u}_l$, our substitution for $\mathbf{u}$ does not impact the scaling parameter, λ, so no modification of the original cost function is needed. A key difference, however, is that we have increased the number of unknowns in the problem from $m + 1$ to $2m + 1$ in order to specify $\mathbf{u}_0$.

Often when this trade-off is made, it is to gain the flexibility to include the magnitude of the control error, $\mathbf{u} - \mathbf{u}_0$, in the cost function. A fortunate side effect is that the outcome of such a cost function minimizes the magnitude of the elements in both $\mathbf{u}^+$ and $\mathbf{u}^-$, driving any constant offset in those terms to zero.

A.2.2   *Dealing with Inequality Constraints*

Equations (A.28) and (A.23) both restate the control in terms of bounded variables, $\mathbf{0} \leq \mathbf{x} \leq \mathbf{h}$. The upper bound, $\mathbf{h}$, represents an additional inequality constraint not found in our standard form, Eq. (A.18).

$$\mathbf{Ax} = \mathbf{b}$$
$$\mathbf{x} \leq \mathbf{h}$$
$$\mathbf{0} \leq \mathbf{x}$$

More generally, we can think of additional constraints that are general linear inequalities and not restricted to simple bounds on the unknown variables.

$$\mathbf{A_{eq}x} = \mathbf{b_{eq}}$$
$$\mathbf{A_u x} \leq \mathbf{b_u}$$
$$\mathbf{0} \leq \mathbf{x} \tag{A.31}$$

The linear inequalities in Eq. (A.31) can be converted into equalities through the introduction of additional, non-negative, variables. These additional variables represent the 'slack' between left-hand side value based on the current unknowns, $\mathbf{A_u x}$, and the bound $\mathbf{b_u}$.

$$\mathbf{A_u x} + \mathbf{y^+} = \mathbf{b_u}$$
$$\mathbf{y^+} \geq \mathbf{0}$$
$$\mathbf{0} \leq \mathbf{x} \tag{A.32}$$

If the inequality in Eq. (A.31) was written as a lower bound, $\mathbf{A_l x} \geq \mathbf{b_l}$, we would subtract an additional variable, $\mathbf{y^-}$, from the right-hand side. In this case $\mathbf{y^-}$ would represent the 'surplus' above the bound.

Returning to the example control allocation program, we remove the upper bounds on our unknown through the addition of slack variables. The control constraints in Eqs. (A.24) and (A.30) become

$$0 \leq \mathbf{x} \leq \mathbf{h} \rightarrow \begin{cases} \mathbf{x} + \mathbf{x^+} = \mathbf{h} \\ \quad \mathbf{x} \geq \mathbf{0} \\ \quad \mathbf{x^+} \geq \mathbf{0} \end{cases} \tag{A.33}$$

$$\begin{aligned} 0 \leq \mathbf{u^+} \leq \mathbf{u}_u - \mathbf{u}_0 \\ 0 \leq \mathbf{u^-} \leq \mathbf{u}_0 - \mathbf{u}_l \end{aligned} \rightarrow \begin{cases} \mathbf{u^+} + \mathbf{x^+} = \mathbf{u}_u - \mathbf{u}_0 \\ \mathbf{u^-} + \mathbf{x^-} = \mathbf{u}_0 - \mathbf{u}_l \\ \quad\quad \mathbf{u^+} \geq \mathbf{0} \\ \quad\quad \mathbf{u^-} \geq \mathbf{0} \\ \quad\quad \mathbf{x^+} \geq \mathbf{0} \\ \quad\quad \mathbf{x^-} \geq \mathbf{0} \end{cases} \tag{A.34}$$

Once again, we note that our substitution has increased the dimensions of the problem. One additional unknown and one additional equality constraint have been added for each inequality that was replaced. For simple bounds, such as those seen in Eqs. (A.34) and (A.33), we will later explore an alternative form of our problem that avoids inflating the problem dimensions. However, the general approach in Eq. (A.32) can still be applied to other bounded parameters.

A.2.3   *Writing a Program for Control Allocation in Standard Form*

The substitutions outlined in Sections A.2.1 and A.2.2 allow us to transform the multi-signed, bounded, effectors in a control allocation problem into a set of non-negative unknowns and some additional equality constraints. Two potential transformations were developed depending on whether a particular problem benefits from the explicit formation of a control error term.

For our direction-preserving program, neither the reference control position, $\mathbf{u}_0$, nor the resulting error, $\mathbf{u} - \mathbf{u}_0$, shows up in the cost function. Thus it is simple to rewrite the effector positions in terms of the lower bound. Starting with Eq. (A.22) and applying the transformation in Eq. (A.33) to the bounded control, Eq. (A.24), we derive the following linear program in standard form.

$$\min_{\left(\mathbf{x}, \lambda, \mathbf{x}^+, \lambda^+\right)^T} -\lambda \tag{A.35}$$

$$\begin{bmatrix} \mathbf{B} & -\mathbf{m}_{des} & \mathbf{0}_{n\times m} & \mathbf{0}_{n\times 1} \\ \mathbf{I}_{m\times m} & \mathbf{0}_{m\times 1} & \mathbf{I}_{m\times m} & \mathbf{0}_{m\times 1} \\ \mathbf{0}_{1\times m} & 1 & \mathbf{0}_{1\times m} & 1 \end{bmatrix} \begin{pmatrix} \mathbf{x} \\ \lambda \\ \mathbf{x}^+ \\ \lambda^+ \end{pmatrix} = \begin{pmatrix} -\mathbf{B}\mathbf{u}_l \\ \mathbf{u}_u \\ 1 \end{pmatrix} \tag{A.36}$$

$$0 \leq \lambda$$

$$0 \leq \lambda^+$$

$$0 \leq \mathbf{x}$$

$$0 \leq \mathbf{x}^+$$

Or, written in terms of the variables in Eq. (A.18),

$$\min_{\mathbf{x}} \mathbf{c}^T \mathbf{x} \mid \mathbf{A}\mathbf{x} = \mathbf{b}, 0 \leq \mathbf{x}$$

$$\mathbf{A} = \begin{bmatrix} \mathbf{B} & -\mathbf{m}_{des} & \mathbf{0}_{n\times m} & \mathbf{0}_{n\times 1} \\ \mathbf{I}_{m\times m} & \mathbf{0}_{m\times 1} & \mathbf{I}_{m\times m} & \mathbf{0}_{m\times 1} \\ \mathbf{0}_{1\times m} & 1 & \mathbf{0}_{1\times m} & 1 \end{bmatrix}$$

$$\mathbf{x} = \begin{pmatrix} \mathbf{u} - \mathbf{u}_l \\ \lambda \\ \mathbf{x}^+ \\ \lambda^+ \end{pmatrix}$$

$$\mathbf{b} = \begin{pmatrix} -\mathbf{B}\mathbf{u}_l \\ \mathbf{u}_u \\ 1 \end{pmatrix}$$

$$\mathbf{c}^T = \left(\mathbf{0}_{m\times 1}, -1, \mathbf{0}_{m\times 1}, 0 \right) \tag{A.37}$$

As written, the program has $m + n + 1$ equality constraints and $2m + 2$ unknowns. Note that, in addition to the slack variable on the effector positions, we require a slack variable on the scaling parameter to enforce its upper bound as well.

The alternative approach above, starting with Eq. (A.28) and removing the upper bounds position using slack variables, can also be applied to our direction-preserving control allocator.

The resulting system is even larger than that above, having $2m + n + 1$ equality constraints and $4m + 2$ unknowns.

$$
\begin{bmatrix}
\mathbf{B} & -\mathbf{B} & -\mathbf{m}_{des} & \mathbf{0}_{n\times m} & \mathbf{0}_{n\times m} & \mathbf{0}_{n\times 1} \\
\mathbf{I}_{m\times m} & \mathbf{0}_{m\times m} & \mathbf{0}_{m\times 1} & \mathbf{I}_{m\times m} & \mathbf{0}_{m\times 1} & \mathbf{0}_{m\times 1} \\
\mathbf{0}_{m\times m} & \mathbf{I}_{m\times m} & \mathbf{0}_{m\times 1} & \mathbf{0}_{m\times m} & \mathbf{I}_{m\times m} & \mathbf{0}_{m\times 1} \\
\mathbf{0}_{1\times m} & \mathbf{0}_{1\times m} & 1 & \mathbf{0}_{1\times m} & \mathbf{0}_{1\times m} & 1
\end{bmatrix}
\begin{pmatrix}
\mathbf{u}^+ \\ \mathbf{u}^- \\ \lambda \\ \mathbf{x}^+ \\ \mathbf{x}^- \\ \lambda^+
\end{pmatrix}
=
\begin{pmatrix}
-\mathbf{B}\mathbf{u}_0 \\ \mathbf{u}_u - \mathbf{u}_0 \\ \mathbf{u}_0 - \mathbf{u}_l \\ 1
\end{pmatrix}
\tag{A.38}
$$

$$0 \le \lambda$$

$$0 \le \mathbf{u}^+$$

$$0 \le \mathbf{u}^-$$

$$0 \le \lambda^+$$

$$0 \le \mathbf{x}^+$$

$$0 \le \mathbf{x}^-$$

Or, once again explicitly substituting into the standard form,

$$\min_{\mathbf{x}} \mathbf{c}^T \mathbf{x} \mid \mathbf{A}\mathbf{x} = \mathbf{b}, 0 \le \mathbf{x} \tag{A.39}$$

$$
\mathbf{A} =
\begin{bmatrix}
\mathbf{B} & -\mathbf{B} & -\mathbf{m}_{des} & \mathbf{0}_{n\times m} & \mathbf{0}_{n\times m} & \mathbf{0}_{n\times 1} \\
\mathbf{I}_{m\times m} & \mathbf{0}_{m\times m} & \mathbf{0}_{m\times 1} & \mathbf{I}_{m\times m} & \mathbf{0}_{m\times 1} & \mathbf{0}_{m\times 1} \\
\mathbf{0}_{m\times m} & \mathbf{I}_{m\times m} & \mathbf{0}_{m\times 1} & \mathbf{0}_{m\times m} & \mathbf{I}_{m\times m} & \mathbf{0}_{m\times 1} \\
\mathbf{0}_{1\times m} & \mathbf{0}_{1\times m} & 1 & \mathbf{0}_{1\times m} & \mathbf{0}_{1\times m} & 1
\end{bmatrix}
$$

$$
\mathbf{x} =
\begin{pmatrix}
\mathbf{u}^+ \\ \mathbf{u}^- \\ \lambda \\ \mathbf{x}^+ \\ \mathbf{x}^- \\ \lambda^+
\end{pmatrix}
$$

$$
\mathbf{b} =
\begin{pmatrix}
-\mathbf{B}\mathbf{u}_0 \\ \mathbf{u}_u - \mathbf{u}_0 \\ \mathbf{u}_0 - \mathbf{u}_l \\ 1
\end{pmatrix}
$$

$$\mathbf{c}^T = \left(\mathbf{0}_{1\times m}, \ \mathbf{0}_{1\times m}, \ -1, \ \mathbf{0}_{1\times m}, \ \mathbf{0}_{1\times m}, \ 0\right)$$

The two programs above were both derived based on the same underlying optimization problem, but have different dimensions. While not directly useful in the formulation above, the additional unknowns in Eq. (A.39) would allow us to pose the cost function or additional constraints as a function of the control error.

In order to avoid confusion with the dimensions of the control allocation problem whose solution is sought, we will, in the discussion of linear programs for the rest of this section, use

k and l to represent the number unknowns and the number of equality constraints, respectively.

$$\mathbf{x}, \mathbf{c} \in \mathfrak{R}^k, \mathbf{b} \in \mathfrak{R}^l, \mathbf{A} \in \mathfrak{R}^{l \times k}, l \leq k$$

In general we expect that the computational requirements, both time and memory, required to arrive at a solution will scale as some function of the problem size. For a specific combination of problem and solver, however, the structure and numerical properties of a larger formulation may enable efficiencies compared to a smaller formulation.

A.2.4   Revised Standard Form with Upper Bound

The addition of slack variables to handle the upper bound on the unknown variables doubles the number of unknowns in our problem as well as adding an additional constraint for every unknown upper bound. Rather than incur this increase in problem size, it would be preferable to adopt a solution strategy that would handle problems in a *revised standard form*.

$$\min_{\mathbf{x}} \mathbf{c}^T \mathbf{x} \mid \mathbf{A}\mathbf{x} = \mathbf{b}, \mathbf{0} \leq \mathbf{x} \leq \mathbf{h} \tag{A.40}$$

Once a program is transformed into the revised form in Eq. (A.40), candidate solutions can be examined. If a given unknown variable, x_i, is at its upper (or lower) bound, a simple change in variables, $\hat{x}_i = h_i - x_i$, can be applied, reversing the bounds for that element. A new program in the revised standard form can be written,

$$\min_{\hat{\mathbf{x}}} \hat{\mathbf{c}}^T \mathbf{x} \mid \hat{\mathbf{A}}\hat{\mathbf{x}} = \hat{\mathbf{b}}, \mathbf{0} \leq \hat{\mathbf{x}} \leq \hat{\mathbf{h}} \tag{A.41}$$

where the parameters $\hat{\mathbf{c}}, \hat{\mathbf{b}}, \hat{\mathbf{h}},$ and $\hat{\mathbf{A}}$ are the same as the original quantities in Eq. (A.40), with the corresponding changes to the reversed ith component.

$$\hat{x}_i = h_i - x_i$$
$$\hat{\mathbf{A}}_{ci} = -\mathbf{A}_{ci}$$
$$\hat{c}_i = c_i$$
$$\hat{\mathbf{b}} = \mathbf{b} + h_i \hat{\mathbf{A}}_{ci} \tag{A.42}$$

This change in variables forms the background for an ad-hoc modification that can sometimes be applied to numerical algorithms intended for application to programs in the standard form, (A.18). Additional logic is added to the algorithm to identify when a variable has reached its upper bound and to apply Eq. (A.42). The solution to a program with upper bounds can thus proceed with only the non-negative inequality constraints being active.

Dealing with physical systems with real effectors, we find that most of the unknown quantities in a control allocation problem have natural upper and lower limits. Thus the revised form for linear program given in Eq. (A.18) is convenient for most of our problems. Unless specifically noted, this is the form adopted for linear programs throughout the rest of this appendix.

The representation of the constraints for our two direction-preserving formulations in this format were previously developed in Section A.2.1. The effect of Eqs (A.22) and (A.24) is to replace the control vector with that about its lower bounds, giving a problem with n constraints and $m + 1$ unknowns.

$$\min_{\mathbf{x}} \mathbf{c}^T \mathbf{x} \mid \mathbf{A}\mathbf{x} = \mathbf{b}, 0 \leq \mathbf{x}$$

$$\mathbf{A} = \begin{bmatrix} \mathbf{B} & -\mathbf{m}_{des} \end{bmatrix}$$

$$\mathbf{x} = \begin{pmatrix} \mathbf{u} - \mathbf{u}_l \\ \lambda \end{pmatrix}$$

$$\mathbf{h} = \begin{pmatrix} \mathbf{u}_u - \mathbf{u}_l \\ 1 \end{pmatrix}$$

$$\mathbf{b} = \begin{pmatrix} -\mathbf{B}\mathbf{u}_l \end{pmatrix}$$

$$\mathbf{c}^T = \begin{pmatrix} \mathbf{0}_{m \times 1}, & -1 \end{pmatrix} \tag{A.43}$$

Representing the control with respect to the errors from a preferred position, the problem has $2m + 1$ unknowns, but it still only has n equality constraints, illustrating the advantage of using this revised form.

$$\min_{\mathbf{x}} \mathbf{c}^T \mathbf{x} \mid \mathbf{A}\mathbf{x} = \mathbf{b}, 0 \leq \mathbf{x}$$

$$\mathbf{A} = \begin{bmatrix} \mathbf{B} & -\mathbf{B} & -\mathbf{m}_{des} \end{bmatrix}$$

$$\mathbf{x} = \begin{pmatrix} \mathbf{u}^+ \\ \mathbf{u}^- \\ \lambda \end{pmatrix}$$

$$\mathbf{h} = \begin{pmatrix} \mathbf{u}_u - \mathbf{u}_0 \\ \mathbf{u}_0 - \mathbf{u}_l \\ 1 \end{pmatrix}$$

$$\mathbf{b} = \begin{pmatrix} -\mathbf{B}\mathbf{u}_0 \end{pmatrix}$$

$$\mathbf{c}^T = \begin{pmatrix} \mathbf{0}_{1 \times m}, & \mathbf{0}_{1 \times m}, & -1 \end{pmatrix} \tag{A.44}$$

Example A.2

Moment Error Minimization in Revised Standard Form Consider the minimum ℓ_1-norm of the moment error presented in Eq. (A.10). We can write the individual moment errors in each axis as additional variables used to enforce the equality constraint, $\mathbf{B}\mathbf{u} = \mathbf{m}_{des} + \mathbf{m}_s$. These auxiliary variables capture the error between the commanded and achieved moment. The cost function has a minimum value of zero when the added variables are driven to zero.

$$\min_{\mathbf{u}} |w_m^T \mathbf{m}_s| \tag{A.45}$$

$$\mathbf{B}\mathbf{u} - \mathbf{m}_s = \mathbf{m}_{des}$$

$$\mathbf{u}_l \leq \mathbf{u} \leq \mathbf{u}_u \tag{A.46}$$

Using a similar transformation to that in Eq. (A.25), the added variables can be made non-negative. We replace **u** using Eq. (A.35) and the resulting linear program is written in a form similar to that seen above.

$$\min_{\mathbf{u}} |w_m^T \mathbf{m}_s| = \min_{\mathbf{u}} w_m^T \mathbf{m}_s^+ + w_m^T \mathbf{m}_s^-$$

$$\mathbf{Bu} = \mathbf{m}_{des} + \mathbf{m}_s$$

$$\mathbf{B}(\mathbf{x} + \mathbf{u}_l) = \mathbf{m}_{des} + \mathbf{m}_s^+ - \mathbf{m}_s^-$$

$$\begin{bmatrix} \mathbf{B} & -\mathbf{I_n} & \mathbf{I_n} \end{bmatrix} \begin{pmatrix} \mathbf{x} \\ \mathbf{m}_s^+ \\ \mathbf{m}_s^- \end{pmatrix} = \mathbf{m}_{des} - \mathbf{Bu}_l 0 \le \mathbf{x} \le \mathbf{u}_u - \mathbf{u}_l$$

$$0 \le \mathbf{m}_s^+$$

$$0 \le \mathbf{m}_s^- \tag{A.47}$$

Setting it up in the standard form for our solvers we see:

$$c^T = \begin{pmatrix} \mathbf{0}_m & \mathbf{w}_{\mathbf{m}n} & \mathbf{w}_{\mathbf{m}n} \end{pmatrix} \tag{A.48}$$

$$\mathbf{A} = \begin{bmatrix} \mathbf{B} & -\mathbf{I_n} & \mathbf{I_n} \end{bmatrix}$$

$$\mathbf{x}^T = \begin{pmatrix} \mathbf{u} - \mathbf{u}_l & \mathbf{m}_s^+ & \mathbf{m}_s^- \end{pmatrix}$$

$$\mathbf{h}^T = \begin{pmatrix} \mathbf{u}_u - \mathbf{u}_l & - & - \end{pmatrix}$$

$$\mathbf{b}^T = \begin{pmatrix} \mathbf{m}_{des} - \mathbf{Bu}_l \end{pmatrix} \tag{A.49}$$

Written in this form, the unknowns capturing the moment error do not have a natural upper bound. If we use a solver that guarantees the cost function is non-increasing, we can set the upper bound for each element of the error to be the initial cost $|w_m^T(\mathbf{Bu}_{init} - \mathbf{m}_{des})|$.

A.3 Properties of Linear Program Solutions

By presenting a linear program in one of the standard forms above, Eqs. (A.18) or (A.40), we have separated the constraints into a simple set of bounds and a linear system of equations,

$$\mathbf{Ax} = \mathbf{b} \tag{A.50}$$

$$\mathbf{A} \in \Re^{l \times k}, \mathbf{x} \in \Re^k, \mathbf{b} \in \Re^l$$

$$0 \le \mathbf{x} \le \mathbf{h} \tag{A.51}$$

In Eq. (A.51) we have included the upper bound, $\mathbf{x} \le \mathbf{h}$. If our program is written in the traditional standard form, Eq. (A.18), the upper bound does not apply and Eq. (A.51) becomes

$$\mathbf{x} \ge \mathbf{0}$$

Vectors satisfying the constraints in Eqs. (A.50) and (A.51) are potential solutions to the linear program. It is instructive to consider the properties of the system of constraints to better understand both the optimal solution and the methods used to find it.

Temporarily ignoring the bounds on our unknown variables and just considering the equality constraints stated in Eq. (A.50), we can make some observations about the solutions to our linear program. For an arbitrary choice of **A** and **b**, it is not guaranteed that Eq. (A.50) even has a solution.

In order to ensure that the constraints in Eq. (A.50) have a solution, we will assume that the matrix **A** is *full rank*. This assumption implies that two conditions are true about our system. Physical analogies relate both of these conditions to the underlying geometry discussed in Chapter 5.

The first is that we will assume that $l < k$; in other words we have more unknowns than equality constraints. The constraints in our control allocation formulations typically originate with a relationship, $\mathbf{Bu} = \mathbf{m}_{des}$, mapping a number of redundant control effectors into a smaller number of moments. As seen above, additional constraints and variables may be added in the process of transforming our program into the standard form. Additional equality constraints do not threaten the condition, $l \leq k$, if they correspond to additional unknowns (as an example consider the slack variables and additional constraints added in Eqs. (A.33) and (A.34)).

The opposite of the above condition—a program with fewer unknowns than equality constraints—is reminiscent of an allocation problem where there are fewer effectors than moment commands. In that case, the allocator could only achieve the desired command if the elements of $\mathbf{m}_{des}$ showed up in a certain ratio. Similarly, if we have more unknowns than constraints, we restrict the values of **b** where a valid solution to Eq. (A.50) exists.

Our full-rank assumption also implies that the rows of **A** are linearly independent. If any row can be written as a linear combination of the others, then, once again, the values of **b** for which $\mathbf{Ax} = \mathbf{b}$ has a solution are restricted. In any programs where the elements of **b** have the same linear relationships as the rows of **A**, then at least one of the constraints is redundant and can be eliminated. If **b** does not display the same relationship then the constraints are mutually exclusive and there is no solution. Bypassing the possibility that we added redundant, or inconsistent, constraints in the process of transforming the problem, this second condition can arise if our suite of control effectors does not offer truly independent control of the moments about each axis.

A.3.1   Basic Solutions

Consider the matrix describing the linear equality constraints, **A**. If we, somewhat arbitrarily, partition **A** into two sub-matrices: a square matrix, $\tilde{\mathbf{B}}$, and the remaining columns in $\tilde{\mathbf{D}}$,

$$\mathbf{A} = \begin{bmatrix} \tilde{\mathbf{B}} & \tilde{\mathbf{D}} \end{bmatrix}, \ \mathbf{A} \in \mathfrak{R}^{l \times k}, \tilde{\mathbf{B}} \in \mathfrak{R}^{l \times l}, \tilde{\mathbf{D}} \in \mathfrak{R}^{l \times k - l}, l \leq k \qquad (A.52)$$

While $\tilde{\mathbf{B}}$ is shown as the first l columns of **A**, it is trivial to rearrange the columns so that the partition could contain any specific set. If the matrix **A** is full rank, then a set of l linearly independent column vectors exist that can be used to form $\tilde{\mathbf{B}}$. This linearly independent set of vectors in $\mathfrak{R}^l$ form a *basis* for $\mathfrak{R}^l$ (Nef 1967).

Any other vector in $\mathfrak{R}^l$, in particular **b**, can be written as a linear combination of the vectors in this set. Defining $\tilde{\mathbf{B}}$ as the square matrix whose columns are these l vectors, **b** can be written,

$$\tilde{\mathbf{B}}\mathbf{x}_{\mathbf{B}} = \mathbf{b}. \qquad (A.53)$$

Solving this system for $\mathbf{x_B}$ yields a unique solution for l variables. These *basic variables* correspond to the columns of $\mathbf{A}$ that were originally selected as part of $\tilde{\mathbf{B}}$. The variables corresponding to the remaining columns in $\mathbf{A}$, those in $\tilde{\mathbf{D}}$, are *non-basic variables*.

A solution to the original equality constraints, $\mathbf{x}$, can be formed by setting all elements corresponding to the basic variables to their values in $\mathbf{x_B}$ and setting the non-basic elements, $\mathbf{x_D}$, equal to zero.

$$\mathbf{Ax} = \tilde{\mathbf{B}}\mathbf{x_B} + \tilde{\mathbf{D}}\mathbf{x_D} = \tilde{\mathbf{B}}\mathbf{x_B} + \tilde{\mathbf{D}}(0, \dots, 0)^T = \tilde{\mathbf{B}}\mathbf{x_B} = \mathbf{b} \tag{A.54}$$

This solution is called a *basic solution* to the linear program with the equality constraints in Eq. (A.50).

Recall that in Eq. (A.42), we demonstrated how to take a program in the revised standard form, where one of the solution variables was at its upper bound, and develop a related program with the corresponding variable equal to zero.

$$\hat{\mathbf{A}}\hat{\mathbf{x}} = \hat{\mathbf{b}}$$

$$\hat{x}_i = h_i - x_i = h_i - h_i = 0$$

$$\hat{\mathbf{A}}_{ci} = -\mathbf{A}_{ci}$$

$$\hat{\mathbf{b}} = \mathbf{b} + h_i \hat{\mathbf{A}}_{ci}$$

Based on the existence of this approach we extend our definition of a *basic solution* to allow the non-basic variables to be at their upper bound when confronted with a problem in the revised standard form.

We will consider a *basic solution* to our program to have at most l variables not at one of their bounds. Since each combination of l independent column vectors corresponds to a particular basic solution, there are a finite number of such solutions.

A.3.2   Degenerate Basic Solutions

In general, all of the basic variables, $\mathbf{x_B}$, are not guaranteed to be non-zero (or not equal to their upper bound). If a given solution has more than $k - l$ variables at one of their bounds, we will term it a *degenerate basic solution*. The basis set for a given degenerate solution is not unique. In constructing $\tilde{\mathbf{B}}$, any column of $\mathbf{A}$ corresponding to an element of $\mathbf{x_B}$ at its bound may be arbitrarily swapped for any non-basic column in $\tilde{\mathbf{D}}$ without changing the solution vector.

Consider, for example, the equality constraints for the direction-preserving linear program setup in Eq. (A.43).

$$\mathbf{A} = \begin{bmatrix} \mathbf{B} & -\mathbf{m}_{des} \end{bmatrix} \tag{A.55}$$

There are n constraints and $m + 1$ unknowns, so we expect a basic solution to have at most n unknowns not at one of their bounds. Recall that the unknown vector is made up of the effector positions and a scaling parameter, λ.

$$\mathbf{x} = \begin{pmatrix} \mathbf{u} - \mathbf{u}_l \\ \lambda \end{pmatrix} \tag{A.56}$$

For a three-moment command outside of Φ, and thus unattainable, the scaling parameter λ is less than one, leaving just two additional basic variables corresponding to two of the control effectors. Just as with the direct allocation case presented in Section 6.6, we expect these two effectors to define a facet of Ω. However, in the case where $\mathbf{m}_{des}$ is aligned with the image of an edge of Ω, the corresponding basic solution only involves a single non-saturated control. This basic solution is degenerate.

A.3.3   Basic Feasible Solutions

If a given solution satisfies the constraints on the problem, $\mathbf{Ax} = \mathbf{b}$, and either $\mathbf{x} \geq \mathbf{0}$ or $\mathbf{0} \leq \mathbf{x} \leq \mathbf{h}$ (the latter applying to problems in the revised standard form), it is a *feasible solution* to the original problem. By simple extension, if a basic solution is also feasible then it is termed a *basic feasible solution*.

Geometrically, if the variables in our problem are bounded, the set of feasible solution vectors, $\mathscr{X}$, is a convex k-dimensional polytope. The extreme points on the polytope all correspond to basic feasible solutions to the constraints; any feasible solution, $\mathbf{x} \in \mathscr{X}$, can be written as a linear combination of these extreme points.

Taking arbitrary values for the cost function, $\mathbf{y} = \mathbf{c}^T\mathbf{x}$, a series of parallel surfaces of at most $k - 1$ dimensions are generated. For a certain range of costs, these surfaces intersect the feasible polytope. The minimum cost, corresponding to the lowest-valued surface that is tangent to the polytope, must contain at least one extreme point.

The geometry of the extreme points of the feasible sets illustrates why basic feasible solutions are of interest to us. The relationship between these points and our desired optimal solution are stated in what is sometimes referred to as the Fundamental Theorem of Linear Programming (Luenberger 1984).

Given a linear program in the form of Eq. (A.18) where $\mathbf{A} \in \mathfrak{R}^{l \times k}$ and rank($\mathbf{A}$) = l:

  (i) If there is a feasible solution, there is a basic feasible solution.
 (ii) If there is an optimal feasible solution, there is an optimal basic feasible solution.

The proof of this theorem can be found in any standard linear programming text.

The implication for control allocation is that, for a given program, we know the optimal effector combination will occur at a basic feasible solution. Starting with the knowledge of the structure of the program, and the fact that, at the optimal solution, at least $k - l$ unknowns will be at one of their bounds, we can make assumptions about how the control effectors will behave under a particular allocator.

The number of basic feasible solutions is finite, limited by the possible combinations of k potential basic vectors taken l at a time. Thus we can envision an iterative solution that searches the extreme points of the feasible set until the optimal solution is found. Fundamentally this approach underlies many linear programming algorithms, including the influential set of algorithms based on Dantzig's original primal-simplex approach.

A.4   Allocating Feasible Commands

Up to this point we have ignored the first condition in Section A.1.2, Condition (A.5), namely how to choose an optimum control solution in the interior of Φ. In general, we can develop a solution to optimize any linear secondary cost function while enforcing a constraint on the moment achieved, $\mathbf{Bu} = \mathbf{m}_{des}$. Assuming the command is attainable, the linear programming solution is feasible. Thus, the equality constraints are guaranteed to be satisfied and the program is built to optimize the objective function in the null space of the B matrix.

Just as in the post-processed, restoring solutions discussed in Chapter 7, we consider two classes of methods for specifying a preferred solution. The first assumes that a desirable effector combination is known and the allocator seeks a solution approaching that solution (measured by minimizing some error metric). The designer has a choice of the preferred solution itself, as well as the specific error metric to be minimized, allowing this sort of allocator to accomplish a number of different objectives.

In the second case, the quality of the solution is defined by some function of the control positions. Prior to invoking the allocator, the effector positions to optimize this function are unknown. However, we do assume that a gradient expressing the linear nature of the objective function is available. If the objective function is truly linear over the range of controls, the minimum can be directly found; more typically, a locally linear approximation about the current operating point is considered, along with some range of allowable motion in the current time-step.

A.4.1 Minimizing Error to a Preferred Solution

A simple secondary objective is to minimize the difference between the solution and a known, preferred control vector (see Section 6.3). Without specifying how such a solution is known, the basic concept is to minimize some norm of the quantity $\mathbf{u} - \mathbf{u}_p$

$$\min_{\mathbf{u}} |\mathbf{u} - \mathbf{u}_p|_n \tag{A.57}$$

$$\mathbf{Bu} = \mathbf{m}_{des}$$

$$\mathbf{u}_l <= \mathbf{u} <= \mathbf{u}_u \tag{A.58}$$

The norm in Eq. (A.57) is carefully chosen to keep the program linear. In the discussion of the moment error in Eq. (A.11), two such norms were introduced. These are also available as design choices for handling the control effector norm above. These design options are to minimize a weighted sum of the magnitude of the individual surface errors, the ℓ_1 norm, and minimizing the maximum individual error, using the ℓ_∞ norm.

In either case, it is convenient to write the unknowns applying the substitution given in Eq. (A.25) relative to the preferred solution, $\mathbf{u} - \mathbf{u}_p = \mathbf{u}^+ - \mathbf{u}^-$.

$$\mathbf{u}_i - \mathbf{u}_{p_i} = \left\{ \begin{array}{l} \mathbf{u}_i^+ \ if \ \mathbf{u}_i - \mathbf{u}_{p_i} \geq 0 \\ \mathbf{u}_i^- \ if \ \mathbf{u}_i - \mathbf{u}_{p_i} < 0 \end{array} \right\} \ for \ i = 1...m \tag{A.59}$$

$$\mathbf{u}_i - \mathbf{u}_{p_i} = \mathbf{u}^+ - \mathbf{u}^-$$

$$\mathbf{u}^+ \geq \mathbf{0}$$

$$\mathbf{u}^- \geq \mathbf{0} \tag{A.60}$$

Substituting into the constraint equation,

$$\mathbf{Bu} = \mathbf{B}(\mathbf{u}^+ - \mathbf{u}^- + \mathbf{u}_p) = \mathbf{m}_{des} \tag{A.61}$$

the beginnings of the parameters for a linear program are created. The final form of the program will depend on the choice of objective function.

$$\Rightarrow \mathbf{B}\mathbf{u}^+ - \mathbf{B}\mathbf{u}^- = \begin{bmatrix} \mathbf{B} & -\mathbf{B} \end{bmatrix} \begin{pmatrix} \mathbf{u}^+ \\ \mathbf{u}^- \end{pmatrix} = \mathbf{m}_{des} - \mathbf{B}\mathbf{u}_p \tag{A.62}$$

$$\Rightarrow \mathbf{A} = \begin{bmatrix} \mathbf{B} & -\mathbf{B} \end{bmatrix}$$

$$\mathbf{x} = \begin{pmatrix} \mathbf{u}^+ \\ \mathbf{u}^- \end{pmatrix}$$

$$\mathbf{h} = \begin{pmatrix} \mathbf{u}_u - \mathbf{u}_p \\ \mathbf{u}_p - \mathbf{u}_l \end{pmatrix} \tag{A.63}$$

Minimum Total Control Error
Using the weighted sum of the absolute value of the individual error terms, we can construct a cost function that is a measure of the total error as,

$$\min_{\mathbf{u}} |\mathbf{w}_{\mathbf{u}}^T(\mathbf{u} - \mathbf{u}_p)| = \min_{\begin{pmatrix} \mathbf{u}^+ \\ \mathbf{u}^- \end{pmatrix}} |\mathbf{w}_{\mathbf{u}}^T(\mathbf{u}^+ - \mathbf{u}^-)| \tag{A.64}$$

$$= \min_{\begin{pmatrix} \mathbf{u}^+ \\ \mathbf{u}^- \end{pmatrix}} (\mathbf{w}_{\mathbf{u}}^T, \mathbf{w}_{\mathbf{u}}^T) \begin{pmatrix} \mathbf{u}^+ \\ \mathbf{u}^- \end{pmatrix} \tag{A.65}$$

From here, the parameters of the linear program can be written with the vector of decision variables $\mathbf{x} = \left(\mathbf{u}^{+T}, \mathbf{u}^{-T} \right)^T$,

$$\mathbf{A} = [\mathbf{B} - \mathbf{B}]$$

$$\mathbf{b} = \mathbf{m}_{des} - \mathbf{B}\mathbf{u}_p$$

$$\mathbf{c}^T = \left(\mathbf{w}_{\mathbf{u}}^T, \mathbf{w}_{\mathbf{u}}^T \right)$$

$$\mathbf{h} = \begin{pmatrix} \mathbf{u}_u - \mathbf{u}_p \\ \mathbf{u}_p - \mathbf{u}_l \end{pmatrix} \tag{A.66}$$

In Eq. (A.66), the preferred control vector, $\mathbf{u}_p$, and the weights on the individual controls, $\mathbf{w}_{\mathbf{u}}$, become design parameters. Some options for $\mathbf{u}_p$ were discussed in Section 6.3.

The choice of design parameters can also be used to emphasize different objectives. Researchers have explored a number of options for weighting schemes, including normalizing the magnitude of effectors or emphasizing individual effector errors that have a larger impact on some secondary objective such as drag, bending loads, or reaction control system and actuator usage. The weights can also be used to force some effectors to follow their preferred position, essentially removing them from the solution, while leaving them available for extreme commands.

Example A.3

Minimizing Combined Effector Deflections For the original example from Eqs. (A.12)–(A.14), we wish to find a control solution that minimizes the sum of the effector deflections away from their zero positions.

The original moment demand in Eq. (A.12) is attainable, so we can utilize the linear program defined in Eq. (A.66) with a preferred control position of $\mathbf{u}_p = \begin{pmatrix} 0, & 0, & 0, & 0, & 0, & 0, & 0 \end{pmatrix}^T$ and all of the control errors equally weighted $\mathbf{w_u} = \begin{pmatrix} 1, & 1, & 1, & 1, & 1, & 1, & 1 \end{pmatrix}^T$. The resulting linear program can be solved with the with the linear programming code of our choice. From the discussion in Section A.3.3, we expect at least four of the controls to be either at their preferred positions or at their maximum or minimum bounds.

Using MATLAB®'s linprog the process of finding the solution looks like:

```
%% Problem setup
md = [2.8493;-.2942;.5726]; %Desired moment
%Control effectiveness
B = [0.7073   -0.7073   -3.4956   -3.0013    3.0013    3.4956    2.1103;...
     1.1204    1.1204   -0.7919   -1.2614   -1.2614   -0.7919    0.0035;...
    -0.3309    0.3309   -0.1507   -0.3088    0.3088    0.1507   -1.2680];
%Effector limits
ul = [-0.9599; -0.9599; -0.5236; -0.5236; -0.5236; -0.5236; -0.5236];
uu = [ 0.4363;  0.4363;  0.5236;  0.5236;  0.5236;  0.5236;  0.5236];

%Control error minimization
wu = ones(7,1);
up = zeros(7,1);
c = [wu;wu];
Aeq= [B -B];
h = [uu-up ; up-ul];
beq = md-B*up;
opt=optimoptions('linprog','algorithm','simplex');
xc = linprog(c,[],[],Aeq,beq, zeros(size(h)),h,zeros(size(h)),opt);

u   = x(1:7)-xc(8:14)+up; %Final effector positions
```

The resulting effector positions that minimize the sum of the absolute deflections are:

$$\mathbf{u}^T = \begin{pmatrix} 0, & 0, & 0, & -0.39, & 0.5236, & 0.1577, & -0.2103 \end{pmatrix} \tag{A.67}$$

The optimal solution does drive three controls, both canards and the right outboard elevon, to their preferred solution, but in this case the left inboard elevon is also driven to its positive stop.

Example A.4

Minimizing Rudder Usage Continuing the previous example, we will add an additional design constraint of minimizing rudder deflections. We proceed by modifying the control error weights, $\mathbf{w_u}$, such that any rudder error will dominate the total error. In the MATLAB® code above we modify the problem setup,

```
...
%Control error minimization
wu = ones(7,1);
wu(7) = 10;
up = zeros(7,1);
c = [wu;wu];
Aeq= [B -B];
h = [uu-up ; up-ul];
...
```

The new solution is,

$$\mathbf{u}^T = \begin{pmatrix} -0.4715, & 0.2532, & 0, & -0.5236, & 0.5236, & 0.0626, & 0 \end{pmatrix}$$

We have achieved the new goal of driving the rudder to zero, but in this case only the rudder and right outboard elevon are at the preferred position. Now there are two controls that have been driven to their stops.

We can verify that the allocator still achieves the maximum moment capability by increasing the magnitude of the moment command (being careful not to ask for an unattainable moment that would make this program infeasible). Recall from the earlier use of this example, Example A.1, that doubling the moment command placed it directly on the boundary of the AMS. Solving the above program with the added scaling, md = 2*md, gives the result

$$\mathbf{u}^T = \begin{pmatrix} -0.2619 & -0.2617 & -0.5236 & -0.5236 & 0.5236 & 0.5236 & -0.5236 \end{pmatrix}$$

driving the rudder to its stop to achieve the desired moment.

It may seem counter-intuitive, but seeking the minimum combined error using the ℓ_1-norm results in some effectors deflected to their limits unless all but n can be driven to the preferred solution. This result follows from the construction of the linear program, and the application of the fundamental theorem.

The optimal solution to the linear program will be a basic feasible solution. A basic feasible solution to the program in Eq. (A.66) has n columns of $\mathbf{A}$ in the basic set. Thus we only expect at most n effectors to not be at their limit or the preferred position. No two elements of $\mathbf{x}$ representing the same effector in $\mathbf{u}^+$ and $\mathbf{u}^-$ will ever be simultaneously non-zero. For a given effector, the position is given as the difference of two variables,

$$u_i = u_i^+ - u_i^- = (u_i^+ - const) - (u_i^+ - const)$$
$$= \tilde{u}_i^+ + \tilde{u}_i^+ \tag{A.68}$$

and the individual elements of $\mathbf{u}^+$ and $\mathbf{u}^-$ can include arbitrary, corresponding, constants. However in the cost function the two variables show up with the same sign.

$$y_i = w_{ui}u_i^+ + w_{ui}u_i^- = w_{ui}(u_i^+ - const) + w_{ui}(u_i^- - const)$$
$$= w_{ui}(u_i^+ + u_i^-) - 2 * w_{ui} * const$$
$$\tilde{u}_i^+ = (u_i^+ - const) \geq 0 \rightarrow const \leq u_i^+$$
$$\tilde{u}_i^+ = (u_i^- - const) \geq 0 \rightarrow const \leq u_i^- \tag{A.69}$$

The contribution to the cost for a given effector, y_i is minimized when the value of *const* reaches its maximum value. Thus, corresponding elements of $\mathbf{u}^+$ and $\mathbf{u}^-$ will never be simultaneously non-zero in the optimal solution and a single effector cannot contribute more than one of our basic variables.

A.4.2   Minimizing Maximum Errors

Utilizing some control effectors at their limits may not be desirable, even if the total absolute deflection is less. Instead, it may be desirable to limit the maximum single deflection. Bodson and Frost (2011) discuss control effort and suggest that instead of the minimum ℓ_1-norm solution, the minimum ℓ_∞-norm may be a better choice for some problems.

The ℓ_∞-norm of a vector is the maximum of the absolute value of the vector's elements. By separating the positive and negative elements of the control error vector as before, the ℓ_∞-norm is just the maximum element of either $\mathbf{u}^+$ or $\mathbf{u}^-$,

$$|\mathbf{u} - \mathbf{u}_p|_\infty = \max_{i=1\ldots m} |\mathbf{u}_i - \mathbf{u}_{p_i}| = \max_{i=1\ldots m} (\mathbf{u}_i^+, \mathbf{u}_i^-) \tag{A.70}$$

Following the notation in Bodson and Frost (2011), we introduce a new variable u^* to represent the ∞-norm in the cost function. From Eq. (A.70), we know that u^* is greater than or equal to any element in $\mathbf{u}^+$ or $\mathbf{u}^-$, thus we add an additional set of inequality constraints to our problem.

$$u^* - \mathbf{u}_i^+ \geq 0 \ (i = 1\ldots m)$$
$$u^* - \mathbf{u}_i^- \geq 0 \ (i = 1\ldots m) \tag{A.71}$$

Following the usual procedure, these new inequalities are converted to equalities through the addition of surplus variables,

$$u^* - \mathbf{u}_i^+ - \delta\mathbf{u}_i^+ = 0 \ (i = 1\ldots m)$$
$$u^* - \mathbf{u}_i^- - \delta\mathbf{u}_i^+ = 0 \ (i = 1\ldots m) \tag{A.72}$$

The entire linear system then becomes

$$
\begin{bmatrix}
\mathbf{B} & -\mathbf{B} & \mathbf{0}_{n \times m} & \mathbf{0}_{n \times m} & \mathbf{0}_{n \times 1} \\
\mathbf{I}_{m \times m} & \mathbf{0}_{m \times m} & \mathbf{I}_{m \times m} & \mathbf{0}_{m \times m} & -\mathbf{1}_{m \times 1} \\
\mathbf{0}_{m \times m} & \mathbf{I}_{m \times m} & \mathbf{0}_{m \times m} & \mathbf{I}_{m \times m} & -\mathbf{1}_{m \times 1}
\end{bmatrix}
\begin{pmatrix}
\mathbf{u}^+ \\ \mathbf{u}^- \\ \delta\mathbf{u}^+ \\ \delta\mathbf{u}^- \\ u^*
\end{pmatrix}
=
\begin{pmatrix}
\mathbf{m}_{des} - \mathbf{Bu}_p \\ \mathbf{0}_{m \times 1} \\ \mathbf{0}_{m \times 1}
\end{pmatrix}
\tag{A.73}
$$

$$
\begin{pmatrix}
\mathbf{u}_u - \mathbf{u}_p \\
\mathbf{u}_p - \mathbf{u}_l \\
\mathbf{u}_u - \mathbf{u}_p \\
\mathbf{u}_p - \mathbf{u}_l \\
\max(|\mathbf{u}_u - \mathbf{u}_p|_\infty, |\mathbf{u}_p - \mathbf{u}_l|_\infty)
\end{pmatrix}
\geq
\begin{pmatrix}
\mathbf{u}^+ \\ \mathbf{u}^- \\ \delta\mathbf{u}^+ \\ \delta\mathbf{u}^- \\ u^*
\end{pmatrix}
\geq
\begin{pmatrix}
\mathbf{0} \\ \mathbf{0} \\ \mathbf{0} \\ \mathbf{0} \\ \mathbf{0}
\end{pmatrix}
\tag{A.74}
$$

Based on the constraints above, u^* is an upper bound on the ∞-norm of $\mathbf{u} - \mathbf{u}_p$. Any value of u^* greater than the desired norm will have all of the terms in Eq. (A.71) as strict inequalities. If u^* appears in the cost function with a positive coefficient, its contribution to the cost can be reduced to the point where one or more of the inequalities is equal to zero, and $u^* = |\mathbf{u} - \mathbf{u}_p|_\infty$, without changing any actual controls.

The final part of our linear program to minimize the ∞-norm is based on the cost function,

$$
\min_{u} \begin{pmatrix} \mathbf{0}_{6m \times 1} & 1 \end{pmatrix}
\begin{pmatrix}
\mathbf{u}^+ \\ \mathbf{u}^- \\ \delta\mathbf{u}^+ \\ \delta\mathbf{u}^- \\ u^*
\end{pmatrix}
\tag{A.75}
$$

A vector of weights, $\mathbf{w_u}$, on the individual elements of the control error can be introduced. These coefficients can be used to normalize the usage of effectors that may have vastly different effectiveness or deflection ranges. Alternatively, the weights can be used to prioritize the usage of certain effectors.

$$
u^* - \mathbf{w_{u}^{+}}_i \mathbf{u}_i^+ - \delta\mathbf{u}_i^+ = 0 \; (i = 1...m)
$$
$$
u^* - \mathbf{w_{u}^{-}}_i \mathbf{u}_i^- - \delta\mathbf{u}_i^+ = 0 \; (i = 1...m)
\tag{A.76}
$$

The end result is a modification to the constraints in Eq. (A.73), similar to that in Bodson and Frost (2011), allowing the positive and negative error components to be weighted separately.

$$
\begin{bmatrix}
\mathbf{B} & -\mathbf{B} & \mathbf{0}_{n \times m} & \mathbf{0}_{n \times m} & \mathbf{0}_{n \times 1} \\
\mathbf{W_u^+} & \mathbf{0}_{m \times m} & \mathbf{I}_{m \times m} & \mathbf{0}_{m \times m} & -\mathbf{1}_{m \times 1} \\
\mathbf{0}_{m \times m} & \mathbf{W_u^-} & \mathbf{0}_{m \times m} & \mathbf{I}_{m \times m} & -\mathbf{1}_{m \times 1}
\end{bmatrix}
\begin{pmatrix}
\mathbf{u}^+ \\ \mathbf{u}^- \\ \delta\mathbf{u}^+ \\ \delta\mathbf{u}^- \\ u^*
\end{pmatrix}
=
\begin{pmatrix}
\mathbf{m}_{des} - \mathbf{Bu}_p \\ \mathbf{0}_{m \times 1} \\ \mathbf{0}_{m \times 1}
\end{pmatrix}
\tag{A.77}
$$

$$
\begin{pmatrix}
\mathbf{u}_u - \mathbf{u}_p \\
\mathbf{u}_p - \mathbf{u}_l \\
\mathbf{u}_u - \mathbf{u}_p \\
\mathbf{u}_p - \mathbf{u}_l \\
\max(|\mathbf{u}_u - \mathbf{u}_p|_\infty, |\mathbf{u}_p - \mathbf{u}_l|_\infty)
\end{pmatrix}
\geq
\begin{pmatrix}
\mathbf{u}^+ \\ \mathbf{u}^- \\ \delta\mathbf{u}^+ \\ \delta\mathbf{u}^- \\ u^*
\end{pmatrix}
\geq
\begin{pmatrix}
\mathbf{0} \\ \mathbf{0} \\ \mathbf{0} \\ \mathbf{0} \\ \mathbf{0}
\end{pmatrix}
\tag{A.78}
$$

Expressed as the input parameters for the standard linear programming format used so far:

$$\mathbf{A} = \begin{bmatrix} \mathbf{B} & -\mathbf{B} & \mathbf{0}_{n\times m} & \mathbf{0}_{n\times m} & \mathbf{0}_{n\times 1} \\ \mathbf{W}_{\mathbf{u}}^{+} & \mathbf{0}_{m\times m} & \mathbf{I}_{m\times m} & \mathbf{0}_{m\times m} & -\mathbf{1}_{m\times 1} \\ \mathbf{0}_{m\times m} & \mathbf{W}_{\mathbf{u}}^{-} & \mathbf{0}_{m\times m} & \mathbf{I}_{m\times m} & -\mathbf{1}_{m\times 1} \end{bmatrix}$$

$$\mathbf{b} = \begin{pmatrix} \mathbf{m}_{des} - \mathbf{B}\mathbf{u}_{p} \\ \mathbf{0}_{m\times 1} \\ \mathbf{0}_{m\times 1} \end{pmatrix}$$

$$\mathbf{c}^{T} = \begin{pmatrix} \mathbf{u}^{+} & \mathbf{u}^{-} & \delta\mathbf{u}^{+} & \delta\mathbf{u}^{-} & u^{*} \end{pmatrix}$$

$$\mathbf{h} = \begin{pmatrix} \mathbf{u}_{u} - \mathbf{u}_{p} \\ \mathbf{u}_{p} - \mathbf{u}_{l} \\ \mathbf{u}_{u} - \mathbf{u}_{p} \\ \mathbf{u}_{p} - \mathbf{u}_{l} \\ \max(|\mathbf{u}_{u} - \mathbf{u}_{p}|_{\infty}, |\mathbf{u}_{p} - \mathbf{u}_{l}|_{\infty}) \end{pmatrix}$$

$$\mathbf{x}^{T} = \begin{pmatrix} \mathbf{u}^{+} & \mathbf{u}^{-} & \delta\mathbf{u}^{+} & \delta\mathbf{u}^{-} & u^{*} \end{pmatrix} \tag{A.79}$$

Example A.5

Minimizing Maximum Control Deflections One again we can look at the solution to our original example with equal weights and preferred deflections of zero. The formulation of the linear program is more complicated given the additional constraints and variables, but following the parameters given in Eq. (A.79), we can set up the problem for solution with a numerical solver

```
[n,m] =size(B);
Aeq = [B -B zeros(n,m) zeros(n,m) zeros(n,1);...
     diag(wu) zeros(m) eye(m) zeros(m) -ones(m,1); ...
     zeros(m) diag(wu) zeros(m) eye(m) -ones(m,1)];
beq = [md ; zeros(2*m,1)];

xMax = [uu ; -ul; uu; -ul; max(max(abs(uu)),max(abs(ul)))];
xMin = [zeros(4*m+1,1)];
ct = [zeros(1,4*m) 1];

opt=optimoptions('linprog','algorithm','simplex');
x = linprog(ct',A,b,Aeq,beq, xMin,xMax,zeros(4*m+1,1),opt);
ui= x(1:m) - x((1:m) + m)
```

As expected, the result has more balanced control usage than the 1-norm solution

$$\mathbf{u}^{T} = \begin{pmatrix} -0.1309, -0.1309, -0.2618, -0.2618, 0.2618, -0.2618 \end{pmatrix} \tag{A.80}$$

In contrast to the previous example, in this case none of the controls are driven to their preferred solution. However, we also have no extreme deflections, with an effector at its bounds. In fact, the solution above is the same as the direct allocation solution seen in Example 6.7. Indeed, for the case looking to minimize the maximum unweighted control deflection, the solution scaled from the boundary of the Φ is the optimal solution.

If we follow the previous example and attempt to minimize the usage of a specific effector, once again setting the weight on the rudder to 10, the optimal solution reduces the rudder usage at the expense of the other effectors.

$$\mathbf{u}^T = \left(-0.3694, 0.3694, 0.0094, -0.3694, 0.3694, 0.3620, -0.0369\right) \tag{A.81}$$

Before, increasing the weight on the rudder was sufficient to force it to its preferred position. In this case the weight just provides the ratio of the rudder error to the largest error.

A.4.3 Optimizing Linear Secondary Objectives

If a preferred solution is not known ahead of time, but the objective being minimized (or maximized) can be expressed as a linear function of the controls, that function may be incorporated into the linear program directly.

$$\min_{\mathbf{u}} F(\mathbf{u}) \tag{A.82}$$

$$F(\mathbf{u}) = f_0 + f_1\mathbf{u}_1 + \cdots + f_m\mathbf{u}_m \tag{A.83}$$

$$\min_{x} F(\mathbf{x} + \mathbf{u}_l) = \min_{\mathbf{x}}(F(\mathbf{x}) + F(\mathbf{u}_l)) \tag{A.84}$$

The addition or subtraction of a constant in the cost function does not effect the location of the optimal value for $\mathbf{x}$. As a result, we eliminate the f_0 and $F(\mathbf{u}_l)$ terms from the cost function, writing $\mathbf{c}^T = (f_1, \dots, f_n)$.

In Chapter 7, a discrete formulation of the control allocation problem was discussed. In addition to handling rate-limited effectors, this formulation allows for optimization of secondary objectives along the projection of gradient vector into the null-space of $\mathbf{B}$. The gradient vector can also be used to similar effect, defining the objective function for our linear program.

If the objective is truly linear over the entirety of the feasible set, Ω, then, combined with a constraint on the achieved moment, $\mathbf{Bu} = \mathbf{m}_{des}$, the linear program will find its optimal value at the actual, constrained minimum of F. If the linear assumption does not hold, a local gradient is evaluated about a reference condition close to the current control solution. By limiting the size of the step the effectors can take in a single frame, the solution approaches the minimum over time. The dynamics of the actuation system may be relied upon to limit the travel in any one step, or the discrete approach taken in Chapter 7 can be implemented, adjusting $\mathbf{u}$ to account for a first-order approximation of the $\Delta\Omega$ the actuator dynamics allow over a single step.

Consider, for example, a linear model of the drag on the vehicle. Ignoring interactions between the surfaces

$$D = f(\mathbf{u}_{ref} + \Delta\mathbf{u}) \approx D(\mathbf{u}_{ref}) + \frac{\partial D}{\partial\mathbf{u}_1}\big|_{\mathbf{u}_{ref}}\Delta\mathbf{u}_1 + \cdots + \frac{\partial D}{\partial\mathbf{u}_m}\big|_{\mathbf{u}_{ref}}\Delta\mathbf{u}_m \tag{A.85}$$

$$D(\mathbf{u}_{ref} + \Delta\mathbf{u}) \approx D(\mathbf{u}_{ref}) + (\nabla D(\mathbf{u}_{ref}))^T\Delta\mathbf{u} \tag{A.86}$$

Using the substitution $\mathbf{x} = \mathbf{u} - \mathbf{u}_l$ and $\mathbf{u} = \mathbf{u}_{ref} + \Delta\mathbf{u}$, then

$$D(\mathbf{u}) = D(\mathbf{u}_{ref}) + (\nabla D(\mathbf{u}_{ref}))^T(\mathbf{u} - \mathbf{u}_{ref})$$

$$= D(\mathbf{u}_{ref}) + (\nabla D(\mathbf{u}_{ref}))^T(\mathbf{x} + \mathbf{u}_l - \mathbf{u}_{ref})$$

$$= D(\mathbf{u}_{ref}) + (\nabla D(\mathbf{u}_{ref}))^T(\mathbf{u}_l - \mathbf{u}_{ref}) +$$

$$(\nabla D(\mathbf{u}_{ref}))^T(\mathbf{x}) \tag{A.87}$$

Over a given time-step, the first two terms are constant and can be ignored when finding the local minimum. The resulting objective function for our program will look like:

$$\min_{\mathbf{x}} w^T\mathbf{x} \tag{A.88}$$

$$w^T = \left(\frac{\partial D}{\partial \mathbf{u}_1}|_{\mathbf{u}_{ref}} \quad \cdots \quad \frac{\partial D}{\partial \mathbf{u}_m}|_{\mathbf{u}_{ref}} \right) \tag{A.89}$$

A.5   Building a Control Allocator for Feasible and Infeasible Solutions

Up to this point, the optimal conditions related to choosing among the multitude of solutions for an attainable command and for ensuring that we allocate commands over the entirety of Φ while handling unattainable commands were considered separately. By treating them independently, we were able to look at different ways to structure these programs. A practical linear program for control allocation must combine features that satisfy both conditions.

The toolbox of previously developed approaches to relaxing the equality constraint for unattainable commands can be combined with those for finding the optimal solution for attainable commands. The appropriate choice of algorithms for each case may depend on the problem at hand; different approaches not only define the behavior of the solution, but also will impact the computational requirements and numerical performance with different solvers.

Three linear programming approaches are presented below, and all meet the three optimality conditions we defined in Eqs. (A.5)–(A.7). The first two provide a framework for selecting from, and combining, the programs defined above. The third approach chooses the behavior for feasible and unfeasible commands to algebraically simplify the problem; while perhaps not as flexible as some of the other proposed architectures, it duplicates the output behavior of some earlier allocators that were developed independently of traditional linear programming solvers.

Examples of all three approaches are included in the ADMIRE simulation (described in Appendix B and available at companion website page to this book). Code for the individual implementations (and for an accompanying solver), are included as sub-functions in the file `LPwrap.m`. This wrapper function appears in the Simulink simulation as an interpreted function block in `NDI_CLAW/DynamicInversionControl/ControlAllocation/GetU/LP_6`. In `GetU`, this function is selected with option 6 for the constant `CAmethod`. The selection of one out of the six individual formulations included as part of `LPwrap` is controlled by an input parameter, `LPmethod`, controlled by a constant block in `NDI_CLAW/DynamicInversionControl/ControlAllocation/GetU/MakeINMAT`.

The linear programs are implemented as drop-in replacements in the frame-wise allocator used in the simulation. The default parameter values for the individual allocators can be changed in `LPwrap3`.

A.5.1 Dual Branch

The most straightforward, and most general, approach to combining our allocators for feasible and infeasible commands is to implement them both as separate parts of the allocator. Buffington (1999) describes a multi-branch approach for a proposed, modular control design for the Innovative Control Effectors concept aircraft. The discussion below follows that description, though the approach can be used more generally by choosing alternate other cost functions in either branch.

The so-called dual-branch allocator operates in two passes. The first, termed a command-feasibility branch determines if the commanded moment is in Φ. Any of the allocators that handle infeasible commands outlined in Section A.1 can be used. The cited implementation minimizes a weighted 1-norm of the moment error, equivalent to the formulation in Eq. (A.45). The feasibility branch is,

$$\min_{\mathbf{u}} w_m^T \mathbf{m}_s$$

$$\mathbf{Bu} = \mathbf{m}_{des} + \mathbf{m}_s$$

$$\mathbf{B}(\mathbf{x} - \mathbf{u}_l) = \mathbf{m}_{des} + \mathbf{m}_s^+ - \mathbf{m}_s^-$$

$$\begin{bmatrix} \mathbf{B} & -\mathbf{I}_n & \mathbf{I}_n \end{bmatrix} \begin{pmatrix} \mathbf{x} \\ \mathbf{m}_s^+ \\ \mathbf{m}_s^- \end{pmatrix} = \mathbf{m}_{des} + \mathbf{Bu}_l \qquad (A.90)$$

If the final value of the objective function from the feasibility branch indicates the command is infeasible (say, a non-zero error or $\lambda < 1$, depending on the choice of objective), the control is deemed 'deficient' and the allocator returns the optimized answer on $\partial(\Phi)$. If the objective function indicates the control is feasible, a 'control sufficiency' branch is invoked to determine the optimal solution.

Because the first branch already determined that the command is feasible, the sufficiency branch optimizes a secondary objective, enforcing $\mathbf{Bu} = \mathbf{m}_{des}$. The sufficiency branch in the reference allocator minimizes the weighted error from a preferred control solution similar to Eq. (A.66). The weighting factor and preferred solution were chosen to implement one of several modes depending on flight phase: minimum deflection, minimum drag, minimum wing loading, minimum radar signature, minimum thrust vector, or null-space injection (for system-identification).

The dual-branch approach potentially involves the solution of two linear programs per call. This computational cost is partially mitigated by recognizing that the second program can be terminated early without jeopardizing the moment allocation. Additionally, if a solver requiring an initial basic feasible solution is used, the output of the first branch may be used to initialize the second.

Two different examples of the dual-branch implementation are provided in the simulation. Both follow a similar structure, minimizing the ℓ_1 norm of the moment error to assess feasibility, and minimizing the error to a preferred solution in the second branch. The first, DB_LPCA (LPmethod $= 0$), uses the ℓ_1 norm for the second branch, while DBinf_LPCA (LPmethod $= 1$) minimizes the ℓ_∞ norm of the control error.

Prioritizing Commands

The two branches above are sufficient to meet the optimality conditions we originally defined. However, the general concept of sequentially solving the problem can be extended to offer the ability to directly prioritize moment commands, distinct from our previous efforts to prioritize certain error terms to be minimized. Because we have assumed the mapping from effector to moment achieved is linear, a complex command can be solved by sequentially optimizing a series of components arranged in priority order.

Not only can individual moment components aligned with the axes be prioritized, but, more generally, the vehicle response given unfeasible commands can be modified. Buffington (1997) examined breaking up the commanded moment into a series of component vectors that can be prioritized:

$$\mathbf{m}_{des} = \mathbf{m}_{des_1} + \cdots \mathbf{m}_{des_q} \tag{A.91}$$

where $\mathbf{m}_{des_i}$ may come from different parts of the control system, allowing the designer to choose the order in which the loops are effectively opened as the moment generating capability is saturated.

$$min_{\mathbf{u}} \lambda_p \tag{A.92}$$

$$\mathbf{B}\mathbf{u} = \lambda_1 \mathbf{m}_{des_1} + \cdots \lambda_q \mathbf{m}_{des_q} \mid \lambda_i = \begin{Bmatrix} 1 \text{ if } i < p \\ 0 \text{ if } i > p \end{Bmatrix}, i \in 1 \cdots q \tag{A.93}$$

$$0 \leq \lambda_p \leq 1 \mathbf{u}_l \leq \mathbf{u} \leq \mathbf{u}_u \tag{A.94}$$

Equations (A.93) and (A.92) specify a series of programs as p is allowed to vary from 1 to q. As each one is sequentially solved, it is necessary to check to see if $\lambda_p = 1$, stopping at the first program where the commanded moment cannot be achieved.

A.5.2   Single-branch or Mixed Optimization

Instead of solving two linear programs, a single linear program can be developed to simultaneously optimize a cost function based on achieving the desired moment for feasible commands and a secondary objective for when the command is feasible. The combined cost function is the weighted sum of the original costs.

$$\min_{\mathbf{u}} J_{comb}(\mathbf{u}) = \min_{\mathbf{u}} J_{feas}(\mathbf{u}) + \epsilon J_{suff}(\mathbf{u}) \tag{A.95}$$

$$\epsilon > 0 \tag{A.96}$$

In the cost function above, the weights on the sufficiency component are chosen such that the feasibility component is prioritized. In general, if the costs are of a similar scale and sign, then $\epsilon \ll 1$. If the sufficiency component is allowed to contribute too strongly, the feasible component will no longer govern the solution for infeasible commands and $\mathbf{B}\mathbf{u}$ may not equal $\mathbf{m}_{des}$ for feasible commands.

This form of control allocation is variously referred to as a single-branch or mixed optimization control allocator.

Example A.6

Direction Preserving and Control-error Minimization Our direction-preserving, Eq. (A.44), and control-error minimizing, Eq. (A.57), allocators can be combined into a single allocator similar to that in Buffington (1999),

$$J(\mathbf{u}) = \epsilon \parallel \mathbf{W_u}(\mathbf{u} - \mathbf{u}_p) \parallel_1 - \lambda \tag{A.97}$$

The resulting linear program becomes

$$\mathbf{A} = [\mathbf{B} \ -\mathbf{B} \ -\mathbf{m}_{des}]$$
$$\mathbf{b} = -\mathbf{B}\mathbf{u}_p$$
$$\mathbf{c}^T = (\epsilon \mathbf{w_u}^T, \epsilon \mathbf{w_u}^T, -1)$$
$$\mathbf{h} = \begin{pmatrix} \mathbf{u}_u - \mathbf{u}_p \\ \mathbf{u}_p - \mathbf{u}_l \\ 1 \end{pmatrix} \tag{A.98}$$

Assuming that $\mathbf{m}_{des}$ is three-dimensional, we know that a basic solution to Eq. (A.98) has at most three unknowns that are not at one of their bounds. For an infeasible command, $\lambda < 1$, leaving just two additional unknowns not at one of their bounds. If ϵ is small enough, the effectors will be driven to maximize λ and, for an infeasible command, the two free unknowns will be those effectors defining the facet of the AMS where the solution is found.

An example of the program in Eq. (A.98) is provided in the simulation as `SB_LPCA` (`LPmethod=5`).

Example A.7

Moment Error and Control-error Minimization Choosing to minimize the ℓ_1-norm of the moment error, Eq. (A.48), for infeasible commands, the cost function from Eq. (A.95) is

$$J(\mathbf{u}) = \parallel \mathbf{W_m}(\mathbf{B}\mathbf{u} - \mathbf{m}_{des}) \parallel_1 + \epsilon \parallel \mathbf{W_u}(\mathbf{u} - \mathbf{u}_p) \parallel_1 \tag{A.99}$$

The resulting linear program is similar to that proposed by Bodson (2001),

$$\mathbf{A} = [\mathbf{I_n} \ -\mathbf{I_n} \ -\mathbf{B} \ \mathbf{B}]$$
$$\mathbf{b} = \mathbf{B}\mathbf{u}_p - \mathbf{m}_{des}$$
$$\mathbf{c}^T = (\mathbf{w_m}^T, \mathbf{w_m}^T, \epsilon \mathbf{w_u}^T, \epsilon \mathbf{w_u}^T)$$
$$\mathbf{h} = \begin{pmatrix} \mathbf{e}_{max} \\ \mathbf{e}_{max} \\ \mathbf{u}_u - \mathbf{u}_p \\ \mathbf{u}_p - \mathbf{u}_l \end{pmatrix} \tag{A.100}$$

An example of the program in Eq. (A.100) is provided in the simulation as MO_LPCA (LPmethod=4).

Both Examples A.7 and A.6 optimize feasible solutions by minimizing a weighted sum of the absolute error between the commanded control and some preferred control vector. Bodson and Frost (2011) introduce a linear program that uses the feasible cost function, minimizing the ℓ_∞-norm from Eq. (A.73) in place of the ℓ_1 norm solution. This formulation is more complex, but also shown to have beneficial properties in balancing effector usage and displaying less sensitivity to small changes in conditions.

$$\mathbf{A} = \begin{bmatrix} \mathbf{B} & -\mathbf{B} & \mathbf{I}_{n\times n} & \mathbf{I}_{n\times n} & \mathbf{0}_{n\times m} & \mathbf{0}_{n\times m} & \mathbf{0}_{n\times 1} \\ \mathbf{W_u} & \mathbf{0}_{m\times m} & \mathbf{0}_{m\times n} & \mathbf{0}_{m\times n} & \mathbf{I}_{m\times m} & \mathbf{0}_{m\times m} & -\mathbf{1}_{m\times 1} \\ \mathbf{0}_{m\times m} & \mathbf{W_u} & \mathbf{0}_{m\times n} & \mathbf{0}_{m\times n} & \mathbf{0}_{m\times m} & \mathbf{I}_{m\times m} & -\mathbf{1}_{m\times 1} \end{bmatrix}$$

$$\mathbf{x} = \begin{pmatrix} \mathbf{u}^+ \\ \mathbf{u}^- \\ \mathbf{m_s}^+ \\ \mathbf{m_s}^- \\ \delta\mathbf{u}^+ \\ \delta\mathbf{u}^- \\ u^* \end{pmatrix}$$

$$\mathbf{b} = \begin{pmatrix} \mathbf{m}_{des} - \mathbf{Bu}_p \\ \mathbf{0}_{m\times 1} \\ \mathbf{0}_{m\times 1} \end{pmatrix}$$

$$\mathbf{h} = \begin{pmatrix} \mathbf{u}_u - \mathbf{u}_p \\ \mathbf{u}_p - \mathbf{u}_l \\ \mathbf{e}_{max} \\ \mathbf{e}_{max} \\ \mathbf{u}_u - \mathbf{u}_p \\ \mathbf{u}_p - \mathbf{u}_l \\ \max(|\mathbf{u}_u - \mathbf{u}_p|_\infty, |\mathbf{u}_p - \mathbf{u}_l|_\infty) \end{pmatrix}$$

$$\mathbf{c}^T = (\mathbf{0}_m, \mathbf{0}_m, \mathbf{w_m}^T, \mathbf{w_m}^T, \mathbf{0}_m, \mathbf{0}_m, \epsilon) \tag{A.101}$$

Multiple Secondary Objectives

So far, the discussion of the mixed optimization linear program has assumed that only a single objective function each for infeasible and feasible commands. If multiple objectives are to be balanced for feasible commands, the additional objectives either are combined in the coefficients of the original cost function, or broken out as separate terms.

The additional objectives will have be to be balanced with the original feasibility and sufficiency goals. The designer is confronted with the difficulty of setting the appropriate weighting to simultaneously balance the various objectives. One approach that has been proposed is to use this kind of formulation to implement objectives that may only apply at specific times; consider a structural load relief response that would only be triggered if the vehicle was in danger of failure due to control induced loads (Frost *et al.* 2015).

$$\min_{\mathbf{u}} J_{comb}(\mathbf{u}) = \min_{\mathbf{u}} J_{feas}(\mathbf{u}) + \epsilon J_{suff}(\mathbf{u}) + \sum_{i=1}^{i=i_{max}} \gamma_i J_i(\mathbf{u}) \qquad (A.102)$$

$$\epsilon > 0, \gamma_i \geq 0 \qquad (A.103)$$

Here, in addition to the original scaling factor ϵ, we invoke a series of weights, γ_i, that can be disabled or enabled depending on the importance of the new objective.

A.5.3 Reduced Program Size without Secondary Optimization

The discussion above focuses on setting up a linear program that takes advantage of the additional degrees of freedom allowed by an under-determined, feasible command to optimize some secondary function of the effector positions. The direct allocator (introduced in Section 6.6) and similar algorithms dealt with control redundancy by scaling feasible solutions from the intersection of the desired command direction and $\partial(\Phi)$. The same approach can be adopted when setting up the linear program, accepting a less general approach in order to reduce the size of our linear program.

The direction-preserving linear program in Eq. (A.44) has three-dimensional basic solutions (assuming that we are allocating three moments). We know that a given solution for the direct allocation algorithms presented previously was defined by at most two controls (assuming a three-objective problem; if controls are allocated for more objectives, the number is still one less than the number of objectives). It makes sense that an equivalent linear program should exist having two-dimensional basic solutions and thus only two equality constraints. Below, the approach developed by Bodson (2001) is used to eliminate one of the constraints in Eq. (A.44).

Starting with the constraints from our original direction-preserving formulation from Eq. (A.9), we remove the upper bound on λ and look for a control solution, $\mathbf{u}^*$, in the direction of the desired moment.

$$\mathbf{B}\mathbf{u}^* = \lambda \mathbf{m}_{des}$$

$$0 \leq \lambda$$

$$\mathbf{u}_l \leq \mathbf{u}^* \leq \mathbf{u}_u \qquad (A.104)$$

If $\mathbf{m}_{des} = \mathbf{0}$, then the solution to Eq. (A.9) is trivial: $\mathbf{u} = \mathbf{0}$. If at least one component of the desired moment is non-zero, $m_{d_i} \neq 0$, then the rows of $\mathbf{B}$ and $\mathbf{m}_{des}$ can be reordered such that m_{d_1} is non-zero. Solving to eliminate λ from the constraint corresponding to this non-zero moment,

$$\lambda \mathbf{m}_{des_1} = \mathbf{B}_{r1} \mathbf{u}^* \rightarrow \lambda = \frac{\mathbf{B}_{r1}\mathbf{u}^*}{\mathbf{m}_{des_1}}, \mathbf{m}_{des_1} \neq 0 \qquad . \quad (A.105)$$

We can thus substitute this result into the other constraints,

$$\mathbf{0} = \lambda \mathbf{m}_{des_i} - \mathbf{B}_{ri}\mathbf{u}^* = \frac{\mathbf{B}_{r1}\mathbf{u}^*}{\mathbf{m}_{des_1}}\mathbf{m}_{des_i} - \mathbf{B}_{ri}\mathbf{u}^*, i = 2...n \qquad (A.106)$$

$$\mathbf{0} = \mathbf{B}_{r1}\mathbf{u}^*\mathbf{m}_{des_i} - \mathbf{B}_{ri}\mathbf{u}^*\mathbf{m}_{des_1} \qquad (A.107)$$

$$\mathbf{0} = (\mathbf{B}_{r1}\mathbf{m}_{des_i} - \mathbf{B}_{ri}\mathbf{m}_{des_1})\mathbf{u}^* \qquad (A.108)$$

Combining all $n - 1$ remaining constraints together, they can be written as:

$$\mathbf{MBu}^* = \mathbf{0} \tag{A.109}$$

where,

$$\mathbf{M} = \begin{pmatrix} y_{d_2} & -y_{d_1} & \cdots & 0 \\ \vdots & 0 & \ddots & 0 \\ y_{d_n} & 0 & \cdots & -y_{d_1} \end{pmatrix} \tag{A.110}$$

This provides $n - 1$ equality constraints that ensure the solution is in the direction of $\mathbf{m}_{des}$. To derive our cost function, we seek to maximize the component of the achieved moment, $\mathbf{Bu} *$, in the direction of $\mathbf{m}_{des}$.

$$\min_{\mathbf{u}^*} \quad -(\mathbf{Bu}^*)^T \mathbf{u}^*$$

$$\mathbf{MBu}^* = \mathbf{0}$$

$$\mathbf{u}_l \leq \mathbf{u}^* \leq \mathbf{u}_u \tag{A.111}$$

We can then apply the transformations to recenter the problem and pose it in the standard form of Equation (A.40) with

$$\mathbf{A} = \mathbf{MB} \tag{A.112}$$

$$\mathbf{b} = -\mathbf{Au}_l \tag{A.113}$$

$$\mathbf{c}^T = -\mathbf{B}^T \mathbf{m}_{des}$$

$$\mathbf{h} = \mathbf{u}_u - \mathbf{u}_l \tag{A.114}$$

The result of this program is the solution corresponding to the moment on the boundary of Φ, $\mathbf{m}_{des}^*$. To find the actual desired moment, the scaling factor, λ, eliminated above, is recomputed. First, the lower limit is added as before, $\mathbf{u}^* = \mathbf{x} + \mathbf{u}_l$. Then we compute the scaling factor,

$$\lambda = \frac{\mathbf{m}_{des}^T \mathbf{Bu}^*}{\mathbf{m}_{des}^T \mathbf{m}_{des}} \tag{A.115}$$

If $\lambda > 1$, then the desired objective is on the interior of Φ and the correct solution is $\mathbf{u} = \mathbf{u}^*/\lambda$. Otherwise, the command is unachievable and $\mathbf{u} = \mathbf{u}^*$.

The allocator represented by Eqs. (A.111)–(A.115) is available as `DPscaled_LPCA`, using `LPmethod = 3`.

A.6 Solvers

The final component that has to be addressed in the implementation of a control allocator through linear programing is that of the numerical routines invoked in the solution. Faced with a control allocator cast as a linear program and desiring a method to solve it, the interested researcher will uncover an array of available software packages that can be used. The code from available linear programming libraries has been utilized to successfully solve control allocation programs by many researchers (Oppenheimer et al., 2010). Indeed, we have already made use of such a canned routine by invoking MATLAB®'s `linprog` function in our examples.

Most generic linear programming libraries are more than capable of handling programs of the scale and format encountered in solving typical control allocation problems. In fact, one potential challenge in implementing a pre-existing library is the degree that any pre-computation to handle generic problems and any optimizations targeted at very large, sparse, systems impacts the capability to rapidly solve a series of small dense problems in a known format. Understanding the underlying algorithms is useful in choosing a software package, optimizing an existing code for use on a specific problem, or in implementing a solution algorithm from scratch.

Common algorithms for linear programming fall in one of two groups: basis exchange methods and interior-point methods. Basis exchange methods, as typified by the classic simplex method and its many variants, traverse a series of basic solutions before arriving at the optimal answer. A newer class of algorithms are interior-point methods that borrow some familiar concepts from non-linear optimization. These algorithms iterate away from the vertices of the feasible set and progress through solutions in the interior of the set, continually moving toward the optimal vertex.

A.6.1   Preprocessing

In the various introductory allocators examined in Section A.5, it was simple to enter the constraints and cost function for the direction-preserving program directly into MATLAB®'s `linprog` command. If we seek the minimum error solution for the same problem, we quickly see that our statement in Eq. (A.11) needs to be transformed into a different form prior to calling our solution code. The inputs to `linprog` are less restrictive than the standard form in Eq. (A.18), the generic toolbox function obscures a good deal of preprocessing that transforms these generic constraints into a more specific form prior to invoking an internal solution algorithm.

Automating the transformation of a problem from a flexible format for the user into the one expected by the selected solution algorithm is one of the preprocessing steps often included in generic linear programming codes. This pre-solve step will also typically look for redundant constraints and variables that can be algebraically removed from the solution. For large programs, it may even be possible to identify partitions of the problem that are decoupled and can be solved separately.

Many traditional applications of linear programing involve large, often sparse sets of constraints and variables; the reduction in problem size and subsequent steps to reorder variables and optimize the factoring of the underlying matrices may have a significant impact on the speed, memory requirements, and robustness of the solution.

By contrast, the formulations of the control allocation problem as linear programs tend to be small and dense, so the scale of the benefit realized by these methods may not be as great. Additionally, because we are interested in allocating controls as part of a real-time control law, the solver will be presented with a series of similar problems of a known form at every step. When presented with a candidate solver, it makes sense to identify those input steps that do not need to occur at every time point.

The more complicated linear programs for control allocation combine terms dealing with different physical quantities (such as control deflections and moments/angular accelerations) in the same vector of decision variables. It is important to consider the approximate scale of the physical units and consider pre-scaling variables to bring them all to a similar order of magnitude – improving the performance of numerical algorithms.

In this text, we have posed allocators directly in the standard form with upper bounds, Eq. (A.40). This form is easily handled by many common simplex-type solvers, such as that included with the simulation on the text website, `simplexuprevsol`. By stating the problem directly in the final form, we can structure the inputs and on-board model for the allocator in a way that allows us to avoid repeating those transformations at every step.

Many interior-point implementations more naturally handle the constraints expressed as linear inequalities. When using such a solver, the input programs can be recast directly into the following form:

$$\min_{\mathbf{x}} \mathbf{c}^T \mathbf{x} \mid \mathbf{A}\mathbf{x} = \mathbf{b}, \mathbf{x} \geq \mathbf{0} \qquad (A.116)$$

Equation (A.116) is sometimes referred to as the *canonical* form for a linear program. Confusingly, it is also common to find it called the *standard* form. This is particularly true in works discussing interior-point methods.[4]

A.6.2   Solution Algorithms

Numerical algorithms to solve linear programs fall into two main categories based on potential solutions visited as they iterate toward an answer. The first set move between basic solutions to the constraints until reaching the optimal basic feasible solution; the iterates in the second category may be non-basic, but they converge, approaching arbitrarily close to the optimal solution. Methods in both categories have been applied to control allocation problems, though below we will detail a single common approach.

The popular simplex method and its variants are a sort of basis exchange method. At a high level, the simplex follows the general approach suggested in Section A.3.3. Starting with an initial basic feasible solution, the algorithm visits a sequence of basic feasible solutions, each one a neighbor of the previous step. The steps are chosen so that the value of the objective function is non-increasing.[5] In this manner, vertices of the polytope bounding the feasible set are visited until the optimal solution is visited. Variants of the primal simplex method differ on how they choose the decision variables to swap in choosing a neighboring basic feasible solution.

Interior-point methods for linear programming are a more recent innovation, offering a better theoretical upper bounds on performance than simplex methods. In practice, the performance difference on relatively small, dense programs (like our control allocation programs) is not large; example allocators set up to compare interior-point solutions for aircraft problems have not shown significant speed improvements over the baseline simplex solvers (Petersen and Bodson 2005). Even so, interior-point solvers are of interest because they have properties that may be valuable compared to simplex solvers.

The iterations in the simplex method can involve basic solutions far from the final solution; additionally, because the algorithm steps along edges of the feasible set, the algorithm may

---

[4] Similarly, *canonical* form shows up variously as that above and the related formulation $\min_{\mathbf{x}_B, \mathbf{x}_N} \mathbf{c}_B^T \mathbf{x}_B + \mathbf{c}_N^T \mathbf{x}_N \mid \bar{\mathbf{A}}_N \mathbf{x}_N + \mathbf{I}\mathbf{x}_B = \mathbf{b} \mid \mathbf{x}_B \geq \mathbf{0}, \mathbf{x}_N \geq \mathbf{0}$. Fortunately the terminology is usually clear from context

[5] Degenerate solutions can lead to cases where the best possible neighboring points yield no reduction in the cost function. In theory, such a situation could result in the algorithm *cycling*, getting stuck repeatedly re-visiting members of the same subset of solutions. In theory, such a situation may result in the algorithm getting stuck cycling among the same set of basis solutions. Anti-cycling codes are simple, and relatively inexpensive to implement, and are good insurance even if problems that cycle are rare in practice. The `simplexuprevsol` code included as part of `LPwrap.m` in the simulation incorporates a simple anti-cycling procedure

take steps of a far from uniform size. The end result is that the simplex may temporarily stall, or iterate with little improvement in the cost function, before continuing. This behavior is problematic in a real-time algorithm if one cannot guarantee that it will reach an optimal solution prior to overrunning the available resources in the current frame. When forced to terminate early, we expect the solution to be better than the initial condition, but it is hard to predict how much better.

Interior-point methods, by contrast, display much more continuous improvement as they iterate. The convergence may slow down as the solution approaches the bounds of the feasible set. As long as a good starting point – away from this slow convergence – is chosen, the iteration appears more conducive to exiting early.

A.6.3   Simplex Method

The simplex method for solving linear programs dates to the 1940s and its development accompanied the formalization of linear programming problems themselves; variants on the simplex are still in wide use. Our interest in using the simplex is two-fold. First, our control allocation problems typically have features that are convenient for the primal simplex method: they are relatively small and dense with more unknowns than equality constraints. Second, and more importantly for our purposes, a basic simplex method is relatively straight-forward to develop.

The simplex method is an iterative algorithm that starts at a basic feasible solution and then proceeds to move along a series of adjacent basic feasible solutions, changing one variable in the basic set each iteration. In this manner the algorithm moves along the edges joining neighboring extreme points until the optimum is found. The presence of degenerate basic solutions means that each iteration does not necessarily correspond to a new vertex. But because the number of basic feasible solutions is finite, if the problem is bounded, eventually we will find the solution.[6]

1. **Initialization** Choose an initial set of basic variables corresponding to a basic feasible solution. If a non-basic variable is at its upper bound, apply Eq. (A.42).
2. **Pricing** Compute the relative cost of each non-basic variable; that is, how much the cost corresponding to the current basis will change if a non-basic variable is added to the solution.
3. **Entering Variable** Choose one of the non-basic variables that will not increase the cost if introduced; if none exists then the current solution is optimal.
4. **Exiting Variable** Identify which bound the basic variables move toward as the new, entering, basic variable is increased. If the entering variable reaches a bound before any of the current basic variables, it will remain a non-basic variable. If one of the current basic unknowns is the first to reach a a bound choose it to leave the basis.
5. **Adjust Bounds** If the exiting variable (or the entering variable if it is remaining non-basic) was driven to its upper bound, apply Eq. (A.42).

---

[6] While the existence of a theoretical upper bound on the number of iterations is a desirable property for an algorithm we wish to execute in real time, in practice this bound scales as an exponential function of the number and is intractable for all but the smallest problems. The continued popularity of simplex algorithms is, in large part, due to the wide gap between the pathological examples, constructed by mathematicians, that visit all possible solutions and the observed performance on real problems (Gill, Murray, and Wright 1981). Simplex solvers have been shown to have good performance on control allocation type problems with both realistic and randomized coefficients (Beck 2002; Petersen and Bodson 2005)

6. **Basis Update** Define a new basic feasible solution exchanging the entering and leaving variables (setting the new non-basic variable to zero and solving for the solution using the new basic set).
7. **Iterate** Using the new solution as the initial condition, iterate from step 4 until the optimal solution is identified.
8. **Assemble Final Solution** Construct the final solution, setting the non-basic variables to zero or their upper bounds based on how many times the procedure in Eq. (A.42) was applied.

The many variations on simplex methods include different options selecting which of the variables enter and leave the basis at each step as well as different factorization methods intended to improve the stability of the solution. The approach followed here is that utilized in the example function (`simplexuprevsol`, included as a sub-function of `LPwrap.m` in the accompanying simulation), and has been kept simple for clarity. The reader interested in implementing their own solver, or just desiring a deeper understanding, is encouraged to seek any of the many available texts on linear programming (Gale, 1960; Luenberger and Ye 2008).

The basic form of the method above was derived assuming that the problem is in the standard form of Eq. (A.18). An ad-hoc modification in steps 1, 4, 5, and 8 allows us to handle upper bounds by reversing the sign on selected unknowns as described previously.

$$\hat{x}_i = h_i - x_i \qquad (A.117)$$

$$\hat{\mathbf{A}}_{ci} = -\mathbf{A}_{ci}$$

$$\hat{c}_i = c_i$$

$$\hat{\mathbf{b}} = \mathbf{b} + h_i \hat{\mathbf{A}}_{ci}$$

Equation (A.117) is applied when whenever a variable is identified that is about to reach its upper bound. This allows the method to proceed as if the problem had no upper bounds. We keep track of a list of which unknowns have been reversed and, upon completion, re-apply Eq. (A.117) to generate the desired answer (assuming of course that subsequent algorithm steps have not already re-reversed them for us).

The method is illustrated in detail in the following example.

Example A.8

Two-dimensional Direction Preserving using the Simplex Method We simplify the example above in Eq. (A.13) to consider a two-dimensional allocator that utilizes only the outboard and inboard elevons, $\mathbf{u}_3$–$\mathbf{u}_6$, to meet a combined pitch/roll acceleration command while preserving direction for unfeasible commands.

$$\mathbf{m}_{\text{des}} = \begin{pmatrix} 6.5 \\ 0.5 \end{pmatrix} \qquad (A.118)$$

Using the formulation presented in Eq. (A.23) we write the problem in standard form as:

$$\min_{\mathbf{x}} \mathbf{c}^T \mathbf{x} \mid \mathbf{A}\mathbf{x} = \mathbf{b}, \mathbf{h} \geq \mathbf{x} \geq \mathbf{0}$$

$$\begin{bmatrix} -3.4956 & -3.0013 & 3.0013 & 3.4956 & -6.5 \\ -0.7919 & -1.2614 & -1.2614 & -0.7919 & -0.5 \end{bmatrix} \mathbf{x} = \begin{pmatrix} 0 \\ -2.1502 \end{pmatrix}$$

$$\mathbf{x} \geq \mathbf{0}$$

$$\min_{\mathbf{x}} y = \begin{pmatrix} 0 & 0 & 0 & 0 & -1 \end{pmatrix} \mathbf{x}$$

$$\mathbf{x} = \begin{pmatrix} \mathbf{u} - \mathbf{u}_l \\ \lambda \end{pmatrix} \tag{A.119}$$

The linear program has two equality constraints and five unknowns: the rank of $\mathbf{A}$ is two and a solution to the equality constraints can be written as a linear combination of two columns of $\mathbf{A}$.

For any basic solution, at least three of the unknowns will be zero. Thus for an infeasible moment command, at least three of the surfaces will be at a lower or an upper bound (recall that we will apply Eq. (A.117)). If the command is feasible, then $\lambda = 1$, freeing an additional surface to move off of its bound. With this knowledge we can begin to see how the basic feasible solutions at each simplex iteration map to the moment (accelerations), Figure A.2, and control, Figure A.3, spaces.

Initialization As an initial control solution we choose,
$\mathbf{u} = (-0.5236 \quad 0.5236 \quad -0.5236 \quad 0.4861)^T$.
Since $\mathbf{u}_2$ is at its upper bound, we will modify the problem as in Eq. (A.42), introducing $\mathbf{E}$ to record any changes in sign we have made.

$$\mathbf{A} = \begin{bmatrix} -3.4956 & 3.0013 & 3.0013 & 3.4956 & -6.5 \\ -0.7919 & 1.2614 & -1.2614 & -0.7919 & -0.5 \end{bmatrix} \tag{A.120}$$

$$\mathbf{b} = \begin{pmatrix} 3.1430 \\ -2.1502 \end{pmatrix}$$

$$\mathbf{E}^T = \begin{pmatrix} + & - & + & + & + \end{pmatrix} \tag{A.121}$$

The corresponding initial solution is $\mathbf{x} = (0, 0, 0, 1.0097, 0.0594)^T$. This point satisfies all of the constraints in Eq. (A.119) and can be used as an initial basic feasible solution to the problem with $\mathbf{x_B} = (\mathbf{x_4}, \mathbf{x_5})^T$ as basic unknowns and $\mathbf{x_D} = (\mathbf{x_1}, \mathbf{x_2}, \mathbf{x_3})^T$ as the set of non-basic unknowns. From, Figure A.2, the corresponding moment, $\mathbf{m} = (0.3864, 0.0297)^T$, denoted as point 0, is in the desired direction, but the scale factor is small $\lambda = 0.0594$.

The cost function at the initial point is partitioned as $\mathbf{c}^T = (\mathbf{c_B} \quad \mathbf{c_D})$, where $\mathbf{c_B}^T = (0 \quad -1)$ and $\mathbf{c_D}^T = (0 \quad 0 \quad 0)$. The resulting cost, $y = \mathbf{c_B}^T \mathbf{x_B} + \mathbf{c_D}^T \mathbf{x_D}$ is equal to -0.594 for this initial condition.

Pricing Based on our partitioning of $\mathbf{A}$ in Eq. (A.54), we want to choose a non-basic variable to enter the basis. In order to continue satisfying the equality constraint, a change to one of the non-basic variables must be accompanied by a corresponding change in the basic variables. For example, if we want to increase x_i from zero we can solve for the required change in $(\mathbf{x_4}, \mathbf{x_5})^T$

$$\tilde{\mathbf{D}} i x_i + \tilde{\mathbf{B}} \begin{pmatrix} \mathbf{x_4} \\ \mathbf{x_5} \end{pmatrix} = \mathbf{0} \rightarrow \tilde{\mathbf{B}} \begin{pmatrix} \mathbf{x_4} \\ \mathbf{x_5} \end{pmatrix} = -\tilde{\mathbf{D}}_1 x_i \tag{A.122}$$

$$\begin{pmatrix} \mathbf{x_4} \\ \mathbf{x_5} \end{pmatrix} = -\tilde{\mathbf{B}}^{-1} \tilde{\mathbf{D}}_i x_i \tag{A.123}$$

The relative change in the cost of the current solution as a new variable is introduced can be computed based on Eq. (A.123). The addition of the new variable changes the cost by $c_i x_i$. The accompanying changes in the basic unknowns from Eq. (A.123) correspond to a change in cost of $\mathbf{c_b}(-\tilde{\mathbf{B}}^{-1} \tilde{\mathbf{D}}_i x_i)$. Thus for every unit change in the new variable, the cost will change by:

$$r_{Di} = c_i - \mathbf{c_b} \tilde{\mathbf{B}}^{-1} \tilde{\mathbf{D}}_i x_i \tag{A.124}$$

Equation (A.124) is used to compute the relative cost for our non-basic variables, x_1, x_2, and x_3.

The first, $\mathbf{x_{D1}} = \mathbf{x_1}$, corresponds to moving the right outside elevon off of its lower limit. Using Eq. (A.123), we compute that, for each degree increase in value, the basic variables (x_4, x_5) must change by $(-0.4930, -0.8029)$. Every degree increase in the right outboard elevon position must be accompanied by a $-0.4930°$ decrease in the left outboard elevon and a corresponding decrease in the scaling factor, λ, increasing the cost function by 0.8029.

$$r_{D1} = 0 - \begin{pmatrix} 0 & -1 \end{pmatrix} \begin{bmatrix} 3.4956 & -6.5 \\ -0.7919 & -0.5 \end{bmatrix}^{-1} \begin{pmatrix} -3.4956 \\ -0.7919 \end{pmatrix}$$

$$= 0 - \begin{pmatrix} 0 & -1 \end{pmatrix} \begin{pmatrix} 0.4930 \\ 0.8029 \end{pmatrix} = 0 - 0.8029$$

$$r_{D1} = 0.8029 \tag{A.125}$$

The process in Eq. (A.125) is repeated for $\tilde{\mathbf{D}}_2 = \mathbf{x_2}$ and $\tilde{\mathbf{D}}_3 = \mathbf{x_3}$,

$$r_{D2} = 0 - \begin{pmatrix} 0 & -1 \end{pmatrix} \begin{bmatrix} 3.4956 & -6.5 \\ -0.7919 & -0.5 \end{bmatrix}^{-1} \begin{pmatrix} 3.0013 \\ 1.2614 \end{pmatrix}$$

$$= 0 - \begin{pmatrix} 0 & -1 \end{pmatrix} \begin{pmatrix} -0.9715 \\ -0.9842 \end{pmatrix} = 0 - 0.9842$$

$$r_{D2} = -0.9842 \tag{A.126}$$

$$r_{D3} = 0 - \begin{pmatrix} 0 & -1 \end{pmatrix} \begin{bmatrix} 3.4956 & -6.5 \\ -0.7919 & -0.5 \end{bmatrix}^{-1} \begin{pmatrix} 3.0013 \\ -1.2614 \end{pmatrix}$$

$$= 0 - \begin{pmatrix} 0 & -1 \end{pmatrix} \begin{pmatrix} 1.4067 \\ 0.2948 \end{pmatrix} = 0 - 0.2948$$

$$r_{D3} = 0.2948 \tag{A.127}$$

Entering Variable From Eqs. (A.125)–(A.127), we see that only x_2, corresponding to the right inboard elevon, will decrease the cost function if it is added to the solution. Recall that we reversed the sign on x_2 in the original solutions, so when it enters the solution it will be moving off of its upper bound.

If multiple non-basic variables had negative entries in $\mathbf{r_D}$, then the largest magnitude would correspond with the quickest reduction in cost.[7] If there were all positive entries in $\mathbf{r_D}$, then any variable introduced into the basic set would increase the cost and the current solution would be the optimum.

Exiting Variable Next we determine the current solution variable to leave the basis and be replaced by x_2. From Eq. (A.123), we can compute how much each of the current basic variables will change as we add x_2

$$\mathbf{y_B^*} = \begin{pmatrix} \mathbf{x_4} \\ \mathbf{x_5} \end{pmatrix} = \tilde{\mathbf{B}}^{-1} \tilde{\mathbf{D}}_2 x_2$$

$$= \begin{bmatrix} 3.4956 & -6.5 \\ -0.7919 & -0.5 \end{bmatrix}^{-1} \begin{pmatrix} 3.0013 \\ 1.2614 \end{pmatrix} x_2$$

$$= \begin{pmatrix} -0.9715 \\ -0.9842 \end{pmatrix} x_2 \tag{A.128}$$

Both entries are negative, indicating that the basic variables increase as x_2 enters the solution. (At this point, if we were not including upper bounds, we would terminate, knowing the problem was unbounded). The negative entries indicate we need to check the corresponding variables against their upper bound when we seek the first variable to be driven to its bound as x_2 is added.

For a given basic variable, we can determine how fast it is driven to its bound by examining the ratio,

$$ratio_i = \begin{cases} \dfrac{y_{0_i}}{y_i^*} & if\, y_i^* > 0 \\ \dfrac{h_i - y_{0_i}}{y_i^*} & if\, y_i^* < r0 \end{cases} \Bigg\} \, i \in (4, 5) \tag{A.129}$$

The result is,

$$\mathbf{ratio} = \begin{pmatrix} 0.0386 & 0.9557 \end{pmatrix} \tag{A.130}$$

The smallest positive ratio corresponds to x_4, the left outboard elevon, indicating that it will be driven to its bound when x_2 has increased to 0.0386. Thus we choose to replace x_4 with x_2 in the basic set. If the increase in x_2 was greater than h_2, then it would be limited before x_4 entered. In this case, we would apply Eq. (A.117) to x_2, moving it to the opposite bound, and continue the next iteration with x_4 still in the basic set.

Adjust Bounds and Update Basis The new basic set is $\mathbf{x_B} = (x_2, x_5)$, and the corresponding non-basic variables are $\mathbf{x_D} = (x_1, x_4, x_5)$. Because of the sign on y_4^*, the sign on x_4 is reversed in Eq. (A.128).

---

[7] In general, any non-basic variable with a negative relative cost can be chosen to enter the cost function and multiple rules for choosing the variable to enter are found in the literature. One such rule, Bland's rule, always chooses the non-basic variable with the lowest index. The simplex solver included with the code accompanying the text invokes this rule in cases where it might cycle at a degenerate basis

$$\mathbf{A} = \begin{bmatrix} -3.4956 & 3.0013 & 3.0013 & -3.4956 & -6.5 \\ -0.7919 & 1.2614 & -1.2614 & 0.7919 & -0.5 \end{bmatrix} \tag{A.131}$$

$$\mathbf{b} = \begin{pmatrix} -0.5176 \\ 0 \end{pmatrix}$$

$$\mathbf{E}^T = (+ \; - \; + \; - \; +) \tag{A.132}$$

Solving for the solution based on this new partitioning of $\mathbf{A}$, $x = (0 \;\; 0.0386 \;\; 0 \;\; 0 \;\; 0.0974)^T$ and the new cost is $y_1 = -0.097$.

Iterate We repeat the process of finding the variable to leave the basis. This time writing Eq. (A.124) in matrix form to solve for the relative costs of all of the non-basic variables simultaneously:

$$r_D = \mathbf{c_D}^T - \mathbf{c_b}^T \tilde{\mathbf{B}}^{-1} \tilde{\mathbf{D}} \quad \mathbf{x_D} = (0.030 \quad 1.013 \quad -1.13) \tag{A.133}$$

The term corresponding to the non-basic variable x_3 represents the largest cost reduction and will enter the basis. Computing the ratio

$$\mathbf{y}*_B = \begin{pmatrix} \mathbf{x_2} \\ \mathbf{x_5} \end{pmatrix} = \tilde{\mathbf{B}}^{-1} \tilde{\mathbf{D}}_1 x_3 = \begin{pmatrix} -1.448 \\ -1.13 \end{pmatrix} x_3 \tag{A.134}$$

Once again both entries are negative and both will be checked against their upper bound.

$$ratio_i = \frac{-y_i^*}{y_{Oi}}, i \in (2, 5)$$

$$\mathbf{ratio} = (0.696 \quad 0.798) \tag{A.135}$$

The minimum positive value corresponds to x_2. It will be replaced by x_3 and the new basis set becomes (x_3, x_9). We refrain from showing $\mathbf{A}$, $\mathbf{b}$, and $\mathbf{c}$, but note that, once again, the column corresponding to the exiting variable is modified. The new solution is $x = (0 \;\; 0 \;\; 0.696 \;\; 0 \;\; 0.0885)^T$ and the corresponding cost is $y_1 = -0.885$, decreasing from the previous step.

We continue to repeat the above steps as outlined in the Table A.1, noticing that on the next iteration all of the relative costs are positive, indicating that introducing any one of the non-basic variables into the solution will increase the cost. Thus we have found the optimum solution.

In the final solution we use $\mathbf{E}$ to rewrite $\mathbf{x}$ as the optimal effector solution, $\mathbf{u} = (-0.5236, -0.5236, 0.1729, 0.5236)$, both right-side elevons are at their lower limit, the left outside elevon is at its upper limit and the edge of $\partial(\Phi)$ is defined by the left inside elevon.

The sequence of steps taken to get to the solution is plotted in Figures A.2 and A.3. In general, the operation of the simplex algorithm is such that it moves through basic solutions along the boundary of the feasible region until it reaches the optimal point. For the direction-preserving control allocator above, each of these basic solutions corresponds to a solution on the boundary of $\partial(\Omega)$ that preserves the direction of the command.

Table A.1 Sequence of Simplex iterations to solve Example A.8

Iter.	$x = \begin{pmatrix} 0 & 0 & 0 & 1.0097 & 0.059 \end{pmatrix}$		To enter
0	$E = \begin{pmatrix} + & - & + & + & + \end{pmatrix}$		
Cost	$x_D = \begin{pmatrix} x_1 & x_2 & x_3 \end{pmatrix}$	$r_d = \begin{pmatrix} 0.80 & -0.98 & 0.29 \end{pmatrix}$	x_2
-0.059	$x_B = \begin{pmatrix} x_4 & x_5 \end{pmatrix}$		
	$y_q = \begin{pmatrix} -0.97 & -0.98 \end{pmatrix}$		To leave
	$ratio = \begin{pmatrix} 0.0386 & 0.9557 \end{pmatrix}$		x_4
Iter.	$x = \begin{pmatrix} 0 & 0.0386 & 0 & 0 & 0.9748 \end{pmatrix}$		To enter
1	$E = \begin{pmatrix} + & - & + & - & + \end{pmatrix}$		
Cost	$x_D = \begin{pmatrix} x_1 & x_4 & x_3 \end{pmatrix}$	$r_d = \begin{pmatrix} 0.303 & 1.013 & -1.1304 \end{pmatrix}$	x_2
-0.097	$x_B = \begin{pmatrix} x_2 & x_5 \end{pmatrix}$		
	$y_q = \begin{pmatrix} -1.448 & -1.1304 \end{pmatrix}$		To leave
	$ratio = \begin{pmatrix} 0.6965 & 0.7984 \end{pmatrix}$		x_3
Iter.	$x = \begin{pmatrix} 0 & 0 & 0.69649 & 0 & 0.885 \end{pmatrix}$		To enter
2	$E = \begin{pmatrix} + & + & + & - & + \end{pmatrix}$		N/A
Cost	$x_D = \begin{pmatrix} x_1 & x_4 & x_2 \end{pmatrix}$	$r_d = \begin{pmatrix} 0.6996 & 0.209554 & 0.78061 \end{pmatrix}$	At optimum
-0.885	$x_B = \begin{pmatrix} x_3 & x_5 \end{pmatrix}$		
			To leave
			N/A

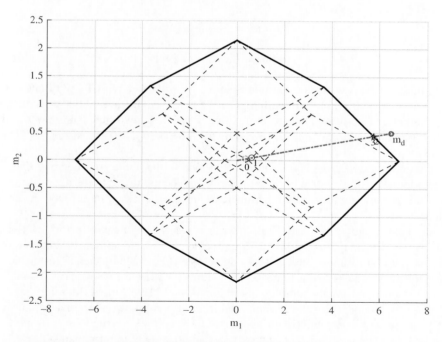

Figure A.2 Progression of angular accelerations showing the convergence to the maximum scaling parameter in the desired moment direction for the simplex iterations to solve Example A.8

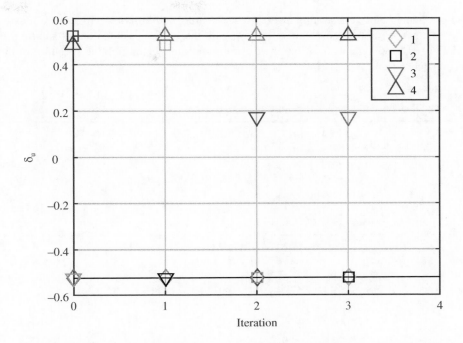

Figure A.3   Control deflections for the simplex iterations to solve example A.8

Example A.9

Results for Two-dimensional Error Minimization using the Simplex Consider the same general problem described in Example A.8, only now we wish to formulate our allocator to minimize the moment error using the approach in Example A.9. In standard form our problem is:

$$\min_{\mathbf{x}} \mathbf{c}^T \mathbf{x} \mid \mathbf{A}\mathbf{x} = \mathbf{b}, \mathbf{h} \geq \mathbf{x} \geq \mathbf{0}$$

$$\begin{bmatrix} -3.4956 & -3.0013 & 3.0013 & 3.4956 & -1 & 0 & 1 & 0 \\ -0.7919 & -1.2614 & -1.2614 & -0.7919 & 0 & -1 & 0 & 1 \end{bmatrix} \mathbf{x}$$

$$= \begin{pmatrix} 0 + 6.5 \\ -2.1502 + 0.5 \end{pmatrix}$$

$$\mathbf{x} \geq \mathbf{0} \tag{A.136}$$

$$\min_{\mathbf{x}} y = \begin{pmatrix} 0 & 0 & 0 & 0 & 1_m & 1_m \end{pmatrix} \mathbf{x}$$

$$\mathbf{x} = \begin{pmatrix} \mathbf{u} - \mathbf{u}_l \\ \mathbf{y}_s^+ \\ \mathbf{y}_s^- \end{pmatrix} \quad \mathbf{h} = \begin{pmatrix} \mathbf{u}_u - \mathbf{u}_l \\ - \\ - \end{pmatrix} \tag{A.137}$$

There are two equality constraints in Eq. (A.136), so a basic solution will have at most two unknowns that are not at their limits. Starting with the initial feasible control solution, $\mathbf{u} = (-0.5236, -0.5236, 0.5236, 0.5236)$, the algorithm progresses through the sequence of iterations presented Table A.2.

Table A.2   Sequence of simplex iterations to solve Example A.9

Iter. 0	$x = (1.047\ \ 1.047\ \ 0\ \ 0\ \ 0\ \ 0\ \ 13.3\ \ 0.5)$ $E = (-\ -\ +\ +\ +\ +\ +\ +)$ $x_D = (x_1\ \ x_2\ \ x_3\ \ x_4\ \ x_5\ \ x_6)$	To Enter
Cost 13.80	$r_d = (-4.29\ \ -4.26\ \ -1.74\ \ -2.70\ \ 2.00\ \ 2.00)$ $x_B = (x_7\ \ x_8)$ $y_q = (3.50\ \ 0.79)$ $ratio = (3.81\ \ 0.63)$	x_1 To Leave x_8
Iter. 1	$x = (0.4158\ \ 1.047\ \ 0\ \ 0\ \ 0\ \ 0\ \ 11.1\ \ 0)$ $E = (-\ -\ +\ +\ +\ +\ +\ +)$ $x_D = (x_8\ \ x_2\ \ x_3\ \ x_4\ \ x_5\ \ x_6)$	To Enter
Cost 11.10	$r_d = (5.41\ \ 2.57\ \ -8.57\ \ -6.99\ \ 2.00\ \ -3.41)$ $x_B = (x_7\ \ x_1)$ $y_q = (8.57\ \ -1.59)$ $ratio = (1.29\ \ 0.26)$	x_3 To Leave x_1 (swap sign)
Iter. 2	$x = (0\ \ 1.047\ \ 0.261\ \ 0\ \ 0\ \ 0\ \ 8.859\ \ 0)$ $E = (+\ -\ +\ +\ +\ +\ +\ +)$ $x_D = (x_8\ \ x_2\ \ x_1\ \ x_4\ \ x_5\ \ x_6)$	To Enter
Cost 8.86	$r_d = (-1.38\ \ -6.00\ \ 5.38\ \ -1.61\ \ 2.00\ \ 3.38)$ $x_B = (x_7\ \ x_3)$ $y_q = (6.00\ \ -1.00)$ $ratio = (1.48\ \ 0.79)$	x_2 To Leave x_3 (swap sign)
Iter. 3	$x = (0\ \ 0.261\ \ 1.047\ \ 0\ \ 0\ \ 0\ \ 4.141\ \ 0)$ $E = (+\ -\ -\ +\ +\ +\ +\ +)$ $x_D = (x_8\ \ x_3\ \ x_1\ \ x_4\ \ x_5\ \ x_6)$	To Enter
Cost 4.14	$r_d = (3.38\ \ 6.00\ \ 1.61\ \ -5.38\ \ 2.00\ \ -1.38)$ $x_B = (x_7\ \ x_2)$ $y_q = (5.38\ \ -0.63)$ $ratio = (0.77\ \ 0.42)$	x_4 To Leave x_2 (swap sign)
Iter. 4	$x = (0\ \ 0\ \ 1.047\ \ 0.4158\ \ 0\ \ 0\ \ 1.904\ \ 0)$ $E = (+\ +\ -\ +\ +\ +\ +\ +)$ $x_D = (x_8\ \ x_3\ \ x_1\ \ x_2\ \ x_5\ \ x_6)$	To enter
Cost 1.90	$r_d = (-3.41\ \ -2.57\ \ 6.99\ \ 8.57\ \ 2.00\ \ 5.41)$ $x_B = (x_7\ \ x_4)$ $y_q = (4.41\ \ -1.26)$ $ratio = (0.43\ \ 0.50)$	x_8 To leave x_7

Table A.2 (*continued*)

Iter.	$x = (0\ 0\ 1.047\ 0.9604\ 0\ 0\ 0\ 0.4312)$	
5	$E = (+\ +\ -\ +\ +\ +\ +\ +)$	
	$x_D = (x_7\ x_3\ x_1\ x_2\ x_5\ x_6)$	To enter
Cost	$r_d = (0.77\ -0.58\ 1.58\ 1.94\ 1.23\ 2.00)$	x_3
0.43	$x_B = (x_8\ x_4)$	
	$y_q = (0.58\ -0.86)$	To leave
	$ratio = (0.74\ 0.10)$	x_4 (swap sign)
Iter.	$x = (0\ 0\ 0.9461\ 1.047\ 0\ 0\ 0\ 0.3724)$	To enter
6	$E = (+\ +\ -\ -\ +\ +\ +\ +)$	N/A
	$x_D = (x_7\ x_4\ x_1\ x_2\ x_5\ x_6)$	
Cost	$r_d = (0.58\ 0.68\ 2.26\ 2.52\ 1.42\ 2.00)$	At optimum
-0.885	$x_B = (x_8\ x_3)$	
		To leave
		N/A

The initial condition chosen was not a particularly close one and the algorithm takes six iterations to find the optimal solution, as shown in Figure A.4. In this case the initial step is to reduce the pitch moment error to zero; the next four iterations all have zero pitch moment error. The iteration only reduces the error term by $0.06\ \text{rad/sec}^2$, illustrating the way the simplex is constrained to step between neighboring vertexes.

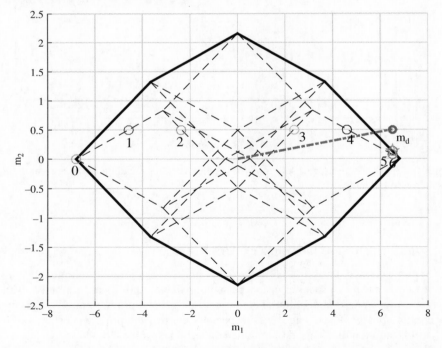

Figure A.4 Simplex iterations to solve Example A.9 showing convergence to the minimum error solution

It is also clear that in solving the problem, individual decision variables are not guaranteed to persist in the basis. In Figure A.5, every effector spends some time as the one effector in the basis (accompanied by an additional variable representing the error). In most cases they only persist in the solution for a single iteration; only the left outboard elevon is in the basic set for consecutive iterations. This example also demonstrates that once a variable leaves the basic set it may return later as part of a different basis; in this case the left inboard elevon reenters the final basis after having been driven out earlier. The final solution $\mathbf{u} = \left(-0.5236, -0.5236, 0.4225, 0.5236 \right)$, has only the left inboard elevon not at one of its limits; the other basic variable in the final solution corresponds to the pitch moment error.

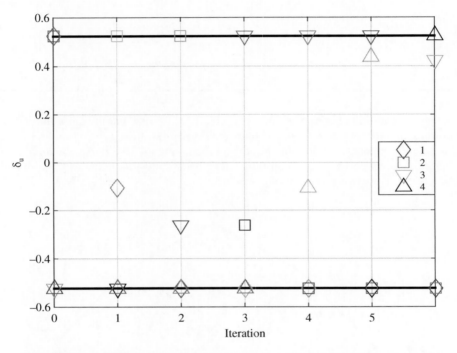

Figure A.5 Sequence of effector solutions for simplex iterations leading to the minimum moment error solution in Example A.9

A.6.4   *Initialization of the Simplex Algorithm*

In the examples above, an initial basic feasible solution was provided. In general this will not be the case, except in specific instances such as the second linear program solved by a dual-branch allocator. To find an initial condition, many linear programming solvers treat the solution in two phases. Phase one solves a specially constructed problem designed to yield a basic feasible solution that is used to initialize the original problem in phase two.

The phase-one program can be setup the addition of slack variables that allowing an arbitrary initial condition to be made feasible.

$$\min_{\tilde{\mathbf{x}}} \tilde{\mathbf{c}}^T \tilde{\mathbf{x}} \tag{A.138}$$

$$[\mathbf{A}\ \mathbf{I}_{m \times m}] \begin{pmatrix} \mathbf{x} \\ \mathbf{y} \end{pmatrix} = \mathbf{b}$$

$$\tilde{\mathbf{x}} = \begin{pmatrix} \mathbf{x} \\ \mathbf{y} \end{pmatrix}$$

$$\mathbf{x} \geq \mathbf{0}$$

$$\mathbf{y} \geq \mathbf{0}$$

$$\tilde{\mathbf{c}}^T = (\mathbf{0}_{1 \times k}, \mathbf{1}_{1 \times l}) \tag{A.139}$$

The initial basic feasible point for this new program is $\tilde{\mathbf{x}}_0 = (\mathbf{0}_{1 \times k}, \mathbf{b})^T$. If a feasible solution to the original problem exists, the minimum cost occurs when all of the slack variables $\mathbf{y}$ have been be driven out of the solution and the cost goes to zero.

If the optimal solution of the phase-one program is non-degenerate, then $\mathbf{y} = \mathbf{0}$ implies that all of the elements of $\mathbf{y}$ are non-basic. The corresponding basic solution in $\mathbf{x}$ represents a basic feasible solution to the original problem.

If the phase one solution is degenerate, one or more of the elements of $\mathbf{y}$ may remain in the basis but take on a zero value. An initial condition for the phase two program can still be found by swapping these elements with a corresponding non-basic element of $\mathbf{x}$, as long as the original $\mathbf{A}$ is full rank (otherwise we would first need to check that the element of $\mathbf{x}$ chosen is linearly independent of those currently in the basis).

The *two-phase* solution above is one way of initializing the program. The same augmented problem in Eq. (A.138) can be used by modifying the cost function in Eq. (A.139) so that it includes the original cost as well as the additional elements.

$$\min_{\tilde{\mathbf{x}}} \tilde{\mathbf{c}}^T \tilde{\mathbf{x}} \tag{A.140}$$

$$\tilde{\mathbf{c}}^T = \begin{pmatrix} \mathbf{c}\ M\mathbf{1}_{1 \times l} \end{pmatrix} \tag{A.141}$$

$$\tilde{\mathbf{x}} = \begin{pmatrix} \mathbf{x} \\ \mathbf{y} \end{pmatrix}$$

The value for the constant, M, must be chosen to be large enough compared to the elements of the existing cost function $\mathbf{c}$ so that $\mathbf{y} = 0$ at the optimal solution; hence the colloquial name of 'Big-M' applied to this algorithm. Big-M approaches are less commonly seen in general practice, but the structure is similar to the single branch or mixed optimization allocator, Eq. (A.100), allowing us to construct a trivial initial basic feasible solution for those problems.

A final option for initializing our solver is to recognize that, at the previous update, we have solved another problem with a similar structure. If we assume that our control effectiveness and commanded moment have not changed dramatically in that small time-step, we might expect the solution to the current problem to be similar to this previous one. If so, we can use the previous solution in a 'warm-start' algorithm to minimize the number of iterations needed to optimize the current solution in a simplex type solver.

Often the final basis from the previous solution will still be feasible. If not we can still solve for the corresponding infeasible basis. There are several ad-hoc algorithms that operate in a similar manner to the primal-simplex that can be used to drive the infeasible vectors back into the feasible set; in general these involve modifying the pricing step in the linear program to consider a measure of infeasibility.

A.7   Afterword

As stated in Chapter 4, the control allocation problem can be written as a system of linear equalities and inequality constraints. Confronted with the potential for multiple solutions for a given moment command, and the desire to find an 'optimal' solution, it is not a significant step to examine whether existing methods for solving constrained optimization problems can be applied. Linear programming is one class of such methods; one that seems particularly suited to the assumption of linear control effectiveness.

The use of linear programming for control allocation was introduced in Section 6.9. The final form of the problem was one of several solution algorithms introduced in Chapter 6 yielding equivalent results; as such it fits well within the framework for frame-wise allocation outlined in Chapter 7. Once the problem is posed as a linear program, the next step is to question whether modifications to that program would allow additional constraints and objectives to be included, in a similar manner to how the frame-wise application allows for additional capability. A quick perusal of the references below, as well as those outlined in Appendix C, will convince the reader that there is indeed a wide variety of ways to state a linear program for control allocation.

Separating the statement of the program from its numerical solution, this appendix expands on the process of developing a linear program for control allocation. We started with an overview of how the problem might be stated generically as a constrained linear optimization problem (Eq. (A.1)). For the purpose of this discussion, the behavior of the control allocator for attainable (Sections A.1.1 and A.4) and unattainable (Section A.1.2) moment commands are considered separately.

For unattainable commands (Section A.1.2), we present two main options, both designed to return moments on the boundary of the attainable set. Direction-preserving allocators behave similarly to the other methods presented in Chapter 6 and may be preferred when it is desirable to minimize out-of-plane moments. Minimizing the moment error is also commonly seen in the literature and, through the choice of norms applied and the weighting applied, provides a mechanism for prioritizing axes.

A standard form for linear programs was presented (Eq. (A.18)), along with a related, revised form (Eq. (A.40)) that simplifies treatment of the bounded effectors found in most control allocation problems. Working through the examples of how to cast the direction-preserving and error-minimization programs into this standard form (Sections A.2.3 and A.2.4, and Example A.2) illustrates methods that can be used, both to cast new programs into this form, as well as to transform a program into the specific form required by a particular solver. Armed with this standard form, basic, degenerate basic, and feasible basic solutions to linear programs were introduced (Section A.3). The properties of these solutions govern what types of optimal solutions the linear program and the underlying control allocation problem will have.

Several approaches to resolving control redundancy for attainable commands are presented. Rather than presenting a series of specific programs, targeted at specific situations, generic forms are presented that minimize different forms of the error from a preferred set of effector positions (Sections A.4.1 and A.4.2). Numerical examples show how the parameters in a specific form can be used to meet a specific goal; more examples can be found in the literature. For example, Buffington (1999) provides a list of options for weighting control errors.

A mechanism for optimizing a linear secondary function, similar to methods for restoring to an unknown solution in Chapter 7, is presented in Section A.4.3. The implementation of this approach is subject to many of the same issues presented there surrounding the potential for chattering and how to define local gradients.

Given an understanding of how to construct a linear program for attainable commands and for unattainable commands, Section A.5 finally brings the two pieces back together into a single linear program that can be used for control allocation. The most straightforward approach, also one of the earlier forms encountered in the references, is to implement two physically separate branches (Section A.5.1). The results of the first branch (designed for unattainable commands) is inspected to determine whether to solve an additional program for attainable commands. More recently, approaches that combine both programs together, weighting them so that achieving the desired moment is prioritized, are commonly seen (Section A.5.2).

Finally, Section A.6 touches on the numerical solution of linear programs. Up to this point the examples presented have been solved using MATLAB®'s built in solver, but any generic linear programming algorithm could have been used. A discussion of some of the common features of generic solvers provides some background for how they might be modified to be integrated into a control allocation algorithm. The details of any such integration are dependent on the specific algorithms used.

One of the potential benefits from casting the control allocation problem as a linear program is the potential to rely on the vast library of previous work that has been done on optimizing the performance and robustness of linear programs. Packaged solvers are readily available, as is extensive documentation of potential algorithms. At first the operation of these algorithms may seem like a black box to the uninitiated. The authors have found that a basic way to gain understanding is to step through the operation of a simplified version of the algorithm in question, at each step relating the internal operation back to the original problem being solved. This approach is presented for a simplified simplex solver in Example A.8. For a similarly accessible discussion of interior-point methods the paper by Petersen and Bodson (2005) is recommended.

Linear programing offers another way to describe desired behavior compared to other approaches of Chapters 6 and 7. This flexibility is accompanied by the same potential challenges faced by many iterative solution techniques developed for use in real-time control: ensuring convergence, numerical performance, approaches to certification and analysis, and on-board computational resources. As time progresses the latter of these becomes easier to deal with, but the first three remain issues to think about as algorithms are implemented.

References

Beck, RE 2002 *Application of Control Allocation Methods to Linear Systems with Four or More Objectives*. PhD Thesis, Virginia Polytechnic Institute & State University .

Bodson, M and Frost, SA 2011 'Load balancing in control allocation,' *AIAA J. Guidance, Control, and Dynamics* **34**(2) 380–387.

Bodson, M 2002 'Evaluation of optimization methods for control allocation,' *AIAA J. Guidance, Control, and Dynamics*, **25** (4), 703–711.

Buffington, J, Chandler, P, and Pachter, M 1998 'Integration of on-line system identification and optimization-based control allocation,' AIAA 98-4487 in *Proceedings of the AIAA Guidance, Navigation, and Control Conference*, pp. 1746–1756.

Buffington, J, Chandler, P, and Pachter, M 1999 'On-line system identification for aircraft with distributed control effectors,' *Int. J. Robust Nonlinear Control*, **9**, 1033–1049

Buffington, JM 1999 'Modular Control Law Design For The Innovative Control Effectors (ICE) Tailless Fighter Aircraft Configuration 101-3.' Final Report Air Vehicles Directorate, Air Force Research Laboratory, Wright-Patterson AFB, Ohio.

Buffington, JM 1997 'Tailless Aircraft Control Allocation.' WL-TM-97-3060, Flight Dynamics Directorate Wright Laboratory, WPAFB, Ohio.

Dantzig, GB 2002 'Linear programming,' *Operations Research,* **50** (1), 42–47.

Frost, S, Bodson, M, Burken, JJ, Jutte, CV, Taylor, BR, and Trinh, KV 2015 'Flight control with optimal cControl allocation incorporating structural load feedback,' *J. Aerospace Information Systems*, **12**, 825–834.

Gale, D 1960 *The Theory of Linear Economic Models.* McGraw-Hill.

Gill, PE, Murray, W, and Wright, MH 1981 *Practical Optimization.* Academic Press.

Luenberger, DG and Ye Y 2008 *Linear and Nonlinear programming*, 3rd edn. Springer.

Luenberger, DG 1984 *Linear and Nonlinear Programming.* Addison-Wesley.

Nef W 1967 *Linear Algebra.* Dover.

Oppenheimer, MW, Doman, DB, and Bolender, MA 2010 'Control allocation,' in *The Control Handbook* 2nd edn. CRC Press, Chapter 8.

Petersen, J and Bodson, M 2005 'Interior-point algorithms for control allocation,' *AIAA J. Guidance, Control, and Dynamics*, **28**(3), 471–480.

B

Flight Simulation

B.1 Introduction

This is the documentation for the flight simulation code available at the companion web site to this book, at www.wiley.com/go/durham/aircraft_control_allocation. The original code is named ADMIRE (Aero-Data Model In a Research Environment). It was created by the Swedish Defence Research Agency (Svensk Försvarets forskningsinstitut, FOI). At the time of this writing version 4.1 of ADMIRE was available at http://www.foi.se/en/Our-Knowledge/Aeronautics/Admire/Downloads/.

Several modifications to the original code have been made, primarily at the flight control system level. Mostly this amounted to the incorporation of a nonlinear dynamic inversion flight control law, and the implementation of various control allocation algorithms referred to in the text.

The high-level source of the modified code in MATLAB® at the companion web site is a Simulink® model named admire_sim_NDI.mdl. This version has been tested with the Student Version of MATLAB® R2014a running on a Macintosh computer.

B.2 Modifications

B.2.1 Three of the top-level blocks have been left almost completely unaltered:

B.2.1.1 Total Computer Delay/Transport Delay Version

This block models time delay within the digital computer that implements the flight control algorithms. In the ADMIRE simulation, this delay is 20 ms, which represents two frames of delay for a computer running at 100 Hz.

B.2.1.2 Saturators, Rate Limiters, and Actuators

This block models the actuator dynamics and incorporates rate and position limiting. The throttle steady state `tss` is not rate limited. Also, the landing gear command, thrust vectoring, and atmospheric disturbances do not go through the delay or actuator blocks.

Aircraft Control Allocation, First Edition. Wayne Durham, Kenneth A. Bordignon and Roger Beck.
© 2017 John Wiley & Sons, Ltd. Published 2017 by John Wiley & Sons, Ltd.
Companion website: www.wiley.com/go/durham/aircraft_control_allocation

B.2.1.3   Aircraft Response

This block models the response of the vehicle to the inputs. The Uncertain Parameters block, which contained 25 constant blocks with variables that are currently set to zero, has been replaced with a single constant block of zeros. This will affect functionality only if users wish to investigate the effects of altering the uncertain parameters. Variable names have been added on the input ports.

B.2.2   Minor modifications consist of the new Pilot and Sensors blocks.

B.2.2.1   Pilot

The Pilot block is set up to handle simple time histories of pilot inputs that will be biased about the values need for trim. This is similar to the original ADMIRE implementation.

B.2.2.2   Sensors

A Sensors block was added at the bottom of the diagram in the feedback path to modify the sensed heading angle. The mod (modulus) command was used to keep the heading in the range of principal values ($0 - 2\pi$ rad) for realism in the waypoint-following part of the control law.

B.3   NDI_CLAW

The major modification consists of a block containing a nonlinear dynamic inversion control law (NDI_CLAW) that completely replaces the original ADMIRE control law. For the naming of new variables, we have tried to follow a convention of Name_units. To limit names to containing only alpha numeric characters and underscores, we use abbreviations such as *mps* for meters per second.

The primary components of NDI_CLAW are the following blocks:

B.3.1   NDI_CLAW/Rate Transition

There is a rate transition block to take the sensor data from the continuous simulation and make it usable by the control law that is modeled as a discrete system. This is done to emulate a digital computer used for the flight control system. A rate of 100 Hz ($dt = 0.01$ s) was chosen because it is a reasonable rate commonly used in modern flight control computers.

B.3.2   NDI_CLAW/PILOT_Mod

This block converts the pilot inputs into usable commands. Currently, this is done with a simple constant scale factor (gain block). For a real aircraft, there likely would be some gradients and command shaping to improve the handling qualities.

B.3.3   NDI_CLAW/INPUT

This block takes the data from the PILOT and SENSORS blocks and puts all the incoming information on a single data bus. This is done to help clean up the diagrams. The block clearly identifies data that comes from sources that are external to the flight computer. The block attempts to emulate what an actual computer would do processing the input data. Several scopes have been added to display data the control law is using. Users may wish to add other scopes, or remove these to make the simulation run faster.

B.3.4   NDI_CLAW/MissionManager

This block generates commands for the autopilot modes of the control law. Note that these commands may or may not be used, depending on the values of other variables in the other parts of the control law. The default configuration will fly a square pattern at different altitudes and velocities over 1000 s of simulation time. This was done to verify that the control law could control the aircraft for the entire envelope of valid data used in the ADMIRE simulation.

Some other constant blocks have been left in the mission manager. These can be used, for example, to stay at the trim condition (x0bare) or go to the first altitude and velocity in the tables. To use them, replace the blocks ALT_table and V_table. To change the altitude and velocity commands, edit the following variables in the initialization file INIT_NDI.m:

Altcmd_t Vector of time (s)
Altcmd_v Vector of altitude commands (m)
Vcmd_t Vector of time (s)
Mcmd_v Vector of Mach number
a Vector of the speed of sound at the commanded altitude (m/s) ($Vcmd_v = Mcmd_v.*a$).

The bank-angle command will do left and right turns of increasing amplitude. This is currently not being used, because the roll response is defaulted to generate bank-angle commands based on waypoints. To change the values of the commanded bank angle, edit the following variables in the initialization file INIT_NDI.m:

PHIcmd_t Vector of time (s)
PHIcmd_v Vector of bank angle commands (°).

B.3.4.1   NDI_CLAW/MissionManager/WaypointNavigation

This block is a simple attempt at an autopilot feature that works with one of the roll axis options to follow waypoints (X and Y earth axes). The output of this block is: the current X and Y command values (Xcmd and Ycmd), the previous commanded values (Xcmdm1 and Ycmdm1 abbreviations of Xcmd minus one and Ycmd minus one), and a maximum bank angle to be used, which is based on the g available at the current flight condition.

There is an attempt to start a turn prior to the waypoint, so that a constant-radius turn will end with the aircraft on the new heading.

Some of the known problems with this navigation system are: insufficient error checking, and incorrect following of waypoints placed too close or requiring very tight turns. The default path is a simple square pattern. The waypoints may be changed by editing the following variables in the initialization file `INIT_NDI.m`

Xwpt X distance (m), north, initially $\begin{bmatrix} 0 & 4000 & 4000 & 0 & 0 \end{bmatrix} * 8$
Ywpt Y distance (m), east, initially $\begin{bmatrix} 0 & 0 & 4000 & 4000 & 0 \end{bmatrix} * 8$

The factor of eight may be adjusted if the closeness of the waypoints results in turns that exceed the capabilities of the airplane.

B.3.5 NDI_CLAW/DynamicInversionControl

This is the top level of the dynamic inversion control law. It is broken into six main subsystems:

B.3.5.1 NDI_CLAW/DynamicInversionControl/Auto_Throttle

The throttle is not commanded by the control allocation. Instead, it was implemented as an energy-based control scheme. It uses commanded altitude and velocity to make an energy command, `Ecmd`. The sensed altitude and velocity are used to calculate the current energy state E. A proportional-integral control law is used to generate throttle commands to drive the energy state to the desired values. The integration is turned off if the error is very large, to prevent the system from overshooting due to integrator build-up. There is a table look-up to reduce the throttle gain when the afterburner is on (`tss` > 0.8) because the afterburner is much more effective than the non-afterburner throttle.

There is also a block `Get_UseSplit`. When `UseSplit` is true, the preferred values change to bias the inboard elevons up and the outboard elevons down. This effectively makes them act as a speed-brake to help decelerate the vehicle.

B.3.5.2 NDI_CLAW/DynamicInversionControl/CMD

The `CMD` (commands) block converts the pilot and mission manager data into commands for the regulator blocks. At the time of this writing, this block is mostly a pass-through that does not have great effect. The altitude command is rate-limited to avoid over-large commands to the auto-throttle. There are unconnected doublet and time history blocks that may be included to observe the aircraft responses to simple commands. To use them, disconnect the input to a gain-of-1 block, and connect the desired input.

B.3.5.3 NDI_CLAW/DynamicInversionControl/REG

These are the regulator blocks that generate the desired accelerations. They are separated into roll, pitch and yaw.

NDI_CLAW/DynamicInversionControl/REG/RollReg

There are three different ways to generate the desired roll acceleration, which are selectable by changing the variable `Rmode` in the initialization file `INIT_NDI.m`. Note that by the logic of the `If` blocks, only the code in the `true` condition runs.

Rmode = 1: Roll-rate Command: This block uses simple proportional control of roll rate. This mode is typical of a pilot flying and using the lateral stick to generate roll-rate commands. There is a command gain that is a function of Mach number. This functionality is typical of high-speed fighter aircraft. Although it is not currently connected, it was included so that users may experiment with different values.

Rmode = 2: Bank-angle Command: This mode emulates an autopilot flying bank-angle commands. There is proportional control of roll rate (similar to `MODE=1`), with an outer loop that has proportional-integral-forward path control of bank angle.

Rmode = 3: Cross-track Command: This mode is used when using the waypoints from the mission manager. It uses proportional control of roll rate and bank angle. The bank-angle command comes from two different blocks. Initially, the bank angle command is the maximum bank angle (with appropriate sign) until the aircraft is on the correct heading. Once the proper heading is achieved, cross-track error and cross-track error rate are used to calculate the required bank angle. This block may create undesired results if the waypoints are not well spaced with an easy-to-follow path. In that case, the airplane may fly in circles or deviate into an undesired location.

NDI_CLAW/DynamicInversionControl/REG/PitchReg

There are three different ways to generate the desired pitch acceleration, which are selectable by changing the variable `Pmode` in the initialization file `INIT_NDI.m`. Note that the way the `If` blocks are implemented, only the code in the true condition runs.

Pmode = 1: Pitch-rate Command: This mode is simple proportional control of pitch rate. This is similar to the roll-rate command system. For low-speed flight, this mode may be preferable to pilots. At higher speeds, pilots would more likely prefer *g*-command for the pitch control inceptor. The current model does not have a *g*-command system.

Pmode = 2: Flight-path Angle Command: This mode performs proportional control on pitch rate and proportional-integral control on flight path angle. Some UAVs have the remote pilot/operator pitch stick command flight path angle.

Pmode = 3: Altitude Command: This mode is used to control altitude with pitch. There is proportional control of pitch rate and flight path angle. Flight path angle command is generated from proportional and integral control on altitude.

NDI_CLAW/DynamicInversionControl/REG/YawReg

This block generates the yaw acceleration command based on a desired sideslip angle. The controller is proportional-integral-forward path on sideslip angle, and proportional on yaw rate. The `n1` (-1) block converts a sideslip rate command into a yaw rate command, $r \approx -\dot{\beta}$. Without yaw thrust vectoring the vehicle is yaw-control limited. Different allocation methods seem to be able to achieve roughly the same maximum sideslip angle, but if one uses more of the AMS the goal can be achieved faster.

B.3.5.4   NDI_CLAW/DynamicInversionControl/AERO_OBM

This block calculates the nominal accelerations, the control effectiveness matrix, and the limits on the aerodynamic effectors. The `AeroLimits` block finds the limits on the aerodynamic effectors based on Mach number per the ADMIRE documentation. There are terms for all 16 inputs to the `ADMIRE_main` program that can act as control effectors with the control law. The 16 input values are:

drc Right canard
dlc Left canard
droe Right outboard elevon
drie Right inboard elevon
dlie Left inboard elevon
dloe Left outboard elevon
dr Rudder
dle Leading edge flaps
ldg Landing gear
tss Thrust command
dty Yaw thrust vectoring
dtz Pitch thrust vectoring
u_dist Forward velocity disturbance
v_dist Side velocity disturbance
w_dist Vertical velocity disturbance
p_dist Roll rate disturbance

Both linear and nonlinear versions of the OBM have been implemented.

Linear model version
The model was linearized in a flight condition with nominal trim effectors $\mathbf{u}_{nom}$:

$$\mathbf{u}_{cmd} = \mathbf{u}_{nom} + B^{-1}(\dot{\mathbf{x}}_{des} - \dot{\mathbf{x}}_{nom})$$
$$\dot{\mathbf{x}}_{nom} = A\mathbf{x} + B\mathbf{u}_{nom}$$

(B.1)

The A and B matrices come from the ADMIRE linearization tool.

Nonlinear model version
The nonlinear model replaces the linearized A and B matrices with table look-ups of the aerodynamic data.

$$\mathbf{u}_{cmd} = \mathbf{u}_{nom} + B^{-1}(\dot{\mathbf{x}}_{des} - \dot{\mathbf{x}}_{nom})$$
$$\dot{\mathbf{x}}_{nom} = \mathbf{f}(\mathbf{x}, \mathbf{u}_{nom}) = \mathbf{f}(\mathbf{x}, \mathbf{u}) - \Delta\mathbf{f}$$

(B.2)

The actual accelerations are:

$$\dot{\mathbf{x}} = \mathbf{f}(\mathbf{x}, \ \mathbf{u}) = \begin{Bmatrix} \dot{p} \\ \dot{q} \\ \dot{r} \end{Bmatrix} \tag{B.3}$$

The accelerations were developed in Chapter 2, Eq. (2.12). ADMIRE documentation adopts the more compact notation of the first edition of Stevens and Lewis (1992):

$$
\begin{aligned}
\dot{p} &= (C_1 r + C_2 p)q + C_3 L + C_4 N \\
\dot{q} &= C_5 pr - C_6(p^2 - r^2) + C_7 M \\
\dot{r} &= (C_8 p - C_2 r)q + C_4 L + C_9 N
\end{aligned}
\tag{B.4}
$$

The moments are total moments, aerodynamic and propulsive. The subscripted C terms are calculated in the initialization file `INIT_NDI.m` from

$$
\begin{aligned}
\Gamma &= I_{xx}I_{zz} - I_{xz}^2 \\
\Gamma C_1 &= (I_{yy} - I_{zz}) - I_{xz}^2 \\
\Gamma C_2 &= (I_{xx} - I_{yy} + I_{zz})I_{xz} \\
\Gamma C_3 &= I_{zz} \\
\Gamma C_4 &= I_{xz} \\
C_5 &= (I_{zz} - I_{xx})/I_{yy} \\
C_6 &= I_{xz}/I_{yy} \\
C_7 &= 1/I_{yy} \\
\Gamma C_8 &= I_{xx}(I_{xx} - I_{yy}) + I_{xz}^2 \\
\Gamma C_9 &= I_{xx}
\end{aligned}
\tag{B.5}
$$

The `Aircraft Response` block provides as outputs the rates and moment coefficients. These are used to calculate the body axis accelerations. These accelerations include the effects of the effectors being at their current deflections. If we calculate the control commands as they change from the current positions, then we can use the actual accelerations as the nominal accelerations:

$$\dot{\mathbf{x}}_{nom} = \dot{\mathbf{x}} = \mathbf{f}(\mathbf{x}, \ \mathbf{u}) \tag{B.6}$$

If we calculate the control commands as the change from some nominal position, which is not the current effector deflections, then we need to subtract off a component that is the difference between the total acceleration and the acceleration with the effectors at their nominal locations.

Consider the linear case:

$$
\begin{aligned}
\dot{\mathbf{x}} &= A\mathbf{x} + B\mathbf{u} = \mathbf{f}(\mathbf{x}, \ \mathbf{u}) \\
\dot{\mathbf{x}}_{nom} &= A\mathbf{x} + B\mathbf{u}_{nom} \\
\dot{\mathbf{x}}_{nom} &= A\mathbf{x} + B\mathbf{u} - (B\mathbf{u} - B\mathbf{u}_{nom}) = \mathbf{f}(\mathbf{x}, \ \mathbf{u}) - \Delta\mathbf{f}
\end{aligned}
\tag{B.7}
$$

To estimate $\Delta\mathbf{f}$ for the nonlinear model, we run the ADMIRE nonlinear aerodynamic model with the controls at their current deflections and their nominal deflections and take the difference.

To calculate the B matrix for the nonlinear OBM, we call the aerodynamic model with one effector disturbed by first a small positive and then a small negative deflection from the nominal position and do a simple difference to get a slope. This can be time consuming to do it for all seven aero effectors every frame, so we only update one column of B each frame to speed up the simulation. The B matrix is initialized to the values for the trim condition.

We have implemented B matrix calculations for only the aerodynamic effectors (the first seven in the list above). For the other effectors, the B matrix is initialized at the trim values, but never updated. If we really want to experiment with the other effectors, we should stay near the trim condition until we implement a way to calculate their B-matrix terms as flight conditions change.

B.3.5.5   NDI_CLAW/DynamicInversionControl/Unom

There are various options for nominal effectors, $\mathbf{u}_{nom}$:

1. Set the nominal values to zero:

$$\mathbf{u}_{nom} = \mathbf{0} \tag{B.8}$$

2. Trim the aircraft and use the trimmed values:

$$\mathbf{u}_{nom} = \mathbf{u}_{trim} \tag{B.9}$$

3. Use the last computed command:

$$\mathbf{u}_{nom} = \mathbf{u}_{cmd}z^{-1} \tag{B.10}$$

4. Use the last computed command sent through an estimated effector transfer function:

$$\mathbf{u}_{nom} = \mathbf{u}_{cmd}z^{-1}G_{act}(z) \tag{B.11}$$

5. Use sensors to determine the current values of the effectors:

$$\mathbf{u}_{nom} = \mathbf{u}_{sensed} \tag{B.12}$$

All of these options are available except the last.

B.3.5.6   NDI_CLAW/DynamicInversionControl/ControlAllocation

At the top level, the control allocation is broken into three subsystems: PreProcess, GetU and PostProcess.

NDI_CLAW/DynamicInversionControl/ControlAllocation/PreProcess

This block sets the inputs for the control allocation routines. `GetLimits` will find the current allowable change in effectors based on the nominal effector values. If the variable `UseRL` is set to 1 in the initialization file, then the commands will be limited to move no more than the rate limits allow each frame. The `INDX` vector is used to make effectors part of the 'active' set, which is determined by the control allocation routine. The default setting is that the seven aerodynamic surfaces are 'active' and the nine other effectors are 'inactive'. We recommend leaving the other effectors 'inactive' for now, although one may want to turn on the thrust vectoring.

NDI_CLAW/DynamicInversionControl/ControlAllocation/GetU

This block is where the control allocation problem, as defined in Section 4, is solved using the methods in Chapter 6 and Appendix A. There currently are six different methods to choose from. A method may be selected by changing the value of the variable `CAmethod` in the initialization file `INIT_NDI.m`.

The seven methods are:

MakeSquare (Option 0):   This is the ganging allocation method described in Sections 6.4 and 8.5. A square matrix, as in Eq. (6.1), results. If the values are commanded to exceed their limits, the commands are clipped to the position limits. This option currently will not enforce rate limits on the commands.

WeightedPseudoClipped (Option 1):   This method is a variant of the weighted pseudo-inverse described in Section 6.5.4 and in Eq. (6.35) (page 85). The weighting matrix is defined in the initialization file `INIT_NDI.m`.

If the values are commanded to exceed their limits, the commands are clipped to the position limits. This option currently will not enforce rate limits on the commands.

WeightedPseudoScaled (Option 2):   This method is the same as the previous, except that if the values are commanded to exceed their limits, the commands are uniformly scaled so that all commands are at or within their limits.

DirectAllocation (Option 3):   The direct allocation method was described in Section 6.6.2.

CGI (Option 4):   This method implements the scaled cascaded generalized inverse described in Section 6.5.5.2. It is the basis for the 'effector blender' algorithm used in the X-35 control law (Bordignon, 2002).

VJA (Option 5):   This is an implementation of Banks' method described in Section 6.8. The name of the module `VJA` is from Banks' original description of the procedure as the 'Vertex Jumping Algorithm'.

LP (Option 6):   This is an implementation of the linear programming methods described in Appendix A. The variable `LPmethod` is used to select between several algorithms. `LPmethod` should be an integer between 0 and 5:

0 DB_LPCA Dual branch control allocation
- Linear program objective error minimization branch (1-norm)
- Control error minimization (1-norm)

1 DBinf_LPCA Dual branch control allocation
- Linear program objective error minimization (1-norm)
- Control error minimization (inf-norm)

2 DP_LPCA Direction preserving control allocation linear program

3 DPscaled_LPCA Direction preserving control allocation linear program
- Reduced formulation (solution scaled from boundary)

4 MO_LPCA Mixed optimization (single branch) control allocation linear program
- Objective error minimizing
- Control error minimizing

5 SB_LPCA Single branch control allocation linear program
- Direction preserving
- Control error minimizing

NDI_CLAW/DynamicInversionControl/ControlAllocation/PostProcess

This block adds the nominal effector values, adds a component in the null space to drive towards a preferred solution, and merges the 'active' and 'inactive' effector commands to get the 16 total commands that feed the ADMIRE simulation. This is the null-space restoring described in Section 7.4.4.

If the reader wants to investigate what happens if there is no null space restoring, delete the line coming out of the `DriveToPreferred` block.

References

Aerodata Model in Research Environment (ADMIRE), Ver. 3.4h, Swedish Defence Research Agency (FOI), Stockholm, Sweden, 2003.

Stevens, BL and Lewis, FL 1992 *Aircraft Control and Simulation,* 1st edn. John Wiley & Sons, pp. 80–81.

Bordignon, K and Bessolo, J 2002 'Control allocation for the X-35B,' AIAA 2002-6020 in *2002 Biennial International Powered Lift Conference and Exhibit, 5-7 November 2002, Williamsburg, Virginia.*

C

Annotated Bibliography

References

Acosta, DM, Yildiz, Y, Craun, RW, Beard, SD, Leonard, MW, Hardy, GH, and Weinstein, M 2015
'Piloted evaluation of a control allocation technique to recover from pilot-induced oscillations,'
J. Aircraft, **52** (1), 130–140.

This paper describes the maturation of a control allocation technique designed to assist pilots
in recovery from pilot-induced oscillations. The control allocation technique to recover from
pilot-induced oscillations is designed to enable next-generation high-efficiency aircraft designs.
Energy-efficient next-generation aircraft require feedback control strategies that will enable low-
ering the actuator rate limit requirements for optimal airframe design. A common issue on aircraft
with actuator rate limitations is they are susceptible to pilot-induced oscillations caused by the
phase lag between the pilot inputs and control surface response. The control allocation technique
to recover from pilot-induced oscillations uses real-time optimization for control allocation to eliminate phase lag in the system caused by control surface rate limiting. System impacts of the control
allocator were assessed through a piloted simulation evaluation of a nonlinear aircraft model in
the NASA Ames Research Centers Vertical Motion Simulator. Results indicate that the control
allocation technique to recover from pilot-induced oscillations helps reduce oscillatory behavior introduced by control surface rate limiting, including the pilot-induced oscillation tendencies
reported by pilots.

Adams, RJ, Buffington, JM, Sparks, AG, and Banda, SS 1994 *Robust Multivariable Flight Control*.
Springer-Verlag.

Manual flight control system design for fighter aircraft is one of the most demanding problems in
automatic control. Fighter aircraft dynamics generally have highly coupled uncertain and nonlinear
dynamics. Multivariable control design techniques offer a solution to this problem. *Robust Multivariable Flight Control* provides the background, theory and examples for full envelope manual
flight control system design. It gives a versatile framework for the application of advanced multivariable control theory to aircraft control problems. Two design case studies are presented for the manual flight control of lateral/directional axes of the VISTA-F-16 test vehicle and an F-18 trust vectoring system. They demonstrate the interplay between theory and the physical features of the systems.

Aerodata Model in Research Environment (ADMIRE), Ver. 3.4h, Swedish Defence Research Agency (FOI), Stockholm, Sweden, 2003.

This document describes the nonlinear aircraft simulation model ADMIRE. It describes the main aircraft model, the flight control system, actuators, the sensor models and the uncertainty parameters with respective limits. This document also contains a description on how to properly install and run the model. The ADMIRE describes a generic small single seated, single engine fighter aircraft with a delta-canard configuration, implemented in MATLAB®/Simulink® Release 13. The model envelope is up to Mach 1.2 and 6000 m altitude. The model is augmented with a longitudinal flight control system (FCS) that controls the pitch rate at low speed and the load factor at higher speeds, and a lateral controller that controls the wind vector roll rate and the angle of sideslip. The longitudinal FCS also contains a very rudimentary speed controller. The model has thrust vectoring capability, although this is not used in the present FCS. For the purpose of the robustness analysis, the model is extended with the possibility to change some predefined uncertainty parameters within prescribed limits. The uncertainty parameters consist of configuration parameter-, aerodynamic-, actuator- and sensor (air data)-uncertainties. The model can be trimmed and linearized within the entire model envelope.

Alwi, H and Edwards, C 2008 'Fault tolerant control using sliding modes with on-line control allocation,' *Automatica*, **44**, 1859–1866.

This paper proposes an on-line sliding mode control allocation scheme for fault tolerant control. The effectiveness level of the actuators is used by the control allocation scheme to redistribute the control signals to the remaining actuators when a fault or failure occurs. The paper provides an analysis of the sliding mode control allocation scheme and determines the nonlinear gain required to maintain sliding. The on-line sliding mode control allocation scheme shows that faults and even certain total actuator failures can be handled directly without reconfiguring the controller. The simulation results show good performance when tested on different fault and failure scenarios.

Azam, M and Singh, S 1994 'Invertibility and trajectory control for nonlinear maneuvers of aircraft,' *AIAA J. Guidance, Control, and Dynamics*, **17** (1), 192–200.

This paper presents an application of the inversion theory to the design of nonlinear control systems for simultaneous lateral and longitudinal maneuvers of aircraft. First, a control law for the inner loop is derived for the independent control of the angular velocity components of the aircraft along roll, pitch, and yaw axes using aileron, elevator, and rudder. Then by a judicious choice of angular velocity command signals, independent trajectory control of the sets of output variables (angle of attack, roll, and sideslip angles), (roll rate, angle of attack, and yaw angle), or (pitch, roll, and yaw angles) is accomplished. These angular velocity command signals are generated in the outer loops using state feedback and the reference angle of attack, pitch, yaw, and roll angle trajectories. Simulation results are presented to show that in the closed-loop system, various lateral and longitudinal maneuvers can be performed in spite of the presence of uncertainty in the stability derivatives.

Beck, RE 2002 *Application of Control Allocation Methods to Linear Systems with Four or More Objectives,* PhD Thesis, Virginia Polytechnic Institute & State University.

Methods for allocating redundant controls for systems with four or more objectives are studied. Previous research into aircraft control allocation has focused on allocating control effectors to provide commands for three rotational degrees of freedom. Redundant control systems have the capability to allocate commands for a larger number of objectives. For aircraft, direct force commands can be applied in addition to moment commands.

When controls are limited, constraints must be placed on the objectives which can be achieved. Methods for meeting commands in the entire set of of achievable objectives have been developed.

The Bisecting Edge Search Algorithm has been presented as a computationally efficient method for allocating controls in the three objective problem. Linear programming techniques are also frequently presented. This research focuses on an effort to extend the Bisecting Edge Search Algorithm to handle higher numbers of objectives. A recursive algorithm for allocating controls for four or more objectives is proposed. The recursive algorithm is designed to be similar to the three objective allocator and to require computational effort which scales linearly with the controls.

The control allocation problem can be formulated as a linear program. Some background on linear programming is presented. Methods based on five formulations are presented. The recursive allocator and linear programming solutions are implemented. Numerical results illustrate how the average and worst case performance scales with the problem size. The recursive allocator is found to scale linearly with the number of controls. As the number of objectives increases, the computational time grows much faster. The linear programmingsolutions are also seen to scale linearly in the controls for problems with many more controls than objectives.

In online applications, computational resources are limited. Even if an allocator performs well in the average case, there still may not be sufficient time to find the worst case solution. If the optimal solution cannot be guaranteed within the available time, some method for early termination should be provided. Estimation of solutions from current information in the allocators is discussed. For the recursive implementation, this estimation is seen to provide nearly optimal performance.

Bemporad, A and Morari, M 1998 'Control of systems integrating logic, dynamics, and constraints,' *Automatica*, **35**, 407–427.

This paper proposes a framework for modeling and controlling systems described by interdependent physical laws, logic rules, and operating constraints, denoted as mixed logical dynamical (MLD) systems. These are described by linear dynamic equations subject to linear inequalities involving real and integer variables. MLD systems include linear hybrid systems, finite state machines, some classes of discrete event systems, constrained linear systems, and nonlinear systems which can be approximated by piecewise linear functions. A predictive control scheme is proposed which is able to stabilize MLD systems on desired reference trajectories while fulfilling operating constraints, and possibly take into account previous qualitative knowledge in the form of heuristic rules. Due to the presence of integer variables, the resulting on-line optimization procedures are solved through mixed integer quadratic programming (MIQP), for which efficient solvers have been recently developed. Some examples and a simulation case study on a complex gas supply system are reported.

Benosman, M, Liao, F, Lum, KY and Wang, JL 2009 'Nonlinear control allocation for non-minimum phase systems,' *IEEE Trans. on Automatic Control*, **17**, 394–404.

In this brief, we propose a control allocation method for a particular class of uncertain over-actuated affine nonlinear systems, with unstable internal dynamics. Dynamic inversion technique is used for the commanded output to track a smooth output reference trajectory. The corresponding control allocation law has to guarantee the boundedness of the states, including the internal dynamics, and satisfy control constraints. The proposed method is based on a Lyapunov design approach with finite-time convergence to a given invariant set. The derived control allocation is in the form of a dynamic update law which, together with a sliding mode control law, guarantees boundedness of the output tracking error as well as of the internal dynamics. The effectiveness of the control law is tested on a numerical model of the non-minimum phase planar vertical take-off and landing (PVTOL) system.

Bodson, M 2002 'Evaluation of optimization methods for control allocation,' *AIAA J. Guidance, Control, and Dynamics*, **25** (4), 703–711.

The performance and computational requirements of optimization methods for control allocation are evaluated. Two control allocation problems are formulated: a direct allocation method that preserves the directionality of the moment and a mixed optimization method that minimizes the error between the desired and the achieved moments as well as the control effort. The constrained optimization problems are transformed into linear programs so that they can be solved using well-tried linear programming techniques such as the simplex algorithm. A variety of techniques that can be applied for the solution of the control allocation problem in order to accelerate computations are discussed. Performance and computational requirements are evaluated using aircraft models with different numbers of actuators and with different properties. In addition to the two optimization methods, three algorithms with low computational requirements are also implemented for comparison: a redistributed pseudoinverse technique, a quadratic programming algorithm, and a fixed-point method. The major conclusion is that constrained optimization can be performed with computational requirements that fall within an order of magnitude of those of simpler methods. The performance gains of optimization methods, measured in terms of the error between the desired and achieved moments, are found to be small on the average but sometimes significant. A variety of issues that affect the implementation of the various algorithms in a flight-control system are discussed.

Bodson, M and Frost, SA 2011 'Load balancing in control allocation,' *AIAA J. Guidance, Control, and Dynamics*, **34** (2), 380–387.

Next-generation aircraft with a large number of actuators will require advanced control allocation methods to compute the actuator commands needed to follow desired trajectories while respecting system constraints. Previously, algorithms were proposed to minimize the ℓ_1 or ℓ_2 norms of the tracking error and of the actuator deflections. This paper discusses the alternative choice of the ℓ_1 norm, or the ℓ_∞ norm. Minimization of the control effort translates into the minimization of the maximum actuator deflection (minmax optimization). This paper shows how the problem can be solved effectively by converting it into a linear program and solving it using a simplex algorithm. Properties of the algorithm are also investigated through examples. In particular, the minmax criterion results in a type of load balancing, where the load is the desired command and the algorithm balances this load among various actuators. The solution using the ℓ_1 norm also results in better robustness to failures and lower sensitivity to nonlinearities in illustrative examples. This paper also discusses the extension of the results to a normalized ℓ_1 norm, where the norm of the actuator deflections are scaled by the actuator limits. Minimization of the control effort then translates into the minimization of the maximum actuator deflection as a percentage of its range of motion.

Bodson, M and Pohlchuck, E 1998 'Command limiting in reconfigurable flight control,' *AIAA J. Guidance, Control, and Dynamics*, **21** (4), 639–646.

Limits on the motion and on the rate of motion of the actuators driving the control surfaces of aircraft significantly affect the performance of flight control systems. After a failure or damage to the aircraft, the constraints become even more restrictive because of the loss of control power. There is also often an increase in cross couplings between the axes and, for a period of time, a significant uncertainty about the moments generated by the individual control surfaces. A model reference adaptive control algorithm is considered for flight control reconfiguration. The tracking performance of the algorithm deteriorates drastically for large maneuvers if actuator saturation is not accounted for. Four methods of command limiting are proposed to handle the problem, which are based on a scaling of the control inputs, a relaxation of the control requirements, a scaling of the reference inputs, and a least-squares approximation of the commanded accelerations. Simulations demonstrate the effectiveness of the algorithms in the reconfigurable flight control application. Even the simplest method is found to considerably improve the responses, and, surprisingly, the performance of all four methods is similar despite their widely different concepts and complexity levels.

In some cases, degraded transient responses are observed, which are attributed to the uncertainty in the aircraft parameters following a failure.

Bolender, M 2004 'Nonlinear control allocation using piecewise linear functions,' *AIAA J. Guidance, Control, and Dynamics*, **27** (6), 1017–1027.

A novel method is presented for the solution of the control allocation problem where the control variable rates or moments are nonlinear functions of control position. Historically, control allocation has been performed under the assumption that a linear relationship exists between the control induced moments and the control effector displacements. However, aerodynamic databases are discrete valued and almost always stored in multidimensional lookup tables, where it is assumed that the data are connected by piecewise linear functions. The approach that is presented utilizes this piecewise linear assumption for the control effector moment data. This assumption allows the control allocation problem to be cast as a piecewise linear program that can account for nonlinearities in the moment/effector relationships, as well as to enforce position constraints on the effectors. The piecewise linear program is then recast as a mixed-integer linear program. It is shown that this formulation accurately solves the control allocation problem when compared to the aerodynamic model. It is shown that the control effector commands for a reentry vehicle by the use of the piecewise linear control allocation method are markedly improved when compared to the performance of more traditional control allocation approaches that use a linear relationship between the control moments and the effectors. The technique is also applied to determine those flight conditions (angle of attack and Mach number) at which the reentry vehicle can be trimmed for the purpose of providing constraint estimates to trajectory reshaping algorithms.

Bolling, JG 1997 *Implementation of Constrained Control Allocation Techniques using an Aerodynamic Model of an F-15 Aircraft*, MS Thesis, Virginia Polytechnic Institute & State University.

Control allocation as it pertains to aerospace vehicles, describes the way in which control surfaces on the outside of an aircraft are deflected when the pilot moves the control stick inside the cockpit. Previously, control allocation was performed by a series of cables and push rods, which connected the 3 classical control surfaces (ailerons, elevators, and rudder), to the 3 cockpit controls (longitudinal stick, lateral stick, and rudder pedals). In modern tactical aircraft however, it is not uncommon to find as many as 10 or more control surfaces which, instead of being moved by mechanical linkages, are connected together by complex electrical and/or hydraulic circuits. Because of the large number of effectors, there can no longer be a one-to-one correspondence between surface deflections on the outside of the cockpit to pilot controls on the inside. In addition, these exterior control surfaces have limits which restrict the distance that they can move as well as the speed at at which they can move. The purpose of Constrained Control Allocation is to deflect the numerous control surfaces in response to pilot commands in the most efficient combinations, while keeping in mind that they can only move so far and so fast. The implementation issues of Constrained Control Allocation techniques are discussed, and an aerodynamic model of a highly modified F-15 aircraft is used to demonstrate the various aspects of Constrained Control Allocation.

Bordignon, KA 1996 *Constrained Control Allocation for Systems with Redundant Control Effectors*, PhD Thesis, Virginia Polytechnic Institute & State University.

Control allocation is examined for linear time-invariant problems that have more controls than degrees of freedom. The controls are part of a physical system and are subject to limits on their maximum positions. A control allocation scheme commands control deflections in response to some desired output. The ability of a control allocation scheme to produce the desired output without violating the physical position constraints is used to compare allocation schemes.

Methods are developed for computing the range of output for which a given scheme will allocate admissible controls. This range of output is expressed as a volume in the n- dimensional output space. The allocation schemes which are detailed include traditional allocation methods such as Generalized Inverse solutions as well as more recently developed methods such as Daisy Chaining, Cascading Generalized Inverses,Null-Space Intersection methods, and Direct Allocation.

Non-linear time-varying problems are analyzed and a method of control allocation is developed that uses Direct Allocation applied to locally linear problems to allocate the controls. This method allocates controls that do not violate the position limits or the rate limits for all the desired outputs that the controls are capable of producing. The errors produced by the non-linearities are examined and compared with the errors produced by globally linear methods.

The ability to use the redundancy of the controls to optimize some function of the controls is explored and detailed. Additionally, a method to reconfigure the controls in the event of a control failure is described and examined. Detailed examples are included throughout, primarily applying the control allocation methods to an F-18 fighter with seven independent moment generators controlling three independent moments and the F-18 High Angle of Attack Research Vehicle (HARV) with ten independent moment generators.

Bordignon, K and Bessolo, J 2002 'Control allocation for the X-35B,' AIAA 2002-6020, in *2002 Biennial International Powered Lift Conference and Exhibit, 5–7 November 2002, Williamsburg, Virginia.*

This paper discusses the methodology and algorithms used for control allocation on the Short Take-Off and Vertical Landing (STOVL) variant of Lockheed Martins Joint Strike Fighter (the X-35B). Control allocation for STOVL vehicles is particularly challenging due to the stringent performance requirements for safe, low-speed flight and the combined use of both aerodynamic and propulsive control effectors. The numerous control effectors' capabilities are a function of many variables including flight condition and engine power setting. To meet these challenging requirements, the X-35B control laws used a control allocation methodology that included a cascaded generalized inverse algorithm, and an onboard model of the aircraft aerodynamic and propulsive characteristics and control effectiveness. To improve the safety and reliability of the vehicle, the methodology needed the ability to handle effector failure scenarios and to avoid structural coupling. The topics covered include: 1) handling of redundant effectors; 2) effector redistribution in the presence of effector limits (position limits and rate limits); 3) effector redistribution in the presence of gain limiting; 4) failure accommodation; and 5) Onboard Model characteristics.

Bordignon, KA and Durham, W 1995 'Closed-form solutions to constrained control allocation problem,' *AIAA J. Guidance, Control, and Dynamics,* **18** (5), 1000–1007.

This paper describes the results of recent research into the problem of allocating several flight control effectors to generate moments acting on a flight vehicle. The results focus on the use of various generalized inverse solutions and a hybrid solution utilizing daisy chaining. In this analysis, the number of controls is greater than the number of moments being controlled, and the ranges of the controls are constrained to certain limits. The control effectors are assumed to be individually linear in their effects throughout their ranges of motion and independent of one another in their effects. A standard of comparison is developed based on the volume of moments or moment coefficients a given method can yield using admissible control deflections. Details of the calculation of the various volumes are presented. Results are presented for a sample problem involving 10 flight control effectors. The effectivenesses of the various allocation schemes are contrasted during an aggressive roll about the velocity vector at low dynamic pressure. The performance of three specially derived generalized inverses, a daisy-chaining solution, and direct control allocation are compared.

Buffington, JM 1997 'Tailless Aircraft Control Allocation' WL-TM-97-3060 Flight Dynamics Directorate, Wright Laboratory WPAFB, Ohio.

This paper presents a flight controller for a tailless aircraft with a large suite of conventional and unconventional control effectors. The controller structure is modular to take advantages of individual technologies from the areas of plant parameter estimation, control allocation, and robust feedback control. Linear models generated off-line provide plant parameter estimates for control. Dynamic inversion control provides direct satisfaction of flying qualities requirements in the presence of uncertainties. The focus of this paper, however, is control allocation. Control allocation is posed as constrained parameter optimization to minimize an objective that is a function of the control surface deflections. The control law is decomposed into a sequence of prioritized partitions, and additional optimization variables scale the control partitions to provide optimal command limiting which prevent actuator saturation. Analysis shows that appropriate prioritization of dynamic inversion control laws provides graceful command and loop response degradation for unachievable commands. Preliminary simulation results show that command variable response remains decoupled for achievable commands while other command limiting methods may result in unacceptable coupled response.

Buffington, J 1999 'Modular Control Law Design for the Innovative Control Effectors (ICE) Tailless Fighter Aircraft Configuration 101-3' US Air Force Research Lab. Report AFRL-VA-WP-TR-1999-3057, Wright-Patterson AFB, Ohio.

A modular flight control system is developed for a tailless fighter aircraft with innovative control effectors. Dynamic inversion control synthesis is used to develop a full envelope flight control law. Minor dynamic inversion command variable revisions are required due to the tailless nature of the configuration studied to achieve nominal stability and performance. Structured singular value and simulation analysis shows that robust stability is achieved and robust performance is slightly deficient due to modeling errors. A multi-branch linear programming-based method is developed and used for allocation of redundant limited control effectors.

Buffington, JM and Enns, DF 1996 'Lyapunov stability analysis of daisy chain control allocation,' *AIAA J. Guidance, Control, and Dynamics*, **19**(6), 1226–1230.

A demonstration that feedback control of systems with redundant controls can be reduced to feedback control of systems without redundant controls and control allocation is presented. It is shown that control allocation can introduce unstable zero dynamics into the system, which is important if input/output inversion control techniques are utilized. The daisy chain control allocation technique for systems with redundant groups of controls is also presented. Sufficient conditions are given to ensure that the daisy chain control allocation does not introduce unstable zero dynamics into the system. Aircraft flight control examples are given to demonstrate the derived results.

Buffington, JM, Enns, DF, and Teel, AR 1998 'Control allocation and zero dynamics,' *AIAA J. Guidance, Control, and Dynamics*, **21** (3), 458–464.

Closed-loop stability for dynamic inversion controllers depends on the stability of the zero dynamics. The zero dynamics, however, depend on a generally nonlinear control allocation function that optimally distributes redundant controls. Therefore, closed-loop stability depends on the control allocation function. A sufficient condition is provided for globally asymptotically stable zero dynamics with a class of admissible nonlinear control allocation functions. It is shown that many common control allocation functions belong to the class of functions that are covered by the aforementioned zero dynamics stability condition. Aircraft flight control examples are given to demonstrate the utility of the results.

Buffington, J, Chandler, P, and Pachter, M 1999 'On-line system identification for aircraft with distributed control effectors,' *Int. J. Robust Nonlinear Control*, **9**, 1033–1049

An algorithm is presented for the identification of aircraft stability derivatives and distributed control derivatives, in real time. Feedback control correlates the effectors' displacement with the aerodynamic angles, while the most commonly used control allocation algorithms correlate the effectors. The result is that valid derivative estimation is not possible. This paper addresses the effector identification problem by including decorrelating excitation into the control allocation cost function while still satisfying the desired control moment, and therefore does not introduce any residual perturbations into the motion variables. A two-step identification algorithm is used where the stability derivative and a generalized control derivative are identified in the first step. Results are shown for a stability axis roll maneuver with the stability and control derivatives being identified for five differential lateral directional effectors.

Burken, JJ, Lu, P, Wu, Z and Bahm, C 2001 'Two reconfigurable flight control design methods: robust servomechanisms and control allocation,' *AIAA J. Guidance, Control, and Dynamics*, **24** (3), 482–493.

Two methods are discussed for design of reconfigurable flight-control systems when one or more control surfaces are jammed. The first is a robust servomechanism control approach, which is a generalization of the classical proportional-plus-integral control to multi-input/multi-output systems. The second proposed method is a control allocation approach based on a quadratic programming formulation. The formulation is formally analyzed, and a globally convergent fixed-point iteration algorithm is used to make onboard implementation of this method feasible. The two methods are applied to reconfigurable entry flight control design for the X-33 vehicle. Nonlinear six-degree-of-freedom simulations demonstrate simultaneous tracking of angle-of-attack and roll-angle commands during control surface failures. The control-allocation method appears to offer more uniform and good performance at the expense of modestly higher computation requirement.

Cadzow, JA 1971 'Algorithm for the minimum-effort problem,' *IEEE Trans. on Automatic Control*, **16** (1), 60–63.

Give a consistent set of m linear equations in n unknown variables, a minimum-effort solution is defined to be a solution of that set of equations whose maximum component's magnitude is the smallest possible. An algorithmic procedure for obtaining a minimum-effort solution is developed. Its development is based on the duality principle from functional analysis. Possible applications of such an algorithm for typical digital control problems is presented in the introductory section. In such situations, it is frequently desirable to effect a given control task while using minimum control amplitude.

Cadzow, JA 1973 'A finite algorithm for the minimum ℓ_∞ solution to a system of consistent linear equations,' *SIAM J. Numerical Analysis*, **10** (4), 607–617.

A column exchange algorithm is presented for the determination of a minimum ℓ_∞ solution to a system of consistent linear equations. The algorithm is based on first solving the associated dual problem as specified by a well-known theorem from functional analysis and then generating the required solution by means of an alignment criterion. The procedure has been shown to have a rapid speed of convergence on all examples treated to date.

Cadzow, JA 1974 'An efficient algorithmic procedure for obtaining a minimum ℓ_∞-norm solution to a system of consistent linear equations,' *SIAM J. Numerical Analysis*, **11** (6), 1151–1165.

Herein is described an algorithmic procedure for determining a minimum l_∞-norm solution to the system of consistent linear equations $[Ax = y,]$ where A is a $m \times n$ matrix of rank m, y is a known $m \times 1$ vector and x is an unknown $n \times 1$ vector. The algorithm's development is based on some fundamental concepts from functional analysis. Its computational efficiency is shown to easily exceed that of the linear programming formulization of the same problem.

Casavola, A and Garone, E 2010 'Fault-tolerant adaptive control allocation schemes for overactuated systems,' *Int. J. Robust and Nonlinear Control*, **20**, 1958–1980.

This paper presents a fault-tolerant adaptive control allocation scheme for overactuated systems subject to loss of effectiveness actuator faults. The main idea is to use an 'ad hoc' online parameters estimator, coupled with a control allocation algorithm, in order to perform online control reconfiguration whenever necessary. Time-windowed and recursive versions of the algorithm are proposed for nonlinear discrete-time systems and their properties analyzed. Two final examples have been considered to show the effectiveness of the proposed scheme. The first considers a simple linear system with redundant actuators and it is mainly used to exemplify the main properties and potentialities of the scheme. In the second, a realistic marine vessel scenario under propeller and thruster faults is treated in full details.

Cui, L and Yang, Y 2011 'Disturbance rejection and robust least-squares control allocation in flight control system,' *AIAA J. Guidance, Control, and Dynamics*, **34** (6), 1632–1643.

The problem of disturbance rejection and control allocation with an uncertain control effectiveness matrix is investigated in this paper for a flight control system. An $\mathcal{H}_2/\mathcal{H}_\infty$ feedback controller is designed to produce the three axis moments and simultaneously suppress disturbance noise. A feedforward controller is used to track the reference signals. Under the condition of uncertainty included in the control effectiveness matrix, a robust least-squares scheme is employed to deal with the problem of distributing the three axis moments to the corresponding control surfaces. The proposed robust least-squares control allocation is studied for both unstructured and structured uncertainties. To illustrate the effectiveness of the proposed scheme, a simulation for an experimental satellite launch vehicle model is conducted. Comparisons of robust least-squares control allocation and pseudoinverse control allocation are presented. Results show that a disturbance is rejected and robust least-squares control allocation is effectively robust to uncertain control effectiveness matrix.

Doman, DB and Ngo, AD 2002 'Dynamic inversion-based adaptive/reconfigurable control of the X-33 on ascent,' *AIAA J. Guidance, Control, and Dynamics*, **25** (2), 275–284.

A quaternion-based attitude control system is developed for the X-33 in the ascent flight phase. A nonlinear control law commands body-axis rotation rates that align the angular velocity vector with an Euler axis defining the axis of rotation that will rotate the body-axis system into a desired-axis system. The magnitudes of the commanded body rates are determined by the magnitude of the rotation error. The commanded body rates form the input to a dynamic inversion-based adaptive/reconfigurable control law. The indirect adaptive control portion of the control law uses online system identification to estimate the current control effectiveness matrix to update a control allocation module. The control allocation nominally operates in a minimum deflection mode; however, if a fault is detected, it can operate in a null-space injection mode that excites and decorrelates the effectors without degrading the vehicle response to enable online system identification. The overall system is designed to provide fault and damage tolerance for the X-33 on ascent. The baseline control law is based on a full envelope design philosophy and eliminates trajectory-dependent gain scheduling that is typically found on this type of vehicle. Results are shown to demonstrate the feasibility of the approach.

Doman, DB, Gamble, BJ, and Ngo, AD 2009 'Quantized control allocation of reaction control jets and aerodynamic control surfaces,' *AIAA J. Guidance, Control, and Dynamics*, **32** (1), 13–24.

A mixed-integer linear programming approach to mixing continuous and pulsed control effectors is proposed. The method is aimed at applications involving reentry vehicles that are transitioning from exoatmospheric flight to endoatmospheric flight. In this flight phase, aerodynamic surfaces are weak and easily saturated, and vehicles typically rely on pulsed reaction control jets for attitude control. Control laws for these jets have historically been designed using single-axis phase-plane analysis, which has proven to be sufficient for many applications where multi-axis coupling is insignificant and when failures have not been encountered. Here, we propose using a mixed-integer linear programming technique to blend continuous control effectors and pulsed jets to generate moments commanded by linear or nonlinear control laws. When coupled with fault detection and isolation logic, the control effectors can be reconfigured to minimize the impact of control effector failures or damage. When the continuous effectors can provide the desired moments, standard linear programming methods can be used to mix the effectors; however, when the pulsed effectors must be used to augment the aerodynamic surfaces, mixed-integer linear programming techniques are used to determine the optimal combination of jets to fire. The reaction jet control allocator acts as a nonuniform quantizer that applies a moment vector to the vehicle, which approximates the desired moment generated by a continuous control law. Lyapunov theory is applied to develop a method for determining the region of attraction associated with a quantized vehicle attitude control system.

Doman, DB, Oppenheimer, MW, and Rone, W 2015 'Selective self-locking actuator and control allocation approach for thermal load minimization,' *AIAA J. Guidance, Control, and Dynamics*, **38** (6), 1110–1117.

The management of heat and energy on military aircraft has become increasingly challenging due to the proliferation of onboard sensors, electronics, and weapons systems. The problem has been compounded by a reduction in the ability of aircraft to dissipate waste thermal energy to the atmosphere as manufacturers have transitioned from structures composed of high-conductivity metals to low-conductivity composites [reference cited]. Future military aircraft may also be subjected to large transient thermal loads produced by onboard directed-energy weapons, which will further increase the challenge [references cited]. Aircraft actuation systems are also contributing more to thermal loads as systems have moved from centralized hydraulic systems to decentralized electrohydrostatic actuator systems or electromechanical actuators (EMAs) [reference cited]. With a myriad of internal heat sources and the reduced capacity to dissipate heat, it is important to study methods for addressing aircraft thermal management. Research in this area has been divided into two primary categories, namely, detailed modeling of the thermal behavior of the aircraft [references cited], and the development of more efficient mechanisms for removing waste thermal energy from the aircraft [reference cited]. In this exposition, a new EMA concept and control strategy is proposed that minimizes thermal loads produced by a flight control system and its actuator suite. The waste thermal energy produced by EMAs equipped with selflocking (SL), back-drivable (BD), and selectively self-locking (SSL) power screws is compared when paired with two different control allocation strategies. The results show that an SSL EMA coupled with a minimum motion control allocation preference vector can significantly reduce waste thermal energy generated by EMAs during cruise and nonaggressive maneuvering. However, under aggressive maneuvers, the benefit of using an SSL EMA is significantly reduced because the electrical power required to overcome control surface inertial loads dwarfs the power required to hold steady aerodynamic loads. Thus, the reduction in waste thermal energy due to the loadholding capacity of the SSL EMA is a much smaller percentage of the total waste thermal energy. Nevertheless, aircraft typically fly aggressive maneuvers for only a small fraction of the time that they operate, and the cumulative savings in waste thermal energy achieved by using an SSL actuator and a thermal load-minimizing control allocation strategy can be significant over the course of an entire flight.

Durham, W 1993 'Constrained control allocation,' *AIAA J. Guidance, Control, and Dynamics*, **16** (4), 717–725.

This paper addresses the problem of the allocation of several airplane flight controls to the generation of specified body-axis moments. The number of controls is greater than the number of moments being controlled, and the ranges of the controls are constrained to certain limits. They are assumed to be individually linear in their effect throughout their ranges of motion and independent of one another in their effects. The geometries of the subset of the constrained controls and of its image in moment space are examined. A direct method of allocating these several controls is presented that guarantees the maximum possible moment can be generated within the constraints of the controls. It is shown that no single generalized inverse can yield these maximum moments everywhere without violating some control constraint. A method is presented for the determination of a generalized inverse that satisfies given specifications which are arbitrary but restricted in number. We then pose and solve a minimization problem that yields the generalized inverse that best approximates the exact solutions. The results are illustrated at each step by an example problem involving three controls and two moments.

Durham, W 1994a 'Constrained control allocation: Three-moment problem,' *AIAA J. Guidance, Control, and Dynamics*, **17** (2), 330–336.

This paper presents a method for the solution of the constrained control allocation problem for the case of three moments. The control allocation problem is to find the 'best' combination of several flight control effectors for the generation of specified body-axis moments. The number of controls is greater than the number of moments being controlled, and the ranges of the controls are constrained to certain limits. The controls are assumed to be individually linear in their effect throughout their ranges of motion and complete in the sense that they generate moments in arbitrary combinations. The best combination of controls is taken to be an apportioning of the controls that yields the greatest total moment in a specified ratio of moments without exceeding any control constraint. The method of solving the allocation problem is presented as an algorithm and is demonstrated for a problem of seven aerodynamic controls on an F-18 airplane.

Durham, W 1994b 'Attainable moments for the constrained control allocation problem,' *AIAA J. Guidance, Control, and Dynamics*, **17** (6), 1371–1373.

Modern tactical aircraft are being designed with many more than the classical three sets of control effectors (ailerons, elevator, and rudder). The next generation of highly maneuverable airplanes are projected to have as many as 20 primary flight control effectors. These controls will all be constrained to certain limits, determined by the physical geometry of the control actuators or in some cases by aerodynamic considerations. The effective allocation, or blending, of these controls to achieve specific objectives is the control allocation problem. The geometry of the constrained control allocation problem was developed in [references cited] we described a means of determining the subset of attainable moments, yielding a description of the boundary that contained the necessary information for the determination of controls in the allocation problem. The method presented in [reference cited], although offering the advantage of generality, was admittedly complicated and difficult to implement. For the control allocation problems of particular interest, such generality is not required, and a simpler method of determining the attainable moment subset is available. That method is the subject of this Note. The problem statement and nomenclature used in this note may be found in [references cited].

Durham, W 1995 'Control stick logic in high-angle-of-attack maneuvering,' *AIAA J. Guidance, Control, and Dynamics*, **18** (5), 1092–1097.

The relationships between pilot control stick inputs and control effector deflections are examined. Specifically, we address multiply redundant control effector arrangements and command-driven control laws. During high-angle-of-attack, low-dynamic-pressure maneuvering, there is both a control power and control coordination problem. Control effector deflections are not one to one with pilot inputs, and the maximum capabilities of effectors to respond to pilot inputs varies dynamically with the state of the airplane. The problem is analyzed in the context of a generic control law that continuously regulates sideslip. A means is presented to relate the fixed control effector limits to the dynamically varying control response limits. This information may be used to re-establish the one-to-one correspondence of pilot inputs to control capabilities.

Durham, W 1996 'Dynamic inversion and model-following control' AIAA Paper 96-3690 in *AIAA Guidance, Navigation and Control Conference*.

The similarities and differences of dynamic inversion control and model-following control laws are examined. For the forms of these control laws assumed in this paper it is shown that dynamic inversion may be considered a special case of model-following. For any given dynamic inversion control law there is a model-following control law that achieves exactly the same response and therefore is in every way equivalent to it. This same model-following control law may be modified in its error dynamics without changing the desired response implied by the dynamic inversion law. The modification in error dynamics may be used to improve the tracking of the desired response in the presence of modeling errors.

Durham, W 1997 'Minimum drag control allocation,' *AIAA J. Guidance, Control, and Dynamics*, **20** (1), 190–193.

In [reference cited], we described a method of control allocation based on the instantaneous rate limits of the control effectors. The chief drawback to this method was the fact that the current positions were dependent on the path (in moment space) followed and would generally result in nonzero deflections in response to zero moment demands. The problem was alleviated by continuously applying unused rate capabilities to drive the solution toward one with the desired characteristics via the null space of the control effectiveness matrix.

We have long advocated the idea of including forces as well as moments in the effects of the controls that would, among other benefits, permit the determination of minimum control generated drag during cruise or maneuvering. Thus, to include just drag, we have not the three-dimensional attainable moment subset (ΔAMS), but the four-dimensional attainable objective subset (ΔAOS), whose coordinates are the three moments plus drag. The moments required of the control effectors are determined by the control law and may be considered to be specified. Because we are in the ΔAOS we actually calculate specified changes in desired moments. The actual change in drag, however, is not specified but is to be chosen as the greatest negative change attainable. Thus, the allocation problem is not to find controls that generate four specified objectives, but that generate three of those objectives and minimize the fourth.

Durham, W 1999 'Efficient, near-optimal control allocation' *AIAA J. Guidance, Control, and Dynamics*, **22** (2), 369–372.

We address the problem of allocating several redundant control effectors in the generation of specified body-axis moments. The effectors are constrained by position limits and are assumed to be linear in their effectiveness. The optimal solution to this problem in terms of generating maximum attainable moments for admissible controls was previously addressed in [references cited]. The computational complexity of the algorithm developed in [reference cited] to obtain such solutions is proportional to the square of the number of controls, which could become problematic in real-time applications. Other methods that generate solutions for maximum attainable moments, e.g., null space intersections [references cited] are even more computationally complex. This Note describes a computationally simple and efficient method to obtain near-optimal solutions. The method is based on prior knowledge of the controls' effectiveness and limits and on precalculation of several generalized inverses based on those data.

Durham, W 2001 'Computationally efficient control allocation' *AIAA J. Guidance, Control, and Dynamics*, **24** (3), 519–524.

The details of a computationally efficient method for calculating near-optimal solutions to the three-objective, linear, control-allocation problem are described. The control-allocation problem is that of distributing the effort of redundant control effectors to achieve some desired set of objectives. The optimal solution is that which exploits the collective maximum capability of the effectors within their individual physical limits. Computational efficiency is measured by the number of floating-point operations required for solution. The method presented returned optimal solutions in more than 90% of the cases examined; nonoptimal solutions returned by the method were typically much less than 1% different from optimal. The computational requirements of the method presented varied linearly with increasing numbers of controls and were much lower than those of previously described facet-searching methods, which increase in proportion to the square of the number of controls.

Durham, W 2013 *Aircraft Flight Dynamics and Control* 1st edn. John Wiley & Sons.
Aircraft flight dynamics and control, including numerical examples of dynamic inversion and control allocation.

Durham, W and Bordignon, K 1996 'Multiple control effector rate limiting,' *AIAA J. Guidance, Control, and Dynamics*, **19** (1), 30–37.
The effect of the choice of control allocation scheme upon individual control effector rate demands is examined. Three previously reported and one new variation of control allocation schemes are described and compared. The new allocation scheme exploits the maximum attainable moment rates of a given control effector configuration and is called moment-rate allocation. The bases of comparison are single-axis sinusoidal and triangular sawtooth, and multiaxis helical time-varying moment demands placed upon the allocation schemes. It is shown that 1) the choice of allocation scheme greatly affects the onset of effective rate limiting for a particular time-varying input but not consistently for all inputs; 2) widely varying results are obtained depending on whether the input is single or multiaxis and on the amplitude and shape of the input; 3) none of the observed behavior (except as regards generalized inverses) was easily predictable for arbitrary time-varying inputs; and 4) from the point of view of moment and moment-rate generating capabilities, moment-rate allocation clearly yielded best results.

Feemster, MG and Esposito, JM 2011 'Comprehensive framework for tracking control and thrust allocation for a highly overactuated autonomous surface vessel,' *J. Field Robotics*, **28**, 80–100.
In this paper, we present a comprehensive trajectory tracking framework for cooperative manipulation scenarios involving marine surface ships. Our experimental platform is a small boat equipped with six thrusters, but the technique presented here can be applied to a multi-ship manipulation scenario such as a group of autonomous tugboats transporting a disabled ship or unactuated barge. The primary challenges of this undertaking are: (1) the actuators are unidirectional and experience saturation; (2) the hydrodynamics of the system are difficult to characterize; and (3) obtaining acceptable performance under field conditions (i.e., GPS errors, wind, waves, etc) is arduous.
To address these issues, we present a frame work that includes trajectory generation, tracking control, and force allocation that, despite actuator limitations, results in asymptotically convergent trajectory tracking. In addition, the controller employs an adaptive feedback law to compensate for unknown difficult to measure hydrodynamic parameters. Field trials are conducted utilizing a 3 meter vessel in a nearby creek.

Frost, SA, Bodson, M, Burken, JJ, Jutte, CV, Taylor, BR, and Trinh, KV 'Flight control with optimal control allocation incorporating structural load feedback,' *J. Aerospace Information Systems*, **12**, 825–834.
Advances in sensors and avionics computation power suggest real-time structural load measurements could be used in flight control systems for improved safety and performance. A conventional transport flight control system determines the moments necessary to meet the pilots command while rejecting disturbances and maintaining stability of the aircraft. Control allocation is the problem of

converting these desired moments into control effector commands. In this paper, a framework is proposed to incorporate real-time structural load feedback and structural load constraints in the control allocator. Constrained optimal control allocation can be used to achieve desired moments without exceeding specified limits on monitored load points. Furthermore, certain criteria can be minimized, such as loads on certain parts of the aircraft. Flight safety issues can be addressed by using system health monitoring information to change control allocation constraints during flight. The framework to incorporate structural loads in the flight control system and an optimal control allocation algorithm are described and demonstrated on a nonlinear simulation of a generic transport aircraft with flight dynamics and static structural loads.

Gaulocher, SL, Roos, C, and Cumer, C 2007 'Aircraft load alleviation during maneuvers using optimal control surface combinations,' *AIAA J. Guidance, Control, and Dynamics*, **30** (2), 591–600.

Control laws in aeronautics are designed to ensure, above all, good handling qualities. However, during extreme maneuvers, which have to be taken into account for aircraft certification, a number of critical structural load limits cannot be guaranteed by this baseline controller. To avoid some modifications of the control law, a solution consists of judiciously exploiting the redundancy of the control surfaces. The aim of this paper is first to find an optimal strategy of the control surface use, which leaves the initial flight behavior unmodified but alleviates a structural load during a selected maneuver. Model predictive control theory solves this offline control allocation problem under actuator saturation constraints. In addition, an identification procedure is proposed to synthesize a new mixing unit that can reproduce this optimal strategy. This methodology is applied to a flexible transport aircraft to alleviate the bending moment at the external wing during a sudden and strong roll maneuver.

Glaze, ML 1998 *The Design and Implementation of a GUI-based Control Allocation Toolbox in the MATLAB® Environment*. MS Thesis, Virginia Polytechnic Institute & State University.

One of the primary considerations in control system design is positioning the control effectors to achieve some desired effect. In the case of aerospace vehicles, the desired effect is the generation of body-axis moments (i.e., roll, pitch, yaw) in response to pilot inputs. Classically, an aircraft was assumed to have three primary control effectors, each responsible for independently controlling one of the three rotational degrees of freedom; the ailerons for roll, the rudder for yaw, and the elevator for pitch. Modern aircraft, however, often utilize these classical effectors in combination with one or more advanced effectors, e.g., flaps, canards, and thrust vectoring, to control the same types of rotational motion. Tactical aircraft provide relevant examples; the F-15 ACTIVE operates with 12 effectors, the F-18 HARV with ten, and the F-16 with five independent control effectors.

The addition of advanced, redundant control effectors, though offering increased aircraft maneuverability, inherently increases the complexity of control system design. In mathematical terms, most modern aircraft are represented as under-determined systems because there are more unknowns (i.e., control effectors) than there are equations (i.e., rotational degrees of freedom). Unlike the uniquely determined classical system, the under-determined system has an infinite number of possible solutions; for the case considered here, there exist an infinite number of possible control configurations for the same set of desired moments. With so many possible combinations, there is no longer an obvious answer to the question of how to allocate controls in response to prescribed demands.

To effectively address the problem of managing multiple, redundant control effectors, a control allocator is introduced to the system. Generally speaking, control allocation is considered as any method that can be used to determine how the controls of a system should be positioned to achieve some desired effect. As discussed in [reference cited], a diverse group of allocation methods exist, from the straightforward direct allocation scheme, to the daisy chaining approach, to the computationally simple generalized inverse algorithms. Each algorithm has advantages and disadvantages with respect to others; it is the designer who determines which is the most suitable in a given situation.

In the interest of determining 'optimal' performance, control allocation has been the subject of much research in the past, and is the foundation for the work presented in this thesis. Unlike previous theoretical research in this area, however, the main focus here is the comparative value of the allocation results in assessing 'optimal' method performance.

Grogan, RL 1994 *A Thesis on the Application of Neural Network Computing to the Constrained Flight Control Allocation Problem.* MS Thesis, Virginia Polytechnic Institute & State University.

The feasibility of utilizing a neural network to solve the constrained flight control allocation problem is investigated for the purposes of developing guidelines for the selection of a neural network structure as a function of the control allocation problem parameters. The control allocation problem of finding the combination of several flight controls that generate a desired body axis moment without violating any control constraint is considered. Since the number of controls, which are assumed to be individually linear and constrained to specified ranges, is in general greater than the number of moments being controlled, the problem is nontrivial. Parallel investigations in direct and generalized inverse solutions have yielded a software tool (namely CAT, for Control Allocation Toolbox) to provide neural network training, testing, and comparison data. A modified back propagation neural network architecture is utilized to train a neural network to emulate the direct allocation scheme implemented in CAT, which is optimal in terms of having the ability to attain all possible moments with respect to a given control surface configuration. Experimentally verified heuristic arguments are employed to develop guidelines for the selection of neural network configuration and parameters with respect to a general control allocation problem. The control allocation problem is shown to be well suited for a neural network solution. Specifically, a six hidden neuron neural network is shown to have the ability to train efficiently, form an effective neural network representation of the subset of attainable moments, and independently discover the internal relationships between moments and controls. The performance of the neural network control allocator, trained on the basis of the developed guidelines, is examined for the reallocation of a seven control surface configuration representative of the F/A-18 HARV in a test maneuver flown using the original control laws of an existing flight simulator. The trained neural network is found to have good overall generalization performance, although limitations arise from the ability to obtain the resolution of the direct allocation scheme at low moment requirements. Lastly, recommendations offered include: (1) a proposed application to other unwieldy control allocation algorithms, with possible accounting for control actuator rate limitations, so that the computational superiority of the neural network could be fully realized; and (2) the exploitation of the adaptive aspects of neural network computing.

Hanger, MB 2011 *Model Predictive Control Allocation,* MS Thesis, Norwegian University of Science and Technology, Department of Engineering Cybernetics.

This thesis develops a control allocation method based on the Model Predictive Control algorithm, to be used on a missile in flight. The resulting Model Predictive Control Allocation (MPCA) method is able to account for actuator constraints and dynamics, setting it aside from most classical methods. A new effector configuration containing two groups of actuators with different dynamic authorities is also proposed. Using this configuration, the MPCA method is compared to the classical methods linear programming and redistributed pseudoinverse in various flight scenarios, highlighting performance differences as well as emphasizing applications of the MPCA method. It is found to be superior to the two classical methods in terms of tracking performance and total cost. Nevertheless, some restrictions and weaknesses were revealed, but countermeasures to these are proposed. The newly developed convex optimization solver CVXGEN is utilized successfully in the method evaluation. Providing solve times in milliseconds even for large problems, CVXGEN makes real-time implementations of the MPCA method feasible.

Harkegard, O 2004 'Dynamic control allocation using constrained quadratic programming,' *AIAA J. Guidance, Control, and Dynamics*, **27** (6), 1028–1034.

Control allocation deals with the problem of distributing a given control demand among an available set of actuators. Most existing methods are static in the sense that the resulting control distribution depends only on the current control demand. In this paper we propose a method for dynamic control allocation, in which the resulting control distribution also depends on the distribution in the previous sampling instant. The method extends regular quadratic-programming control allocation by also penalizing the actuator rates. This leads to a frequency-dependent control distribution, which can be designed to, for example, account for different actuator bandwidths. The control allocation problem is posed as a constrained quadratic program, which provides automatic redistribution of the control effort when one actuator saturates in position or in rate. When no saturations occur, the resulting control distribution coincides with the control demand fed through a linear filter.

Harkegard, O 2004b 'QCAT-quadratic programming control allocation toolbox for Matlab' URL: http://research.harkegard.se/qcat/ (link verified on 4 January 2016).

Harkegard, O and Glad, T 2005 'Resolving actuator redundancy-optimal control vs. control allocation,' *Automatica*, **41**, 137–144.

This paper considers actuator redundancy management for a class of overactuated nonlinear systems. Two tools for distributing the control effort among a redundant set of actuators are optimal control design and control allocation. In this paper, we investigate the relationship between these two design tools when the performance indexes are quadratic in the control input. We show that for a particular class of nonlinear systems, they give exactly the same design freedom in distributing the control effort among the actuators. Linear quadratic optimal control is contained as a special case. A benefit of using a separate control allocator is that actuator constraints can be considered, which is illustrated with a flight control example.

Hausner, MA 1965 *A Vector Space Approach to Geometry* Prentice-Hall; Dover Publications, 2010.

The effects of geometry and linear algebra on each other receive close attention in this examination of geometry's correlation with other branches of math and science. In-depth discussions include a review of systematic geometric motivations in vector space theory and matrix theory; the use of the center of mass in geometry, with an introduction to barycentric coordinates; axiomatic development of determinants in a chapter dealing with area and volume; and a careful consideration of the particle problem.

Jin, J 2005 'Modified pseudoinverse redistribution methods for redundant controls allocation,' *AIAA J. Guidance, Control, and Dynamics*, **28** (5), 1076–1079.

In recent years, control allocation problems have been intensively studied [references cited] following the work of Durham. There are several solution methods: direct control allocation, [references cited] daisy chaining, [reference cited] a linear-programming (LP) method,[references cited] a quadratic-programming (QP) method, [references cited] and a pseudoinverse-redistribution (PIR) method. [references cited] In this Note, a PIR method is discussed.

A pseudoinverse has been used as a selector or a distributor [reference cited]. Shtessel *et al*. applied a pseudoinverse to the control allocation problem for a tailless aircraft [reference cited]. However, it is fixed, and it does not consider the control surfaces' limits. Virnig and Bodden developed a PIR method for a short takeoff and vertical landing aircraft and then demonstrated the algorithm by means of a simulator [reference cited]. The PIR method repeats the process of calculating a pseudoinverse and a control vector by setting the saturated elements of a control vector to their limit values until a solution is obtained or there is no unsaturated element. Bordignon and Bessolo developed a modified method and applied it to the X-35 aircraft's control allocation problem [reference cited]. Although the PIR method sets the saturated control elements to their limit values, the modified method selects only one saturated element and sets it to its limit value during the redistribution process.

The purpose of this Note is to propose other modified methods and to compare them with the conventional PIR method and Bordignon and Bessolo's method. A different selection of one saturated element results in a difference in performance. For this purpose, two different selections are proposed. The performance metrics of concern are the calculation times, the percentage of times that a method converges to an optimal solution, and the characteristics of the errors between the optimal solutions and the redistributed solutions. Numerical examples are presented for comparisons.

Jingping, S, Weiguoa, Z, and Suilaoa, L 2011 'A control allocation method based on equivalent virtual control surfaces,' *Procedia Engineering*, **15**, 1256–1260.

Most existing allocation methods need the control effectiveness matrix to solve the problem, and their allocation accuracy directly depends on the identification precision of the control effectiveness matrix. If the matrix is poorly identified, the designed allocator may not work well. This paper presents an allocation method based on equivalent virtual control surface whose aerodynamic effect is equivalent to the real effectors. The method divides the design of control allocation system into two parts: the longitudinal and lateral, solve the problem and find the allocation law by choosing reasonable equivalent virtual effectors. Simulation results show that the proposed method can make the responses of roll angle and yaw angle follow the command signals well.

Johansen, TA and Fossen, TI 2013 'Control allocation—A survey,' *Automatica*, **49** (5), 1087–1103.

Control allocation problems for marine vessels can be formulated as optimization problems, where the objective typically is to minimize the use of control effort (or power) subject to actuator rate and position constraints, power constraints as well as other operational constraints. In addition, singularity avoidance for vessels with azimuthing thrusters represent a challenging problem since a non-convex nonlinear program must be solved. This is useful to avoid temporarily loss of controllability in some cases. In this paper, a survey of control allocation methods for overactuated vessels are presented.

Johansen, TA, Fossen, TI and Berge, SP 2004 'Constrained nonlinear control allocation with singularity avoidance using sequential quadratic programming,' *IEEE Trans. on Control Systems Technology*, **12**, 211–216.

Control allocation problems can be formulated as optimization problems, where the objective is typically to minimize the use of control effort (or power) subject to actuator rate and position constraints, and other operational constraints. Here we consider the additional objective of singularity avoidance, which is essential to avoid loss of controllability in some applications, leading to a nonconvex nonlinear program. We suggest a sequential quadratic programming approach, solving at each sample a convex quadratic program approximating the nonlinear program. The method is illustrated by simulated maneuvers for a marine vessel equipped with azimuth thrusters. The example indicates reduced power consumption and increased maneuverability as a consequence of the singularity-avoidance.

Johansen, TA, Fossen, TI and Tendel, P 2005 'Efficient optimal constrained control allocation via multi-parametric programming' *AIAA J. Guidance, Control, and Dynamics*, **28** (3), 506–515.

Constrained control allocation is studied, and it is shown how an explicit piecewise linear representation of the optimal solution can be computed numerically using multiparametric quadratic programming. Practical benefits of the approach include simple and efficient real-time implementation that permits software verifiability. Furthermore, it is shown how to handle control deficiency, reconfigurability, and flexibility to incorporate, for example, rate constraints. The algorithm is demonstrated on several overactuated aircraft control configurations, and the computational complexity is compared to other explicit approaches from the literature. The applicability of the method is further demonstrated using overactuated marine vessel dynamic position experiments on a scale model in a basin.

Khatri, AK, Singh, J, and Kumar, N 2013 'Accessible regions for controlled aircraft maneuvering,' *AIAA J. Guidance, Control, and Dynamics*, **36** (6), 1829–1834.

Design of maneuvers for carefree access of an aircraft to its complete flight envelope including post stall regimes is useful not only from a combat strategy point of view, but also for devising recovery strategies from an accident scenario. Maneuvers for an aircraft can be efficiently designed if a priori knowledge of its maneuverability characteristics is available to the control designers. Different types of agility metrics that characterize aircraft maneuvering capabilities have been proposed in literature based on different criteria [references cited]. A recent approach to define maneuverability characteristics is based on computing 'attainable equilibrium sets,' as suggested in [reference cited] and [reference cited]. This approach involves computing a two dimensional (2-D) section of attainable equilibrium sets of a particular maneuver using an inverse trimming formulation. Construction of maneuvers based on attainable equilibrium sets involves accessing desired aircraft states in the attainable equilibrium set from a normal flying condition, such as a level flight trim condition. Computing an attainable equilibrium set for a given aircraft model and developing control algorithms to switch aircraft states between different operating points lying within the accessible region defined by attainable equilibrium sets are thus essential ingredients of aircraft maneuver design. For aircraft models, which are inherently nonlinear due to nonlinear aerodynamics in the post stall regimes, and because of various couplings, use of nonlinear control design techniques based on dynamic inversion (DI) or sliding-mode control (SMC) have been proposed for control prototyping to design maneuvers [references cited]. Using bifurcation theory and continuation methods, Raghavendra *et al.* [reference cited] computed spin solutions for the F18/high-alpha research vehicle (HARV) model, and demonstrated use of the DI controller to recover the aircraft from a flat oscillatory spin motion. Recently, the authors have proposed a systematic approach using bifurcation analysis and continuation methods in conjunction with a SMC technique to design maneuvers for a nonlinear aircraft model [reference cited].

Kishore, WCA, Sen, S, Ray, G, and Ghoshal, TK 2008 'Dynamic control allocation for tracking time-varying control demand,' *AIAA J. Guidance, Control, and Dynamics*, **31**, 1150–1157

Modern aircraft, missiles, and launch vehicles are fitted with more control surfaces/actuators than controlled variables to achieve fast maneuvers. These extra effectors provide a certain degree of redundancy and hence can be used to achieve multiple control objectives. There are several combinations of actuator positions which produce the same virtual control effort demanded by the high level controller and hence achieve desired performance. Control allocation of such overactuated systems is generally formulated as an optimization problem to minimize energy loss, power consumption, and other similar costs, subject to constraints such as actuator saturation in rate and position, and a specified set of dynamics for operation in the linear range. Extensive surveys that compare and identify the limitations of control allocation algorithms are presented in [references cited]. Among these, weighted norm minimization [references cited] provides an explicit solution, but without taking actuator saturation into consideration. This methodology has been extended in [reference cited] to accommodate the saturation constraints by selecting a weight matrix in a linear matrix inequality (LMI) formulation, but it is only suitable for a limited region of control demand space. Linear programming, mixed integer, and quadratic programming approaches have also been applied to solve the allocation problems formulated as weighted norm minimization and convex quadratic optimal allocation subject to linear equality and inequality constraints [references cited].

Given a probability distribution for each of the unknown aerodynamic stability and control parameters, a Monte Carlo simulation of the time response allows for a statistical evaluation of the uncertain aerodynamic parameter effects on the aircraft behavior. Prior to integrating the system equations over the desired time of flight, each fractional variation in the aerodynamic coefficient was computed via a random number generator. After each simulation, flying qualities metrics were evaluated and the binary results were recorded: the aircraft either passed or failed the metric. The process was repeated a sufficient number of times such that the resulting probability of violation was statistically significant based on confidence bounds. Experiments were performed using a dynamic inversion and a model-following control law to command angle-of-attack. The results show that the properly designed model-follower has better performance robustness than a dynamic inverter with the same performance objectives.

Kocurek, N and Durham, W 1997 'Dynamic inversion and model-following flight control: A comparison of performance robustness,' in *22nd Atmospheric Flight Mechanics Conference*.

Given a probability distribution for each of the unknown aerodynamic stability and control parameters, a Monte Carlo simulation of the time response allows for a statistical evaluation of the uncertain aerodynamic parameter effects on the aircraft behavior. Prior to integrating the system equations over the desired time of flight, each fractional variation in the aerodynamic coefficient was computed via a random number generator. After each simulation, flying qualities metrics were evaluated and the binary results were recorded: the aircraft either passed or failed the metric. The process was repeated a sufficient number of times such that the resulting probability of violation was statistically significant based on confidence bounds. Experiments were performed using a dynamic inversion and a model-following control law to command angle-of-attack. The results show that the properly designed model-follower has better performance robustness than a dynamic inverter with the same performance objectives.

Lavretsky, E, Diecker, R, and Brinker, J 2009 'Robust Control Effector Allocation,' United States Patent Application 20090143925.

Method and apparatus is disclosed which allocates the execution of a commanded vehicle maneuver among the vehicle's control effectors capable of affecting such maneuver, with consideration given to the possible nonlinear and/or non-monotonic effects each control effector's displacement may have on the vehicle and on each other's performance.

Leedy, JQ 1998 *Real-time Moment-rate Constrained Control Allocation for Aircraft with a Multiply-redundant Control Suite*. MS Thesis, Virginia Polytechnic Institute & State University.

The problem of aircraft control allocation is that of finding a combination of control positions that cause the resulting aircraft moments to most closely satisfy a given desired moment vector. The problem is easily solved for the case of an aircraft having three control surfaces, each of which primarily imparts moments in each of the three aircraft axes. In this simple case, the solution to the control allocation problem is uniquely determined. However, many current and future aircraft designs employ a larger set of control effectors, resulting in a control redundancy in the sense that more than one combination of control positions can produce the same desired moment. When taking into account both the position and rate constraints of the control effectors, the problem is significantly more complex. Constrained moment-rate control allocation guarantees a control solution that can achieve every possible moment that is physically realizable by the aircraft. Addressed here is the real-time performance of moment-rate constrained control allocation as tested on a desktop simulation. Issues that were deemed interesting or potentially problematic in earlier batch simulation, such as control chattering due to restoring and apparent control wind-up, are investigated and an evaluation is made of the overall feasibility of these algorithms. The purpose of the research is to confirm that the results obtained from batch simulation testing are also valid using maneuvers representative of real-time flight and representative simulation frame sizes, and to uncover potential problems not observed in batch simulation.

Liao, F, Lum, K, Wang, JL, and Benosman, M 2009 'Adaptive control allocation for non-linear systems with internal dynamics,' *IET Control Theory & Applications*, **4**, 909–922

An adaptive control allocation method is presented for a general class of non-linear systems with internal dynamics and unknown parameters. A certainty equivalence indirect adaptive approach is used to estimate the unknown parameters. Based on the estimated parameters, model reference control and control allocation techniques are used to control the non-linear system subject to control constraints and internal dynamics stabilisation. A Lyapunov design approach with the property of convergence to a positively invariant set is proposed. The derived adaptive control allocation is in the form of a dynamic update law, which, together with a stable model reference control, guarantees that the closed-loop non-linear system be input to-state stable.

Liu, Y and Crespo, LG 2012 'Adaptive control allocation in the presence of actuator failures,' *J. Control Science and Engineering*, **2012**, 1–16.

This paper proposes a control allocation framework where a feedback adaptive signal is designed for a group of redundant actuators and then it is adaptively allocated among all group members. In the adaptive control allocation structure, cooperative actuators are grouped and treated as an equivalent control eector. A state feedback adaptive control signal is designed for the equivalent eector and adaptively allocated to the member actuators. Two adaptive control allocation algorithms, guaranteeing closed-loop stability and asymptotic state tracking when partial and total loss of control eectiveness occur, are developed. Proper grouping of the actuators reduces the controller complexity without reducing their ecacy. The implementation and eectiveness of the strategies proposed is demonstrated in detail using several examples.

Lombaerts, T, Schravendijk, M, Ping, C and Mulder, J 2011 'Adaptive nonlinear flight control and control allocation for failure resilience,' in Holzapfel, F and Theil, S (eds) *Advances in Aerospace Guidance, Navigation and Control*, Springer, pp. 41–53.

In this publication, reconfiguring control is implemented by making use of Adaptive Nonlinear Dynamic Inversion (ANDI) for autopilot control. The adaptivity of the control setup is achieved by making use of a real time identified physical model of the damaged aircraft. In failure situations, the damaged aircraft model is identified by the so-called two step method in real time and this model is then provided to the model-based adaptive NDI routine in a modular structure, which allows flight control reconfiguration on-line. Three important modules of this control setup are discussed in this publication, namely aerodynamic model identification, adaptive nonlinear control, and control allocation. Control allocation is especially important when some dynamic distribution of the control commands is needed towards the different input channels. After discussing this modular adaptive controller setup, reconfiguration test results are shown for damaged aircraft models which indicate satisfactory failure handling capabilities of this fault tolerant control setup.

Low, CP 2005 'An efficient algorithm for the minimum cost min-max load terminal assignment problem,' *IEEE Communications Letters*, **9** (11), 1012–1014.

One of the main issues to be addressed in topological design of centralized networks is that of assigning terminals to concentrators in such a way that each terminal is assigned to one (and only one) concentrator and the total number of terminals assigned to any concentrator (which is referred to as load in this paper) does not overload that concentrator, i.e. is within the concentrator's capacity. Under these constraints, an assignment with the lowest possible cost is sought. An assignment of terminals to concentrators which minimizes the maximum load among the concentrators (which qualitatively represents congestion at some hot spots in a network service area) is referred to as a min-max load assignment. In this paper, we consider the problem of finding a min-max load assignment with the-lowest cost. We call this problem the Minimum Cost Min-Max Load Terminal Assignment Problem (MCMLTAP). We present an algorithm for MCMLTAP and prove that the problem is optimally solvable in polynomial time using our proposed algorithm.

Luo, Y, Serrani, A, Yurkovich, S, Doman, DB, and Oppenheimer, MW 2007 'Model-predictive dynamic control allocation scheme for reentry vehicles,' *AIAA J. Guidance, Control, and Dynamics*, **30** (1), 100–113.

Allocation of control authority among redundant control effectors, under hard constraints, is an important component of the inner loop of a reentry vehicle guidance and control system. Whereas existing control allocation schemes generally neglect actuator dynamics, thereby assuming a static relationship between control surface deflections and moments about a three-body axis, in this work a dynamic control allocation scheme is developed that implements a form of model-predictive control. In the approach proposed here, control allocation is posed as a sequential quadratic programming problem with constraints, which can also be cast into a linear complementarity problem and therefore solved in a finite number of iterations. Accounting directly for nonnegligible dynamics of the actuators with hard constraints, the scheme extends existing algorithms by providing asymptotic tracking of time varying input commands for this class of applications. To illustrate the effectiveness of the proposed scheme, a high fidelity simulation for an experimental reusable launch vehicle is used, in which results are compared with those of static control allocation schemes in situations of actuator failures.

Mulkens, JM and Ormerod, AO 1993 'Measurements of aerodynamic rotary stability derivatives using a whirling arm facility,' *J. Aircraft*, **30** (2), 178–183.

This work is part of a program of research in which the high angle-of-attack region is of particular interest. Equipment and methods have been developed to adapt a whirling arm facility for the measurement of the effects of path curvature on two generic combat aircraft configurations. An explanation is given of the merits of using a whirling arm and some of the difficulties are mentioned. The derivatives associated with steady rotation have been assessed at angles of attack up to 30 deg. Both longitudinal and directional tests have been made and comparisons with the results of oscillatory tests are presented. For the directional results, little difference was found. The longitudinal results, however, showed a significant difference at certain high angles of attack. These differences, which were of different signs for the two models tested, have to be attributed to effects associated with the rate of change-of-incidence.

Munro, BC 1992 *Airplane Trajectory Expansion for Dynamics Inversion*. MS Thesis, Virginia Polytechnic Institute & State University.

In aircraft research, there is keen interest in the procedure of determining the set of controls required to perform a maneuver from a definition of the trajectory. This is called the inverse problem. It has been proposed that if a complete set of states and state time derivatives can be derived from a trajectory then a model-following solution can allocate the controls necessary for the maneuver. This paper explores the problem of finding the complete state definition and provides a solution that requires numerical differentiation, fixed point iteration and a Newton's method solution to nonlinear equations. It considers trajectories that are smooth, piecewise smooth, and noise ridden. The resulting formulation was coded into a FORTRAN program. When tested against simple smooth maneuvers, the program output was very successful but demonstrated the limitations imposed by the assumptions and approximations in the development.

Nelson, MD 2001 *A Comparison of Two Methods used to deal with Saturation of Multiple, Redundant Aircraft Control Effectors*. MS Thesis, Virginia Polytechnic Institute & State University.

A comparison of two methods to deal with allocating controls for unattainable moments in an aircraft was performed using a testbed airframe that resembled an F/A-18 with a large control effector suite. The method of preserving the desired moment direction to deal with unattainable moments is currently used in a specific control allocator. A new method of prioritizing the pitch axis is compared to the moment-direction preservation. Realtime piloted simulations are completed to evaluate the characteristics and performance of these methods.

A direct comparison between the method of preserving the moment direction by scaling the control solution vector and prioritizing the pitching moment axis is performed for a specific case. Representative maneuvers are flown with a highly unstable airframe to evaluate the ability to achieve the specific task. Flight performance and pilot interpretation are used to evaluate the two methods.

Pilot comments and performance results favored the method of pitch-axis prioritization. This method provided favorable flight characteristics compared to the alternative method of preserving the moment direction for the specific tasks detailed in this paper.

Oh, J-H, Jamoom, MB, McConley, MW and Feron, E 1999 'Solving control allocation problems using semidefinite programming,' *AIAA J. Guidance, Control, and Dynamics*, **22** (3), 494–497.

The capabilities of modern combat and civilian aircraft keep increasing. In particular, modern-day aircraft have many available control surfaces and thrust vectoring capabilities that offer significant advantages over conventional architectures based on three control surfaces only, including reduced electromagnetic signature, tailless designs, energy-efficient maneuvering, and most importantly, much needed redundancy in case of battle damage. The trend toward the presence of multiple actuators in modern aircraft is likely to continue, with the advent of distributed actuation systems based, for example, on micro electro mechanical systems. The control allocation problem is to nd a harmonious way to manage several actuators together to produce desired effects (usually moments) on the aircraft. The requirement for simplicity enables the reduction of software development costs by reusing existing control architectures as much as possible.

Oppenheimer, MW and Doman, DB 2004 'Methods for compensating for control allocator and actuator interactions,' *AIAA J. Guidance, Control, and Dynamics*, **27** (5), 922–927.

Numerous control allocation algorithms have been developed for aircraft for the purpose of providing commands to suites of control effectors to produce desired moments or accelerations. A number of approaches have been developed that ensure that the commands provided to the effectors are physically realizable. These actuator command signals are feasible in the sense that they do not exceed hardware rate and position limits. Buffington developed a linear programming-based approach that separately considered cases where sufficient control power was available to meet a moment demand and a control deficiency case where the moment deficiency was minimized. Bodson developed a linear programming approach where the sufficiency and deficiency branches were considered simultaneously, which resulted in computational savings. Quadratic programming approaches have also been considered. In the early 1990s, Durham developed a constrained control allocation approach called direct allocation that was based on geometric concepts of attainable moment sets. Page and Steinberg as well as Bodson have presented excellent survey papers that compare and contrast many of the control allocation approaches developed over the last two decades. Recent work in the area has resulted in the development of control allocation algorithms that can accommodate cases where the moments or accelerations produced by the control effectors are nonlinear functions of the effector position.

Oppenheimer, MW, Doman, DB, and Bolender, MA 2010 'Control allocation', in *The Control Handbook*, 2nd edn. CRC Press.

Over the past few decades, much emphasis has been placed on over-actuated systems for air vehicles. Over-actuating an air vehicle provides a certain amount of redundancy for the flight control system, thus potentially allowing for recovery from off-nominal conditions. Due to this redundancy, control allocation algorithms are typically utilized to compute a unique solution to the over-actuated problem. Control allocators compute the commands that are applied to the actuators so that a prescribed set of forces or moments are generated by the control effectors. Usually, control allocation problems are formulated as optimization problems so that all of the available degrees of freedom can be utilized and, when sufficient control power exists, secondary objectives can be achieved. A conventional aircraft utilizes an elevator for pitch control, ailerons for roll control, and a rudder for yaw control. As aircraft designs have advanced, more control effectors (some unconventional) have been placed on the vehicles. In some cases, certain control effectors may be able to exert significant influence upon multiple axes. When a system is equipped with more effectors than controlled variables, the system may be over-actuated. The allocation, blending, or mixing of these control effectors to achieve some desired objectives constitute the control allocation problem. Due to over-actuation and the influence of control surfaces on multiple controlled variables, it can be difficult to determine an appropriate method of how to translate a controlled variable command into a control surface command. Some air vehicle concepts have been designed with 10 or more control effectors and only three controlled variables. As the number of control effectors increases, the determination of ad hoc control allocation schemes becomes more difficult and the need for systematic control allocation algorithms increases. In addition, rate and position limits of the control effectors must be considered in order to achieve a realistic solution. Not only is the mixing of control surface effects critical, but it is also desirable to enable the aircraft to recover from off-nominal conditions, such as a failed control surface, when physically possible. Reconfigurable controllers can adjust control system parameters to adapt to off-nominal conditions [references cited]. In reconfigurable control systems, a control allocation algorithm can be used to perform automatic redistribution of the control power requests among a large number of control effectors, while still obeying the rate and position limits of the actuators.

Orr, JS 2013 *High Efficiency Thrust Vector Control Allocation* PhD Thesis, The University of Alabama in Huntsville.

The design of control mixing algorithms for launch vehicles with multiple vectoring engines yields competing objectives for which no straightforward solution approach exists. The designer seeks to optimally allocate the effector degrees of freedom such that maneuvering capability is maximized subject to constraints on available control authority. In the present application, such algorithms are generally restricted to linear transformations so as to minimize adverse control-structure interaction and maintain compatibility with industry-standard methods for control gain design and stability analysis. Based on the application of the theory of ellipsoids, a complete, scalable, and extensible framework is developed to effect rapid analysis of launch vehicle capability. Furthermore, a control allocation scheme is proposed that simultaneously balances attainment of the maximum maneuvering capability with rejection of internal loads and performance losses resulting from thrust vectoring in the null region of the admissible controls. This novel approach leverages an optimal parametrization of the weighted least squares generalized inverse and exploits the analytic properties of the constraint geometry so as to enable recovery of more than ninety percent of the theoretical capability while maintaining linearity over the majority of the attainable set.

Orr, JS and Slegel, NJ 2014 'High-efficiency thrust vector control allocation,' *AIAA J. Guidance, Control, and Dynamics*, **37** (2), 374–382.

A generalized approach to the allocation of redundant thrust vector slew commands for multi-actuated launch vehicles is presented, where deflection constraints are expressed as omniaxial or elliptical deflection limits in gimbal axes. More importantly than in the aircraft control allocation problem, linear allocators (pseudoinverses) are preferred for large booster applications to facilitate accurate prediction of the control-structure interaction resulting from thrust vectoring effects. However, strictly linear transformations for the allocation of redundant controls cannot, in general, access all of the attainable moments for which there is a set of control effector positions that satisfies the constraints. In this paper, the control allocation efficiency of a certain class of linear allocators subject to multiple quadratic constraints is analyzed, and a novel single-pass control allocation scheme is proposed that augments the pseudoinverse near the boundary of the attainable set. The controls are determined over a substantial volume of the attainable set using only a linear transformation; as such, the algorithm maintains compatibility with frequency domain approaches to the analysis of the vehicle closed-loop elastic stability. Numerical results using a model of a winged reusable booster system illustrate the proposed technique's ability to access a larger fraction of the attainable set than a pseudoinverse alone.

Page, AB and Steinberg, ML 2002 'High-fidelity simulation testing of control allocation methods,' AIAA 2002-4547 in *AIAA Guidance, Navigation, and Control Conference and Exhibit*.
This paper describes high-fidelity simulation testing of some of the more popular advanced control allocation techniques integrated with two separate dynamic inversion control laws. The allocation methods include variations of quadratic programming, linear programming, direct allocation, cascaded generalized inverse, and weighted pseudo-inverse. Results are presented for single and multi-axis pitch and roll maneuvers with and without actuator failures. A velocity vector roll is also considered for the no failure case. Results show that a robust control law can mask differences in the various control allocation routines and lead to similar performance from both optimal and sub-optimal allocation methods. Furthermore, the current results illustrate that the closed-loop performance does not directly follow from the open- loop measures that are widely used in the literature.

Paradiso, JA 1991 'Adaptable method of managing jets and aerosurfaces for aerospace vehicle control,' *AIAA J. Guidance, Control, and Dynamics*, **14** (1), 44–50.
An actuator selection procedure is presented that uses linear programming to optimally specify bounded aerosurface deflections and jet firings in response to differential torque and/or force commands. This method creates a highly adaptable interface to vehicle control logic by automatically providing intrinsic actuator decoupling, dynamic response to actuator reconfiguration, dynamic upper bound and objective specification, and the capability of coordinating hybrid operation with dissimilar actuators. The objective function minimized by the linear programming algorithm is adapted to realize several goals, i.e., discourage large aerosurface deflections, encourage the use of certain aerosurfaces (speedbrake, body flap) as a function of vehicle state, minimize drag, contribute to translational control, and adjust the balance between jet firings and aerosurface activity during hybrid operation. A vehicle model adapted from Space Shuttle aerodynamic data is employed in simulation examples that drive the actuator selection with a six-axis vehicle controller tracking a scheduled re-entry trajectory.

Petersen, J and Bodson, M 2002 'Fast implementation of direct allocation with extension to coplanar controls,' *AIAA J. Guidance, Control, and Dynamics*, **25** (3), 464–473.
The direct allocation method is considered for the control allocation problem. The original method assumed that every three columns of the controls effectiveness matrix were linearly independent. Here, the condition is relaxed, so that systems with coplanar controls can be considered. For fast online execution, an approach using spherical coordinates is also presented, and results of the implementation demonstrate improved performance over a sequential search. Linearized state-space models of a C-17 aircraft and of a tailless aircraft are used in the evaluation.

Petersen, J and Bodson, M 2005 'Interior-point algorithms for control allocation,' *AIAA J. Guidance, Control, and Dynamics*, **28** (3), 471–480.

Linear-programming formulations of control allocation problems are considered, including those associated with direct allocation and mixed 1 -norm objectives. Primal-dual and predictor-corrector path-following interior-point algorithms, that are shown to be well suited for the control-allocation problems, are described in some detail with an emphasis on preferred implementations. The performance of each algorithm is evaluated for computational efficiency and for accuracy using linear models of a C-17 transport and a tailless fighter aircraft. Appropriate choices of stopping tolerances and other algorithm parameters are studied. Comparisons of speed and accuracy are made to the simplex method. Results show that real-time implementation of the algorithms is feasible, without requiring excessive number of computations.

Petersen, J and Bodson, M 2006 'Constrained quadratic programming techniques for control allocation,' *IEEE Trans. on Automatic Control*, **14** (1), 91–98.

The paper considers the objective of optimally specifying redundant control effectors under constraints, a problem commonly referred to as control allocation. The problem is posed as a mixed 2-norm optimization objective and converted to a quadratic programming formulation. The implementation of an interior-point algorithm is presented. Alternative methods including fixed-point and active set methods are used to evaluate the reliability, accuracy and efficiency of a primal-dual interior-point method. While the computational load of the interior-point method is found to be greater for problems of small size, convergence to the optimal solution is also more uniform and predictable. In addition, the properties of the algorithm scale favorably with problem size. Index Terms Control allocation, flight control, interior-point methods, quadratic programming, redundant control effectors.

Pratt, RW (ed) 2000 Flight Control Systems: Practical Issues in Design and Implementation. Institution of Engineering and Technology, Control Engineering Series 57.

A complete reference on modern flight control methods for fixed wing aircraft, this authoritative book includes contributions from an international group of experts in their respective specialized fields. Split into two parts, the first section of the book deals with the fundamentals of flight control systems design, whilst the second concentrates on genuine applications based on modern control methods used in the latest aircraft.

Scalera, KR 1999 *A Comparison of Control Allocation Methods for the F-15 ACTIVE Research Aircraft Utilizing Real-Time Piloted Simulations*. MS Thesis, Virginia Polytechnic Institute & State University.

A comparison of two control allocation methods is performed utilizing the F-15 ACTIVE research vehicle. The control allocator currently implemented on the aircraft is replaced in the simulation with a control allocator that accounts for both control effector positions and rates. Validation of the performance of this Moment Rate Allocation scheme through real-time piloted simulations is desired for an aircraft with a high fidelity control law and a larger control effector suite.

A more computationally efficient search algorithm that alleviates the timing concerns associated with the early work in Direct Allocation is presented. This new search algorithm, deemed the Bisecting, Edge-Search Algorithm, utilizes concepts derived from pure geometry to efficiently determine the intersection of a line with a convex faceted surface.

Control restoring methods, designed to drive control effectors towards a 'desired' configuration with the control power that remains after the satisfaction of the desired moments, are discussed. Minimum-sideforce restoring is presented. In addition, the concept of variable step size restoring algorithms is introduced and shown to yield the best tradeoff between restoring convergence speed and control chatter reduction.

Representative maneuvers are flown to evaluate the control allocator's ability to perform during realistic tasks. An investigation is performed into the capability of the control allocators to reconfigure the control effectors in the event of an identified control failure. More specifically, once the control allocator has been forced to reconfigure the controls, an investigation is undertaken into possible performance degradation to determine whether or not the aircraft will still demonstrate acceptable flying qualities.

A direct comparison of the performance of each of the two control allocators in a reduced global position limits configuration is investigated. Due to the highly redundant control effector suite of the F-15 ACTIVE, the aircraft, utilizing Moment Rate Allocation, still exhibits satisfactory performance in this configuration. The ability of Moment Rate Allocation to utilize the full moment generating capabilities of a suite of controls is demonstrated.

Schierman, JD, Ward, DG, Hull, JR, Gandhi, N, Oppenheimer, MW, and Doman, DB 2004 'Integrated adaptive guidance and control for re-entry vehicles with flight-test results,' *AIAA J. Guidance, Control, and Dynamics*, **27** (6), 975–988.

To enable autonomous operation of future reusable launch vehicles, reconfiguration technologies will be needed to facilitate mission recovery following a major anomalous event. The Air Force's Integrated Adaptive Guidance and Control program developed such a system for Boeing's X-40A, and the total in-flight simulator research aircraft was employed to flight test the algorithms during approach and landing. The inner loop employs a model-following/dynamic-inversion approach with optimal control allocation to account for control-surface failures. Further, the reference-model bandwidth is reduced if the control authority in any one axis is depleted as a result of control effector saturation. A backstepping approach is utilized for the guidance law, with proportional feedback gains that adapt to changes in the reference model bandwidth. The trajectory-reshaping algorithm is known as the optimum-path-to-go methodology. Here, a trajectory database is precomputed off line to cover all variations under consideration. An efficient representation of this database is then interrogated in flight to rapidly find the 'best' reshaped trajectory, based on the current state of the vehicle's control capabilities. The main goal of the flight-test program was to demonstrate the benefits of integrating trajectory reshaping with the essential elements of control reconfiguration and guidance adaptation. The results indicate that for more severe, multiple control failures, control reconfiguration, guidance adaptation, and trajectory reshaping are all needed to recover the mission.

Servidia, PA 2010 'Control allocation for gimballed/fixed thrusters,' *Acta Astronautica*, **66**, 587–594.

Some overactuated control systems use a control distribution law between the controller and the set of actuators, usually called control allocator. Beyond the control allocator, the configuration of actuators may be designed to be able to operate after a single point of failure, for system optimization and/or decentralization objectives. For some type of actuators, a control allocation is used even without redundancy, being a good example the design and operation of thruster configurations. In fact, as the thruster mass flow direction and magnitude only can be changed under certain limits, this must be considered in the feedback implementation. In this work, the thruster configuration design is considered in the fixed (F), single-gimbal (SG) and double-gimbal (DG) thruster cases. The minimum number of thrusters for each case is obtained and for the resulting configurations a specific control allocation is proposed using a nonlinear programming algorithm, under nominal and single-point of failure conditions.

Shaw, RL 1985 *Fighter Combat: Tactics and Maneuvering*. Naval Institute Press.

This book provides a detailed description of one-on-one dog-fights and multi-fighter team work tactics, as well as discussions on aircraft and weapons systems.

Snell, S, Enns, D, and Garrard, W 1992 'Nonlinear inversion flight control for a supermaneuverable aircraft,' *AIAA J. Guidance, Control, and Dynamics*, **15** (4), 976–984.

Nonlinear dynamic inversion affords the control system designer a straightforward means of deriving control laws for nonlinear systems. The control inputs are used to cancel unwanted terms in the equations of motion using negative feedback of these terms. In this paper, we discuss the use of nonlinear dynamic inversion in the design of a flight control system for a supermaneuverable aircraft. First, the dynamics to be controlled are separated into fast and slow variables. The fast variables are the three angular rates and the slow variables are the angle of attack, sideslip angle, and bank angle. A dynamic inversion control law is designed for the fast variables using the aerodynamic control surfaces and thrust vectoring control as inputs. Next, dynamic inversion is applied to the control of the slow states using commands for the fast states as inputs. The dynamic inversion system was compared with a more conventional, gain-scheduled system and was shown to yield better performance in terms of lateral acceleration, sideslip, and control deflections.

Sommerville, D 1929 *An Introduction to the Geometry of n Dimensions,* Methuen; Dover Publishing 1958, pp. 9–10.

Selected representative topics that not only illustrate the extensions of theorems of three-dimensional geometry but that reveal results that are unexpected and where analogy would be a faithless guide.

Stevens, BL and Lewis, FL 2015 *Aircraft Control and Simulation,* 3rd edn. Wiley-Blackwell.

Aircraft Control and Simulation provides comprehensive, expert-led guidance to the topic, accessible to both students and professionals involved in the design and modeling of aerospace vehicles. Updated to include new coverage of Unmanned Aerial Vehicles, this new third edition has been expanded throughout to cover the latest advances in the field.

The material progresses steadily from motion and aerodynamics equations through advanced control methods, using detailed real-world examples with model software details provided. Fundamental principles give way to dynamic analysis, stability evaluation, multivariable control, and more, including geodesy and the gravitational theory behind suborbital aircraft.

Tauke, G and Bordignon, K 2002 'Structural coupling challenges for the X-35B,' AIAA 2002-6004 in *AIAA 2002 Biennial International Powered Lift Conference and Exhibit*.

The Short Takeoff and Vertical Landing variant of Lockheed Martin's Joint Strike Fighter (the X-35B) had several structural coupling margin verification issues. These issues included the use of multi-path non-linear dynamic-inversion control laws and the use of the propulsion as part of the flight control system. The control law challenges included: 1) multi-path analysis, 2) gain limiting with real time gain computation, 3) dealing with the relatively low coupling frequency of a high inertia control surface, 4) dealing with inertial coupling through the pilot and control stick, 5) open and closed loop margin testing. The propulsion system challenges include: 1) non-linear response characteristics including response at one frequency affecting gain at a different frequency, 2) rigid body coupling effects, 3) gear mode interaction, 4) dealing with the relatively low effector frequency response, 5) test methodology. This paper shows how the X-35 team successfully handled the variety of structural coupling challenges associated with the innovative control system.

Tjonnas, J and Johansen, TA 2008a 'Optimizing adaptive control allocation with actuator dynamics,' *Modeling, Identification and Control*, **29** (2), 67–75.

In this work we address the optimizing control allocation problem for an over-actuated nonlinear time- varying system with actuator dynamic where parameters affine in the actuator and effector model may be assumed unknown. Instead of optimizing the control allocation at each time instant, a dynamic approach is considered by constructing actuator reference update-laws that represent an asymptotically optimal allocation search. By using Lyapunov analysis for cascaded set-stable systems, uniform global/local asymptotic stability is guaranteed for the optimal equilibrium sets described by the system, the control allocation update-law and the adaptive update-law, if some persistence of exitation condition holds. Simulations of a scaled-model ship, manoeuvred at low-speed, demonstrate the performance of the proposed allocation scheme.

Tjonnas, J and Johansen, TA 2008b 'Adaptive control allocation,' *Automatica*, **44** (11), 2754–2765. In this work we address the control allocation problem for a nonlinear over-actuated time-varying system where parameters affine in the effector model may be assumed unknown. Instead of optimizing the control allocation at each time instant, a dynamic approach is considered by constructing update-laws that represent asymptotically optimal allocation search and adaptation. Using Lyapunov analysis for cascaded set-stable systems, uniform global/local asymptotic stability is guaranteed for the sets described by the system, the optimal allocation update-law and the adaptive update-law.

Tol, HJ, de Visser, CC, van Kampen, E, and Chu, QP 'Nonlinear multivariate spline-based control allocation for high-performance aircraft,' *AIAA J. Guidance, Control, and Dynamics*, **37** (6), 1840–1862. High-performance flight control systems based on the nonlinear dynamic inversion principle require highly accurate models of aircraft aerodynamics. In general, the accuracy of the internal model determines to what degree the system nonlinearities can be canceled; the more accurate the model, the better the cancellation, and with that, the higher the performance of the controller. In this paper, a new control system is presented that combines nonlinear dynamic inversion with multivariate simplex spline-based control allocation. Three control allocation strategies that use novel expressions for the analytic Jacobian and Hessian of the multivariate spline models are presented. Multivariate simplex splines have a higher approximation power than ordinary polynomial models and are capable of accurately modeling nonlinear aerodynamics over the entire flight envelope of an aircraft. This nonlinear spline based controller is applied to control a high-performance aircraft (F-16) with a large flight envelope. The simulation results indicate that perfect feedback linearization can be achieved throughout the entire flight envelope, leading to a significant increase in tracking performance compared with ordinary polynomial-based nonlinear dynamic inversion.

Wilson, DJ, Riley, DR, and Citurs, KD 1993 'Aircraft Maneuvers for the Evaluation of Flying Qualities and Agility, Vol. 2: Maneuver Descriptions And Selection Guide,' Wright Laboratory Technical Report WL-TR-93-3082. A set of aircraft maneuvers has been developed to augment evaluation maneuvers used currently by the flying qualities and flight test communities. These maneuvers extend evaluation to full, aircraft dynamics throughout the aircraft flight envelope. As a result, a tie has been established between operational use and design parameters without losing control of the aircraft evaluation process. Twenty maneuvers are described as an initial set to examine primarily high-angle-of-attack conditions. Perhaps as important as the maneuvers themselves is the method used to select them. These maneuvers will allow direct measurement of flying qualities throughout the flight envelope instead of merely comparing parameters to specification values.

Wurth, S, Hart, J, and Baxter, J 2002 'X-35B integrated flight propulsion control fault tolerance development,' AIAA 2002-6019 in *AIAA 2002 Biennial International Powered Lift Conference and Exhibit*. The development of the Lockheed Martin Joint Strike Fighter Short Takeoff and Vertical Landing Concept Demonstrator Aircraft (X-35B) Integrated Flight Propulsion Control system required a rigorous and thorough application of an established fly by wire flight control standard. This standard was modified to adapt a legacy twin-engine propulsion system into a single-engine flight-critical application. The modification included establishing acceptable and achievable fault tolerance requirements, rigorous fault tolerance testing, and assessment of all propulsion system fault tolerance risks. This process resulted in a determination that the X-35B IFPC system had acceptable risk for the flight test program.

Index

Aircraft Control Allocation, First Edition. Wayne Durham, Kenneth A. Bordignon and Roger Beck.
© 2017 John Wiley & Sons, Ltd. Published 2017 by John Wiley & Sons, Ltd.
Companion website: www.wiley.com/go/durham/aircraft_control_allocation

Printed and bound by CPI Group (UK) Ltd, Croydon, CR0 4YY

17/04/2025

14658878-0001